Differential Quadrature and Differential Quadrature Based Element Methods

Theory and Applications

Differential Quadrature and Differential Quadrature Based Element Methods

Theory and Applications

Xinwei Wang

AMSTERDAM • BOSTON • HEIDELBERG • LONDON
NEW YORK • OXFORD • PARIS • SAN DIEGO
SAN FRANCISCO • SINGAPORE • SYDNEY • TOKYO

Butterworth-Heinemann is an imprint of Elsevier

Butterworth-Heinemann is an imprint of Elsevier
The Boulevard, Langford Lane, Kidlington, Oxford OX5 1GB, UK
225 Wyman Street, Waltham, MA 02451, USA

Notices
Knowledge and best practice in this field are constantly changing. As new research and experience broaden our understanding, changes in research methods, professional practices, or medical treatment may become necessary.

Practitioners and researchers must always rely on their own experience and knowledge in evaluating and using any information, methods, compounds, or experiments described herein. In using such information or methods they should be mindful of their own safety and the safety of others, including parties for whom they have a professional responsibility.

To the fullest extent of the law, neither the Publisher nor the authors, contributors, or editors, assume any liability for any injury and/or damage to persons or property as a matter of products liability, negligence or otherwise, or from any use or operation of any methods, products, instructions, or ideas contained in the material herein.

British Library Cataloguing-in-Publication Data
A catalogue record for this book is available from the British Library

Library of Congress Cataloging-in-Publication Data
A catalog record for this book is available from the Library of Congress

ISBN: 978-0-12-803081-3

For information on all Butterworth-Heinemann publications
visit our website at http://store.elsevier.com/

Typeset by Thomson Digital

Printed and bound in the United States of America

MATLAB® is a trademark of The MathWorks, Inc. and is used with permission. The MathWorks does not warrant the accuracy of the text or exercises in this book. This book's use or discussion of MATLAB® software or related products does not constitute endorsement or sponsorship by The MathWorks of a particular pedagogical approach or particular use of the MATLAB® software.

Working together
to grow libraries in
developing countries

www.elsevier.com • www.bookaid.org

Contents

Preface

Time is flying. More than 20 years have passed since the differential quadrature method (DQM) was introduced to the author by Professor Charles W. Bert in 1991. The author joined Dr. Bert's differential quadrature (DQ) research group at the University of Oklahoma, Norman, Oklahoma, USA. A year later, he returned to his home country and became a professor at Nanjing University of Aeronautics and Astronautics, Nanjing, China. He has continued to do research on the engineering applications of the DQM and contributed to the development of the DQM.

The DQM can often yield numerically exact results as to the analytical solutions. Due to its attractive features of rapid convergence, high accuracy, and computational efficiency, the DQM is now a well-known method worldwide. The early developments of the DQM and its applications in general engineering up to the year of 1999 are well documented in a book written by Professor Shu (Chang Shu: *Differential Quadrature and Its Application in Engineering*, London: Springer-Verlag Limited, 2000). New developments on the DQ method and its applications to structural mechanics have been made since then. Although the progression of the development and application of the DQM in the area of structural mechanics is clear from the past researches, but these have been scattered over many papers. In addition, a variety of different quadrature formulations by varying the degree of the polynomials, treatment of boundary conditions and employing regular or irregular grid points exist in literature. This has often caused confusion for researchers and engineers and led to a difficulty to make a choice of a DQM or DQ-based element method for solving practical problems.

From time to time, the author receives emails domestically and internationally to request papers or ask questions related to the applications of the DQM as well as to the implementation. Therefore, there is a need to write a book to explain these new developments and their applications in detail as well as to provide FORTRAN programs and MATLAB files to overcome the difficulty in the implementation by using the well-developed methods.

The book is primarily written for scientists and engineers who are interested in applying the DQM and DQ-based element methods to problems in the area of structural mechanics, including static, buckling, vibration, and dynamic problems. The book is also suitable for graduate students majoring in engineering. With the help of the useful information provided in the book, they can use these methods effectively to solve a variety of problems in their research as well as in practice. The book can serve as a reference book for researchers in many fields such as biosciences, transport processes, fluid mechanics, static and dynamic structural mechanics, static aeroelasticity, and lubrication mechanics, where quantitative analysis is needed.

This book is focused to explore the applications of the DQ technique to problems in the area of structural mechanics, including static stress analysis, buckling analysis, and vibration and dynamic analysis. Linear, geometrically nonlinear, and material nonlinear problems are involved. Important aspects are emphasized and discussed in detail. The main body of the book contains two segments. One is theory and the other is applications. To ease the implementation effort, FORTRAN programs and MATLAB files are included in the appendices. The outline of this book is as follows.

Chapter 1 presents the basic principle of the DQM. For completeness considerations, the original method to determine the weighting coefficients is also included. The explicit formulas to compute the weighting coefficients are given, including the one based on the polynomials and the one based on the harmonic functions. The latter is called the harmonic differential quadrature method (HDQM). Various grid distributions are summarized. Error analysis of the DQM is briefly discussed. Local adaptive DQM is given. Special attention is paid to the DQ-based time integration scheme, an unconditional stable numerical integration method. Examples are given to demonstrate its capability over the existing time integration schemes.

Chapter 2 presents the basic principle of the differential quadrature element method (DQEM). Two different ways are described to formulate the weighting coefficients of the DQ beam element. One way uses the Hermite interpolation and the other employs the Lagrange interpolation. For DQ plate element, the approach of the mixed Hermite interpolation with Lagrange interpolation can also be used. Assemblage procedures are given and several examples are worked out in detail for illustrations.

Chapter 3 presents various approaches to apply the multiple boundary conditions. Although all methods work equally well in one-dimensional problem; however, some of them have difficulty in applying the multiple boundary conditions at the corners for two-dimensional problems, such as the rectangular plate problems, since these approaches have four degrees of freedom at corners but only three boundary conditions are available. Examples are given to show the importance of choosing an appropriate way to apply the multiple boundary conditions.

Chapter 4 presents the basic principle of the weak-form quadrature element method (QEM), one of the DQ-based element methods. Different from the strong-form DQEM presented in Chapter 2, the formulations of the QEM are essentially similar to the high-order finite element method (FEM) or the time-domain spectral element method (SEM). However, differences do exist and are demonstrated in details.

Chapters 5–11 present various applications of the DQM and the DQ-based element methods. The applications include in-plane stress analysis, static analysis, linear and nonlinear buckling analysis, and free vibration and dynamic analysis. The importance of using an accurate way to apply the multiple boundary conditions and of choosing the right nonuniform grid distribution is demonstrated.

In appendices, several FORTRAN programs and subroutines are provided to save the readers' programming effort. With simple modifications, these programs can be used to handle different materials, different applied loads as well as different boundary conditions, and to solve a variety of problems in the area of structural mechanics, such as static, buckling, and free vibration analysis. These programs and subroutines have also been converted to the MATLAB files and functions.

Finally, a summary on our research work is given in Appendix XI. Our contribution to the development of the DQM as well as to its applications is briefly described. More results, which are not included in the book, can be found in the listed journal papers.

The book was completed in the State Key Laboratory of Mechanics and Control of Mechanical Structures, Nanjing University of Aeronautics and Astronautics, Nanjing, China.

Xinwei Wang
Nanjing, China

Acknowledgments

It is clearly impossible for me to acknowledge everyone who has made a contribution, in some manner, to publish this book. I owe a major debt to my colleagues and graduate students who made many contributions to the development of the DQM as well as applications of the DQM. They are Professor Renhuai Liu, Yongliang Wang, Beiqing He, Guangming Zhou, Chuwei Zhou, and Xinfeng Wang, Dr. Meilan Tan, Feng Liu, Lifei Gan, Jian Liu, Zhangxian Yuan, and Chunhua Jin, and Mr. Bin Liu, Xudong Shi, Linghai Jiang, Mengsheng Wang, Chengwei Dai, Zhongbin Zhou, Wei Zhang, Chunling Xu, Feng Wang, Guohui Duan, Luyao Ge, and Yu Wang, and Miss Zhe Wu. The materials presented in this book include their contributions.

I am particularly grateful to Professor Charles W. Bert at the University of Oklahoma, USA, who introduced the DQM to me and encouraged me to do research on the DQM continuously. Special thanks go to Dr. Zhangxian Yuan and Mr. Luyao Ge for helping me to convert the FORTRAN programs into MATLAB files, and also to Mr. Luyao Ge and Yu Wang for redrawing some plots.

My sincere thanks and appreciation also go to each of the reviewers for their helpful reviews and valuable comments to improve the book. They include Professor Alfred G. Striz at the University of Oklahoma, USA, Professor Wen Chen at Hohai University, China, and several anonymous reviewers.

The staff at Elsevier have been very helpful and cooperative. I was assisted in the manuscript preparation by Cari Owen, Editorial Project Manager, Science & Technology Books (Engineering). I express my heartfelt thanks to Cari who works with great care and in a friendly and considerate way; to Pauline Wilkinson, the Production Project Manager, who prepared the proofs with great accuracy, and to the other individuals at Elsevier who were involved in the editing and production aspects of the book.

Last but not least, I would like to thank my family for their understanding, encouragement, and support. I devote this book to my wife, Guoying, for her infinite patience and taking most of the household duties, and also to my little Grandson, Chenqi, to ease my (his Grandpa's) guilt from spending too little time to play with him.

The researches were partially supported by the National Natural Science Foundation of China (10972105), the Aeronautical Science Foundation of China (2004ZB52006), Jiangsu Natural Science Foundation (BK99116), National Doctorial Foundation of China (20020287003), PAPD, and the Research Fund of State Key Laboratory of Mechanics and Control of Mechanical Structures (Nanjing University of Aeronautics and Astronautics) (Grant No. 0214G02).

DIFFERENTIAL QUADRATURE METHOD

1.1 INTRODUCTION

With the advance of computer technology, nowadays numerical simulations play an important role in science and engineering. Various numerical methods have been used in numerical analysis and are regarded as powerful tools for solving partial differential equations (PDEs). To name a few, finite element method (FEM) [1], finite difference method (FDM) [2], finite volume method (FVM) [3], and methods of weighted residuals (MWR) [4,5] such as Galerkin method and collocation method. Among all aforementioned methods, the most widely used methods are FEM and FDM. Perhaps due to its flexibility and ability in dealing with complex geometries and boundary conditions, FEM is more widely used in the area of structural mechanics [1].

It is known that none of the aforementioned numerical methods is versatile and can be used to solve all problems efficiently. Each method has its own merits and limitations. Even the most widely used FEM still suffers from difficulty in analyzing problems when phenomena such as singularities, steep changes, stress concentration, and large deformation exist. For example, the computational efficiency of the FEM is lost when the method is used to analyze problems of metal forming and elastoplasticity, high-velocity impact, dynamic crack propagation, explosion, and shock wave. Even with the modern computing machines, analysis for guided wave propagation in three-dimensional solids by the classical FEM would require impractical computational resources (the computational time and memory storage requirements) [6]. Therefore, along with the ever-growing advancement of faster computing machines, the research into the development of new efficient methods for numerical simulations is an ongoing parallel activity [7].

Motivated by the needs of modern science and technology, considerable efforts have been made in the development of new numerical methods, such as various meshless or mesh-free methods [8], wavelet-based numerical methods [9], the differential quadrature method (DQM) and the differential quadrature-based element method [7], the discrete singular convolution (DSC) algorithm [10], and the high-order FEM [11]. A few examples of the requirements of new methods are as follows: (1) simulations of many dynamic systems often require very fast numerical solution of the equations of the system mathematical models [7]; (2) the computer-aided design (CAD) process in which the database often requires large computer storage and the interpolative manipulations for the operating design parameters may be less accurate as well as quite timeconsuming [7]; and (3) structural health monitoring (SHM) applications call for both efficient and powerful numerical tools to predict the behavior of ultrasonic-guided waves since the existing well-known FEM would require impractical computational resources. The aforementioned methods try to resolve the limitations existing in the classical FEM and to fulfil the needs of modern science and technology.

To reduce the enormous computational costs in simulations of wave propagation in solids, higher-order FEMs with polynomial degrees $p > 2$ [11], time-domain spectral element methods (SEM) [12–14], and weak-form quadrature element method (QEM) [15] are proposed. Besides the merit of high rate of convergence, they also possess the advantages existing in the FEM and can be implemented into the commercial software such as ABAQUS, since these methods are essentially the FEMs. Obviously, these methods still have some limitations existing in the classical FEM. For example, they cannot be efficiently used in the large deformation analysis such as analysis of metal forming.

Aimed at resolving the limitations existing in the FEM for analyzing problems of metal forming and elastoplasticity, high-velocity impact, dynamic crack propagation, and explosion, various mesh-free methods [8] are proposed. Due to less mesh dependency, mesh-free methods can eliminate possible mesh distortion and entanglement encountered in FEM to analyze large deformation and explosion problems. For example, the method of smoothed particle hydrodynamic (SPH) [16], a mesh-free method, is proposed and implemented into the commercial finite element software LS-DYNA to complement the deficiency of the FEM in dealing with the explosion and high-velocity impact problems. Extended finite element method (XFEM) [17], another mesh-free method [8], is proposed to analyze the problem of dynamic crack propagations. Since the finite element mesh can remain unchanged during crack propagation with using the XFEM, the hybrid method of FEM together with XFEM is convenient to analyze such problems. It is seen that each method is proposed to overcome certain difficulty existing in FEM and to fulfil certain needs by the modern science and technology. Mesh-free methods have some disadvantages: their approximate functions are much complicated and larger computation effort is usually needed; and dealing with essential boundary conditions is far more complicated than the classical FEM [8].

Since wavelets possess multiresolution and localization properties, various wavelet-based methods are proposed to meet some needs of modern science and technology [9]. Similar to the mesh-free methods, wavelet-based methods can be more efficiently used to analyze problems with singularities, steep changes, and stress concentration. For example, the wavelet Galerkin method (WGM) is proposed and has been successfully used in solving a variety of PDEs in regular and irregular computational domains [18]. The drawback of the WGM is that the method can only handle simple boundary conditions and is complicated when nonhomogeneous boundary conditions are considered in two-dimensional cases [9].

The DSC algorithm [10] is efficient and robust for solving the Fokker–Planck equation describing various physical phenomena. The method employs compactly support wavelet interpolating functions. Banded differential matrices with well-behaved condition numbers are obtained thus the DSC is suitable for large-scale computations. Due to employing exterior grid points for treating boundary conditions, spurious eigenvalues are removed although the matrices are not symmetric. The unique advantage of the DSC is that it can also yield accurate high-order mode frequencies. Perhaps its drawback is not easy to apply the free boundary conditions, especially at free corners of anisotropic rectangular plates. Although with the aid of the iteratively matched boundary (IMB) method [19], this issue may be resolved to a certain degree but is very complicated and the existing problem has not completely solved yet.

Because of clear superiority of the FEM and FDM in engineering applications, some new or hybrid methods are proposed to improve the classical FEM and FDM: the aforementioned higher-order FEMs [11], SEM [12–14], QEM [15], mesh-free least-squares-based FDM [20], wavelet finite element method (WFEM) [21], and wavelet-optimized finite difference method (WOFD) [22]. These methods raise the computational efficiency and extend the application ranges of the classical FEM and FDM.

DQM was originated by Bellman and Casti in the early 1970s [23,24]. The method has a relatively recent origin in the later 1980s [25] and has been gradually emerging as a distinct numerical solution technique for the initial- and/or boundary-value problems of physical and engineering sciences since then [7].

In fact, the DQM can be formulated via the polynomial-based collocation method, one of the popular methods of MWR [26,27]. As a numerical method, the DQM can be applied in the fields of biosciences, transport processes, fluid mechanics, static and dynamic structural mechanics, static aeroelasticity, and lubrication mechanics. It has been shown that the DQM is simple and can yield highly accurate numerical solutions with minimal computational effort. The differential quadrature-based element methods are proposed to overcome some deficiency existing in the conventional DQM and extended the application range of the DQM in dealing with complex geometry and boundary conditions. The developed DQM has seemingly a high potential as an alternative to the classical finite difference and FEMs [7].

Summaries on the new development of the DQM as well as on its applications to structural mechanics problems up to the year of 2000 can be found in Refs. [7, 27–30]. Since 2000, further developments on the DQM have been made and the DQM has been used successfully for solutions of many engineering problems. Although the progression on the development and application of the DQM in the area of structural mechanics is clear from the past researches, these have been scattered over many papers. Besides, a variety of different quadrature formulations by varying the degree of the polynomials, treating boundary conditions, and employing regular or irregular grid points exist in literatures. This has often caused confusion for researchers and engineers and led to a difficulty in making a choice between a DQM and a DQ-based element method for solving practical problems. Improper choice can be very dangerous since the DQM may admit spurious complex eigenvalues, namely, the stability problem whose occurrence is not well understood in general [31]. To resolve this issue, this book presents new developments on the DQM, the strong-form differential quadrature element method (DQEM) and the weak-form differential quadrature element method (QEM) systematically. A variety of applications are demonstrated. Suggestions are made, and some FORTRAN programs and converted MATLAB files are provided. It is hoped that the provided information can help the researchers and engineers to use the DQM for solving practical problems.

In this chapter, the basic principle of DQM is presented. Various existing DQ formulations are summarized, including the original DQM, the modified DQM, the HDQM, the local adaptive differential quadrature method (LaDQM), and the DQ-based time integration scheme. Existing explicit formulas to compute the weighting coefficients are given. Seven grid distributions are summarized and their discrete error is briefly discussed. Numerical examples are given and some recommendations are made.

1.2 INTEGRAL QUADRATURE

There are many integral quadrature methods available, such as trapezoidal rule, Newton–Cotes quadrature, Gaussian quadrature, and Gauss–Lobatto–Legendre (GLL) quadrature. With the advent of computer, Gaussian quadrature has become a well-known numerical integration method. Since it is ideally suited for computers, Gaussian quadrature has been exclusively used in the FEM [1]. On the other hand, the GLL quadrature has become a well-known method in recent years and is widely used in the time-domain SEM [12–14].

The advantage of Gaussian quadrature is its high accuracy. The method is accurate up to a polynomial of degree $(2N - 1)$ with only N abscissas. To achieve the same accuracy, Gaussian quadrature requires the least number of functional evaluations as compared to other quadrature methods. The advantages of GLL quadrature are that its accuracy is up to a polynomial of degree $(2N - 3)$ and that it can result a diagonal mass matrix.

Consider one-dimensional integration of $f(x)$ in the range of $[-1, 1]$. Numerical integration can be written by

$$I = \int_{-1}^{1} f(x)\,dx = \sum_{i=1}^{N} H_i f(x_i) \tag{1.1}$$

where N is the total number of integral points, and H_i and x_i are the weights and abscissas in the quadrature, respectively.

The key to success is the right choice of abscissas. For Gaussian quadrature, the abscissas are the roots of the Nth-order Legendre polynomial. For GLL quadrature, the abscissas are the roots of the first-order derivative of the Nth-order Legendre polynomial together with the two end points ($x_i = \pm 1$). The abscissas are often called as the GLL points [13].

Once the abscissas are given, the weights can be obtained by

$$H_j = \int_{-1}^{1} l_j(x)\,dx \quad (j = 1, 2, ..., N) \tag{1.2}$$

where $l_j(x)$ are the Lagrange interpolation function defined by

$$l_j(x) = \frac{(x-x_1)(x-x_2)...(x-x_{j-1})(x-x_{j+1})...(x-x_N)}{(x_j-x_1)(x_j-x_2)...(x_j-x_{j-1})(x_j-x_{j+1})...(x_j-x_N)} = \prod_{\substack{k=1 \\ k \neq j}}^{N} \frac{x-x_k}{x_j-x_k} \tag{1.3}$$

For GLL quadrature, Eq. (1.1) can be expressed by

$$I = \int_{-1}^{1} f(x)\,dx = \frac{2}{N(N-1)}[f(-1) + f(+1)] + \sum_{i=2}^{N-1} H_i f(x_i) \tag{1.4}$$

The $(N-2)$ weights can be computed by

$$H_j = \frac{2}{N(N-1)\left[P_N(x_j)\right]^2} \quad (j = 2, 3, ..., N-1) \tag{1.5}$$

where $P_{N-1}(x)$ is the Legendre polynomial of degree $(N-1)$.

In the time-domain SEM, GLL quadrature plays an important role in formulations of the element stiffness matrix and mass matrix. Due to a diagonal mass matrix, a crucial reduction of the complexity and of the cost of the numerical time integration is achieved, since the dynamic equation can be explicitly integrated with the usage of the central FDM.

Appendix I lists the abscissas and weights ($2 < N < 22$) in GLL quadrature for readers' reference. If larger N is required, the abscissas and weights can be obtained by using the software Maple (http://www.maplesoft.com). A simple Maple program is also given in Appendix I for readers' reference.

1.3 DIFFERENTIAL QUADRATURE METHOD

Consider one-dimensional function $f(x)$ in the domain of $[-1,1]$. Assume $f(x)$ is continuous and differentiable with respect to x. Analog to Eq. (1.1), one has

$$\left(\frac{df(x)}{dx}\right)_{x=x_i} = \sum_{j=1}^{N} A_{ij} f(x_j) \quad (i = 1, 2, ..., N) \tag{1.6}$$

where N is the total number of grid points, and A_{ij} and x_i are called the weighting coefficients of the first-order derivative with respect to x and grid points in the DQM, respectively.

For higher-order derivatives, similar expressions to Eq. (1.6) exist. For example,

$$\left(\frac{d^2 f(x)}{dx^2} \right)_{x=x_i} = \sum_{j=1}^{N} B_{ij} f(x_j) \qquad (i = 1, 2, ..., N) \tag{1.7a}$$

$$\left(\frac{d^3 f(x)}{dx^3} \right)_{x=x_i} = \sum_{j=1}^{N} C_{ij} f(x_j) \qquad (i = 1, 2, ..., N) \tag{1.7b}$$

$$\left(\frac{d^4 f(x)}{dx^4} \right)_{x=x_i} = \sum_{j=1}^{N} D_{ij} f(x_j) \qquad (i = 1, 2, ..., N) \tag{1.7c}$$

where B_{ij}, C_{ij}, and D_{ij} are called the weighting coefficients of the second-, third-, and fourth-order derivatives, respectively, with respect to x in the DQM.

It is seen that the essence of the DQM is that the derivative of a function $f(x)$ with respect to x at a given discrete point is approximated as a weighted linear sum of the function values at all discrete points, including the two end points.

Similar to the integral quadrature method, the key to success is also the distribution of grid points. In the early days, uniformly distributed grid points were exclusively used in the DQM for solving problems in the area of structural mechanics. Although uniformly distributed grid points work very well for certain problems, however, nonuniformly distributed grid points should be used to ensure the solution accuracy as well as numerical stability in the applications of the DQM. In 1991, Sherbourne and Pandey [32] first reported that the DQM with uniformly distributed grid points gave poor solutions for buckling problems of composite plates. With nonuniform grid points, the DQM can give accurate buckling load. Thus, the choice of the grid points is critical in the applications of DQM. Several distributions of grid points are available in literature [33] and will be summarized in Section 1.6.

Although the DQM can be regarded as a specific class of mixed collocation methods [26], and the roots of the Chebyshev polynomials yield minimum residuals for ordinary differential equations [5], the roots of the Chebyshev polynomials are not the best choice as the grid points in the DQM, since the two end points, that is, $x_i = \pm 1$, must be included in the DQM in order to apply the boundary conditions.

Except the uniformly distributed grid points, the widely used nonuniformly distributed grid points in $[-1,1]$ are given by

$$x_j = -\cos \frac{(j-1)\pi}{(N-1)} \qquad (j = 1, 2, ..., N) \tag{1.8}$$

It is known that Eq. (1.8) consists the extreme points of the Nth-order Chebyshev polynomial together with two end points ($x_i = \pm 1$). In other words, the grid points are the roots of the first-order derivative of the Nth-order Chebyshev polynomial, not the Legendre polynomial, together with two end points ($x_i = \pm 1$). Sometimes, Eq. (1.8) is mistakenly called the GLL points in literatures. Since no simple formula as Eq. (1.8) is available to compute the grid points, GLL points, widely used in the time-domain SEM, are seldom used in the DQM.

Once the grid points are chosen, the remaining task is to determine the weighting coefficients. Unlike the integral quadrature, the determination of the weighting coefficients of the DQM is more flexible. Currently, there are two major ways to determine the weighting coefficients. One way is based on the polynomials, called the DQM, and the other one is based on the harmonic functions, called the HDQM [34].

1.4 DETERMINATION OF WEIGHTING COEFFICIENTS

Without loss of generality, consider one-dimensional function $f(x)$ that is continuous and differentiable with respect to x in the domain $[a,b]$. According to the differential quadrature rule, one has the following form:

$$L\{f(x)\}_{x=x_i} = \sum_{j=1}^{N} E_{ij} f(x_j) \quad (i=1, 2, ..., N)$$ (1.9)

where L is the linear operator, $x_j(j=1,2,...,N)$ are N discrete points in the entire domain including the two end points, that is, $a=x_1 < x_2 < \cdots < x_N = b$, and E_{ij} are the weighting coefficients of the corresponding linear operator, respectively.

To determine the weighting coefficients by using the way based on the polynomials, Eq. (1.9) must be exact for all polynomials of degree less than or equal to $(N-1)$ at all grid points. In other words, the following equation must be satisfied exactly:

$$L\{x^{k-1}\}_{x=x_i} = \sum_{j=1}^{N} E_{ij} x_j^{k-1} \quad (k,i=1, 2, ..., N)$$ (1.10)

Solving the N sets of N linear algebraic equations, Eq. (1.10), yields the $N \times N$ weighting coefficients.

Let $L = \dfrac{d}{dx}$ and denote $f_j = f(x_j)$, one has

$$\left(\frac{df}{dx}\right)_{x=x_i} = f_i' = \sum_{j=1}^{N} A_{ij} f_j \quad (i=1, 2, ..., N)$$ (1.11)

Then Eq. (1.10) becomes

$$\left(\frac{dx^{k-1}}{dx}\right)_{x=x_i} = (k-1)x_i^{k-2} = \sum_{j=1}^{N} A_{ij} x_j^{k-1} \quad (k,i=1, 2, ..., N)$$ (1.12)

In matrix form, Eq. (1.12) can be written as

$$[G]=[A][V]$$ (1.13)

where $[A]$ is the weighting coefficient matrix of the first-order derivatives with respect to x, and

$$[G]=\begin{bmatrix} 0 & 1 & 2x_1 & \cdots & (N-1)x_1^{N-2} \\ 0 & 1 & 2x_2 & \cdots & (N-1)x_2^{N-2} \\ \vdots & \vdots & \vdots & \ddots & \vdots \\ 0 & 1 & 2x_N & \cdots & (N-1)x_N^{N-2} \end{bmatrix}_{N \times N}$$ (1.14)

$$[V] = \begin{bmatrix} 1 & x_1 & \cdots & x_1^{N-1} \\ 1 & x_2 & \cdots & x_2^{N-1} \\ \vdots & \vdots & \ddots & \\ 1 & x_N & \cdots & x_N^{N-1} \end{bmatrix}_{N \times N} \tag{1.15}$$

Since $[V]$ is the Vandermonde matrix, its inverse $[V]^{-1}$ exists. One has

$$[A] = [G][V]^{-1} \tag{1.16}$$

In this way, the $N \times N$ weighting coefficients A_{ij} can be obtained.

Similarly, let $f_i^{[n]} = f^{[n]}(x_i) = \left(\dfrac{d^n f(x)}{dx^n} \right)_{x=x_i}$ $(n = 2, 3, 4)$. One has

$$f_i^{[2]} = \sum_{j=1}^{N} B_{ij} f_j \tag{1.17}$$

$$f_i^{[3]} = \sum_{j=1}^{N} C_{ij} f_j \tag{1.18}$$

$$f_i^{[4]} = \sum_{j=1}^{N} D_{ij} f_j \tag{1.19}$$

The weighting coefficients of higher-order derivatives, B_{ij}, C_{ij}, D_{ij}, can be obtained by a similar way as the determination of the A_{ij}.

Since A_{ij} are known, B_{ij} can be alternatively computed by following equations, namely,

$$\begin{aligned} f_i^{[2]} &= \left(\frac{d^2 f(x)}{dx^2} \right)_{x=x_i} = \left(\frac{d}{dx} \left(\frac{df(x)}{dx} \right) \right)_{x=x_i} = \sum_{k=1}^{N} A_{ik} \left(\frac{df(x)}{dx} \right)_{x=x_k} \\ &= \sum_{k=1}^{N} A_{ik} \left(\sum_{j=1}^{N} A_{kj} f(x_j) \right) = \sum_{j=1}^{N} \sum_{k=1}^{N} A_{ik} A_{kj} f_j = \sum_{k=1}^{N} B_{ij} f_j \quad (i = 1, 2, ..., N) \end{aligned} \tag{1.20}$$

It is obvious that B_{ij} can be conveniently computed without solving the N sets of N linear algebraic equations once A_{ij} are obtained, namely,

$$B_{ij} = \sum_{k=1}^{N} A_{ik} A_{kj} \quad (i, j = 1, 2, ..., N) \tag{1.21}$$

Once A_{ij} and B_{ij} are known, C_{ij} and D_{ij} can be computed by

$$C_{ij} = \sum_{k=1}^{N} A_{ik} B_{kj} = \sum_{k=1}^{N} B_{ik} A_{kj} \quad (i, j = 1, 2, ..., N) \tag{1.22}$$

$$D_{ij} = \sum_{k=1}^{N} A_{ik} C_{kj} = \sum_{k=1}^{N} B_{ik} B_{kj} = \sum_{k=1}^{N} C_{ik} A_{kj} \quad (i, j = 1, 2, ..., N) \tag{1.23}$$

To determine the weighting coefficients by using the way based on the harmonic functions, Eq. (1.9) must be exact for all triangular functions defined by (1.24).

$$1, \sin\frac{\pi(1+x)}{2}, \cos\frac{\pi(1+x)}{2}, \sin\pi(1+x), \cos\pi(1+x), ...,$$
$$\sin\frac{(N-1)\pi(1+x)}{4}, \cos\frac{(N-1)\pi(1+x)}{4} \tag{1.24}$$

where N is an odd number.

In other words, for the HDQM, it requires that when $f(x)$ takes functions defined by Eq. (1.24), Eq. (1.1) should be satisfied exactly.

Follow the similar procedures, the weighting coefficients A_{ij}, B_{ij}, C_{ij} and D_{ij} can be determined.

It should be mentioned that the HDQM is not widely used, since there is no much difference between DQM and HDQM when the total number of grid points is large enough, say $N > 13$.

1.5 EXPLICIT FORMULATION OF WEIGHTING COEFFICIENTS

Numerical experience revealed that due to the round-off errors, the Vandermonde matrix $[V]$ is ill conditioned for $N > 15$ and its inverse cannot be obtained numerically. To overcome this difficulty, another efficient way is used to compute the weighting coefficients explicitly.

Without loss of generality, consider one-dimensional function $f(x)$ that is continuous and differentiable with respect to x in the domain $[a,b]$. It is well known that $f(x)$ can be expressed by using Lagrange interpolation functions once x_j and $f(x_j)(j=1,2,...,N)$ are given, namely [27],

$$f(x) = \sum_{j=1}^{N} l_j(x)f(x_j) = \sum_{j=1}^{N} l_j(x)f_j \tag{1.25}$$

where $a = x_1 < x_2 < \cdots < x_N = b$, and $l_j(x)$ are the Lagrange interpolation functions defined by Eq. (1.3), namely,

$$l_j(x) = \frac{(x-x_1)(x-x_2)...(x-x_{j-1})(x-x_{j+1})...(x-x_N)}{(x_j-x_1)(x_j-x_2)...(x_j-x_{j-1})(x_j-x_{j+1})...(x_j-x_N)} = \prod_{\substack{k=1 \\ k \neq j}}^{N} \frac{x-x_k}{x_j-x_k} \tag{1.26}$$

Differentiating Eq. (1.25) with respect to x yields

$$f'(x) = \sum_{j=1}^{N} l'_j(x)f_j \tag{1.27}$$

Thus, at point x_i one has

$$f'_i = f'(x_i) = \sum_{j=1}^{N} l'_j(x_i)f_j \qquad (i=1,2,...,N) \tag{1.28}$$

Comparing Eq. (1.28) to Eq. (1.11) gives

$$A_{ij} = l'_j(x_i) \quad (i, j = 1, 2, ..., N) \tag{1.29}$$

where,

$$A_{ij} = l'_j(x_i) = \begin{cases} \displaystyle \prod_{\substack{k=1 \\ k \neq i,j}}^{N} (x_i - x_k) \Bigg/ \prod_{\substack{k=1 \\ k \neq j}}^{N} (x_j - x_k) & (i \neq j) \\[4mm] \displaystyle \sum_{\substack{k=1 \\ k \neq i}}^{N} \frac{1}{(x_i - x_k)} & (i = j) \end{cases} \tag{1.30}$$

Once A_{ij} are found, B_{ij}, C_{ij}, D_{ij} can be computed by using Eqs. (1.21)–(1.23). Alternatively, B_{ij}, C_{ij}, D_{ij} can be directly computed by following recursive formulas, namely,

$$l_j^{[k]}(x_i) = \begin{cases} k\left[l_i^{[k-1]}(x_i) l'_j(x_i) - \dfrac{l_j^{[k-1]}(x_i)}{x_i - x_j} \right] & (i \neq j) \\[4mm] \displaystyle -\sum_{\substack{m=1 \\ m \neq i}}^{N} l_m^{[k]}(x_i) & (i = j) \quad (2 \leq k \leq N-1) \end{cases} \tag{1.31}$$

Similar expressions to Eq. (1.30) and Eq. (1.31) can also be found in the book written by Shu [29]. To obtain the explicit formula for computing A_{ij} in the HDQM, the first-order derivative with respect to x, Eq. (1.27) can be used, where the Lagrange interpolation functions are now defined by

$$l_j(x) = \frac{\sin\dfrac{\pi}{8}(x - x_1)\cdots\sin\dfrac{\pi}{8}(x - x_{j-1})\sin\dfrac{\pi}{8}(x - x_{j+1})\cdots\sin\dfrac{\pi}{8}(x - x_N)}{\sin\dfrac{\pi}{8}(x_j - x_1)\cdots\sin\dfrac{\pi}{8}(x_j - x_{j-1})\sin\dfrac{\pi}{8}(x_j - x_{j+1})\cdots\sin\dfrac{\pi}{8}(x_j - x_N)}, \quad x \in [-1,1] \tag{1.32}$$

Thus, one has

$$A_{ij} = \begin{cases} \dfrac{\pi}{8} \displaystyle \prod_{\substack{k=1 \\ k \neq i,j}}^{N} \sin\dfrac{\pi}{8}(x_i - x_k) \Bigg/ \prod_{\substack{k=1 \\ k \neq j}}^{N} \sin\dfrac{\pi}{8}(x_j - x_k) & (i \neq j) \\[5mm] \dfrac{\pi}{8} \displaystyle \sum_{\substack{k=1 \\ k \neq i}}^{N} \dfrac{\cos\dfrac{\pi}{8}(x_i - x_k)}{\sin\dfrac{\pi}{8}(x_i - x_k)} & (i = j) \end{cases} \tag{1.33}$$

Again, once A_{ij} are found, B_{ij}, C_{ij}, D_{ij} can be computed by using Eqs. (1.21)–(1.23). Alternatively, the weighting coefficients B_{ij}, C_{ij}, D_{ij} can be directly computed by the recursive formulae of Eq. (1.31). Note that N can be an even or odd number.

FORTRAN subroutines and MATLAB functions to compute A_{ij} by Eq. (1.30) or Eq. (1.33) are given in several programs included in Appendix II.

1.6 VARIOUS GRID POINTS

Sherbourne and Pandey [32] are the first ones to report that the DQ solution is very sensitive to grid spacing when the DQM is used to analyze the buckling of anisotropic plate. Reliable solution can be only obtained by the DQM with nonuniformly distributed points. Several nonuniform grid points, successfully used in the applications of the DQM, are available in literature [33] and summarized further for reference. The grid points are given in the domain $[-1,1]$ for simplicity and ready to be converted into other solution domains.

$$\text{I : Uniformly distributed grid points : } \qquad x_i = -1 + 2\frac{(i-1)}{N-1} \quad (i=1,2,...,N) \tag{1.34}$$

$$\text{II : } \quad -1, \; x_i = -\cos\frac{(2i-3)\pi}{2N-4} \, (i=2,3,...,N-1), 1 \tag{1.35}$$

$$\text{III : } \quad x_i = -\cos\frac{(i-1)\pi}{N-1} \quad (i=1,2,...,N) \tag{1.36}$$

$$\text{IV : } \quad -1, x_i = -\cos\frac{(2i-1)\pi}{2N} \quad (i=2,3,...,N-1), 1 \tag{1.37}$$

$$\text{V : } \quad -1, x_i = (N-2)\text{Gaussian quadrature points } (i=2,3,...,N-1), \, 1 \tag{1.38}$$

$$\text{VI : } \quad x_i = \frac{-\cos\left[\dfrac{(2i-1)\pi}{2N}\right]}{\cos\left[\dfrac{\pi}{2N}\right]} \quad (i=1,2,...,N) \tag{1.39}$$

$$\text{VII : } \qquad \text{Gauss} - \text{Lobatto} - \text{Legendre (GLL) integration points (see Appendix I)} \tag{1.40}$$

As mentioned earlier, sometimes Grid III is mistakenly called GLL points in literature. Actually Grid III is Chebyshev–Gauss–Lobatto grid distribution, that is, the extreme points of the $(N-1)$th Chebyshev polynomial together with the two end points, but the GLL points, Grid VII, is the extreme points of the $(N-1)$th Legendre polynomial together with the two end points.

A FORTRAN subroutine, called SUBROUTINE GRULE (N,X,W,Y) [35], is provided in Appendix II that can be used to calculate the Grid V numerically. The subroutine has also been converted to the MATLAB function.

1.7 ERROR ANALYSIS

Without loss of generality, consider one-dimensional function $f(x)$ that is continuous and differentiable with respect to x in the domain $[a, b]$. According to Ref. [33], $f(x)$ can be expressed as

$$f(x) = \sum_{j=1}^{N} l_j(x)f_j + \frac{f^{(N)}(\xi)}{N!}\omega_N(x), \qquad \xi \in [a,b] \tag{1.41}$$

where $l_j(x)$ is Lagrange interpolation function defined by Eq. (1.26) and $\omega_N(x) = \prod_{k=1}^{N}(x-x_k)$.

It is seen that Eq. (1.41) is slightly different from Eq. (1.25), since Eq. (1.25) is only approximately valid unless $f(x)$ is an $(N-1)$th-order polynomial. Taking the first-order derivative with respect to x yields

$$f'(x) = \sum_{j=1}^{N} l_j'(x)f_j + E_1 \tag{1.42}$$

In Eq. (1.42), E_1 is the error term of the first-order derivative and given by

$$E_1 = \frac{f^{(N)}(\xi)}{N!}\omega_N'(x) + \frac{\omega_N(x)}{N!}\frac{df^{(N)}(\xi)}{dx} \tag{1.43}$$

Note that at the grid point $\omega_N(x_i) = 0 \quad (i=1, 2, ..., N)$. From Eq. (1.43), the maximum error among all grid points is

$$E_1^{max} = \left|f^{(N)}(\xi)\right|\frac{\max\left|\omega_N'(x_i)\right|}{N!} = \left|f^{(N)}(\xi)\right|E_M \qquad (i=1, 2, ..., N) \tag{1.44}$$

where E_M is called the maximum error coefficient whose expression for various grid points is given in [33].

If $f(x)$ is a polynomial of an order less than or equal to $(N-1)$, $\left|f^{(N)}(\xi)\right| = 0$. In other words, the error E_1 is identically zero at any point. In such cases, the DQ solution is exact and insensitive to the grid spacing.

From Eq. (1.44), it is seen that E_1^{max} would be different for various grid points listed in Section 1.6. In other words, the convergence rate is different for various grid points. This will be illustrated by two simple examples.

EXAMPLE 1.1

Consider a function given by

$$f(x) = \frac{\ln[1+\tan^2(1.5x)]}{3} \qquad x \in [-1,1] \tag{1.45}$$

Its first-order derivative with respect to x can be easily obtained as

$$f'(x) = \tan(1.5x) \quad x \in [-1,1] \tag{1.46}$$

Figure 1.1 illustrates the results of the first-order derivative at various grid points by the DQM with $N=19$. In Fig. 1.1, symbols are DQ data and line is the exact solution. It is seen that DQ data agree well with the exact solution for all seven sets of grid spacing listed in Section 1.6. However, the convergence rate is different and illustrated in Fig. 1.2.

Figure 1.2 shows the variation of the first-order derivative at the end point ($x=1$) with the number of grid points. N varies from 7 to 19. The exact solution is 14.10142. It is obvious that the convergence rate is different for various sets of grid spacing. Among the seven sets of grid spacing, the convergence rate of the DQM with Grid II is the fastest and the one of the DQM with Grid I is the lowest.

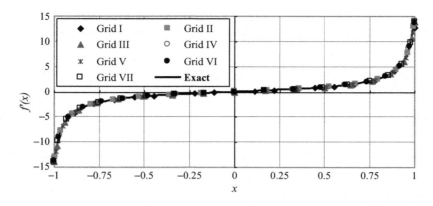

FIGURE 1.1

Comparison of DQ Results with Exact Data ($N = 19$)

EXAMPLE 1.2

Consider a function given by

$$f(x) = \frac{\arctan(5x)}{5} \quad x \in [-1, 1] \tag{1.47}$$

Its first-order derivative with respect to x can be easily obtained as

$$f'(x) = \frac{1}{1 + 25x^2} \quad x \in [-1, 1] \tag{1.48}$$

Figure 1.3 illustrates the results of the first-order derivative at various grid points by the DQM with $N = 19$. In Fig. 1.3, symbols are DQ data and the solid line is the exact solution. It is seen that except

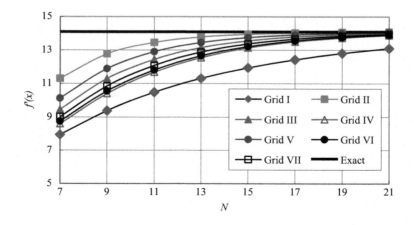

FIGURE 1.2

Convergence of the DQM with Various Grid Spacing at the End Point ($x = 1$)

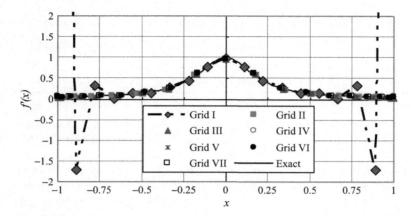

FIGURE 1.3

Comparison of DQ Results with Exact Data ($N = 19$)

the result obtained by the DQM with Grid I, all other DQ data agree well with the exact solution. The DQM with Grid I only yield accurate results in the center region. The results near the ends are very poor, and the Runge's phenomenon occurs.

Figure 1.4 shows the variation of the first-order derivative at the center point ($x = 0$) with the number of grid points. N varies from 7 to 19. The exact solution is 1.0. It is obvious that the convergence rate is different for various grid points. Among the seven sets of grid spacing, the convergent rate of the DQM with Grid I is the fastest and the one of the DQM with Grid II is the lowest, quite different from the one at the end point.

Table 1.1 lists the maximum percentage errors for the two examples. It is seen that the maximum error usually occurs at the end points. However, the maximum error, denoted by the bold numbers in Table 1.1, occurs at the point next to the middle point for Grid II and Grid V. This is quite different from the conclusion made in the open literature.

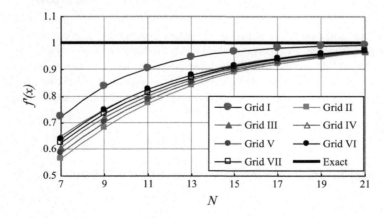

FIGURE 1.4

Convergence of the DQM with Various Grid Spacing at the Center Point ($x = 0$)

Table 1.1 Maximum Absolute Percentage Errors of the First-Order Derivative ($N = 19$)

Grid Spacing	Grid I	Grid II	Grid III	Grid IV	Grid V	Grid VI	Grid VII
Example I	9.153	**2.331**	1.187	2.553	**1.189**	2.306	1.756
Example II	66063	**6.714**	13.923	89.810	**6.538**	71.844	34.570

Table 1.2 lists the percentage errors at the middle point for Example 1.2. It is seen that the convergence rate of the DQM with Grid I is almost an order higher than the ones of the DQM with other grid points.

From data listed in Figs. 1.2 and 1.4 and Tables 1.1 and 1.2, it is seen that the convergence rate is different for different sets of the grid spacing and is different at different grid points for the same grid spacing. Runge's phenomenon may occur if the uniformly distributed grid spacing (Grid I) is used in the DQM; thus, nonuniformly distributed grid spacing should be used in the applications of the DQM for reliability considerations.

The DSC [10,19] with the nonregularized Lagrange's delta sequence kernel, called the DSC-LK [31], can yield very accurate solutions although uniform grid spacing is used in the DSC algorithm. Careful study shows that the weighting coefficients of the DSC-LK are exactly the same as the corresponding weighting coefficients of the DQM at the center point with Grid I. The convergence property at the center point, shown in Fig. 1.4 and Table 1.2, explains why the DSC-LK can yield very accurate solutions.

Example 1.2 demonstrates the importance of selecting suitable grid points in the applications of the DQM. Currently, Grid III is the widely used grid spacing. For certain cases that will be demonstrated in Chapter 8, however, only Grid V can yield reliable solutions.

The DQ error estimations for higher-order derivatives can be treated in a similar way. More information can be found in [33].

1.8 LOCAL ADAPTIVE DIFFERENTIAL QUADRATURE METHOD

The LaDQM, introduced by Wei and his colleagues [36,37], employs both localized interpolating basis functions and exterior grid points for applying boundary conditions. The concept might come from the high-order finite difference. Compared to the DQM, the LaDQM has two good features. One, the matrix structure is banded due to the use of localized interpolating basis functions, and thus it may be more efficient for large-scale computations; second, spurious eigenvalues may be avoided due to using exterior grid points. For dynamic analysis, spurious eigenvalues will cause instability during numerical time integration thus degrade the dynamic response.

For the LaDQM, Eq. (1.6) is modified to

Table 1.2 Percentage Errors of the First-Order Derivative at the Middle Point ($N = 19$)

Grid Spacing	Grid I	Grid II	Grid III	Grid IV	Grid V	Grid VI	Grid VII
Example II ($x = 0$)	0.970	5.293	4.521	3.855	4.886	3.900	4.191

$$\left(\frac{df(x)}{dx}\right)_{x=x_i} = \sum_{j=N_1}^{N_2} A_{ij} f(x_j) \quad (i=1,2,...,N) \tag{1.49}$$

where N_1 and N_2 are adaptive numbers.

N_L and N_R denote the number of exterior points outside the left (x_1) and right (x_N) boundaries, and M the computational band width. There are two ways to compute N_1 and N_2, one way results box-banded (*BB*) matrices and the other results in uniform-banded (*UB*) matrices. The following equation is used to compute N_1 and N_2 [37],

$$\begin{cases} BB: & N_1 = \max(i-M, 1-N_L), \quad N_2 = \min(N_1 + 2M, N+N_R) \\ UB: & N_1 = \max(i-M, 1-N_L), \quad N_2 = \min(i+M, N+N_R) \end{cases} \tag{1.50}$$
$$(i=1,2,...,N;\ 2M+1 \le N)$$

Note that M is adaptive. It can be seen that the LaDQM is reduced to the DQM if the interpolating basis functions are the same as the DQM and may be equivalent to the DSC algorithm if same exterior and interior grid points are used as the DSC.

Similar to the DQM, the weighting coefficient of the first-order derivative, A_{ij}, can be explicitly calculated by

$$A_{ij} = \begin{cases} \displaystyle \prod_{\substack{k=N_1 \\ k\ne i,j}}^{N_2} (x_i - x_k) \Big/ \prod_{\substack{k=N_1 \\ k\ne j}}^{N_2} (x_j - x_k) & (N_1 \le j \le N_2;\ i \ne j) \\[20pt] \displaystyle \sum_{\substack{k=N_1 \\ k\ne i}}^{N_2} \frac{1}{(x_i - x_k)} & (N_1 \le j \le N_2;\ i = j) \\[12pt] 0 & \text{(others)} \end{cases} \tag{1.51}$$

Once A_{ij} are found, B_{ij}, C_{ij}, D_{ij} can be computed by using Eqs. (1.21)–(1.23). Alternatively, B_{ij}, C_{ij}, D_{ij} corresponding to $k = 2, 3, 4$ can be directly computed by the similar recursive formulas as Eq. (1.31), namely,

$$l_j^{[k]}(x_i) = \begin{cases} k\left[l_i^{[k-1]}(x_i) l_j'(x_i) - \dfrac{l_j^{[k-1]}(x_i)}{x_i - x_j} \right] & (N_1 \le j \le N_2;\ i \ne j) \\[16pt] -\displaystyle\sum_{\substack{m=N_1 \\ m\ne i}}^{N_2} l_m^{[k]}(x_i) & (N_1 \le j \le N_2;\ i = j) \\[12pt] 0 & \text{for all other } j \end{cases} \tag{1.52}$$

where $l_j(x)$ is the localized Lagrange interpolation functions defined by

$$l_j(x) = \frac{(x-x_{N_1})(x-x_{N_1+1})...(x-x_{j-1})(x-x_{j+1})...(x-x_{N_2})}{(x_j-x_{N_1})(x_j-x_{N_1+1})...(x_j-x_{j-1})(x_j-x_{j+1})...(x_j-x_{N_2})} = \prod_{\substack{k=N_1 \\ k\ne j}}^{N_2} \frac{x-x_k}{x_j-x_k} \tag{1.53}$$
$$(N_1 \le j \le N_2;\ x_{N_1} \le x \le x_{N_2})$$

For simply supported or clamped ends, N_L or N_R is 1, and for free ends, N_L or N_R is 2 [37].

1.9 DIFFERENTIAL QUADRATURE TIME INTEGRATION SCHEME
1.9.1 THE METHOD OF THE DQ-BASED TIME INTEGRATION

The DQM can be also used in time integration and several schemes have been proposed. Perhaps Chen is the first one to use the DQM for the time integration [30]. Later Fung [38,39] proposed a new DQ-based time integration scheme, which is unconditionally stable, higher-order accurate, and computationally efficient. A year later, Shu et al. [40] proposed a block-marching technique with DQ discretization for initial-value problems. Chen and Tanaka [41] also proposed a DQ-based step-by-step time integration algorithm.

In this section, the scheme proposed by Fung [38,39] will be presented. Since the DQ-based time-integration scheme is reliable, computationally efficient, and also suitable for time integrations over long time duration [42,43].

In the time interval, $u(t)$ is expressed as

$$u(t) = \sum_{j=0}^{N} l_j(t) u_j \qquad t \in [0, \Delta t] \qquad (1.54)$$

where $u_j = u(t_j)$, $(N + 1)$ are the total number of grid points, t_j $(j = 0, 1, 2, ..., N)$ are the grid points in $(0, \Delta t)$, and $l_j(t)$ is the Lagrange interpolation functions. Note that the right end point $(t = \Delta t)$ is not a grid point in Eq. (1.54), that is, $0 = t_0 < t_1 < t_2 < \cdots < t_N < \Delta t$. This is quite different from the DQM in physical domain presented previously in this chapter.

The first-order derivative with respect to time t is given by

$$\frac{du(t)}{dt} = \dot{u}(t) = \sum_{j=0}^{N} \dot{l}_j(t) u_j \qquad t \in [0, \Delta t] \qquad (1.55)$$

where the over dot denotes the first-order derivative with respect to time t. From Eq. (1.55) one obtains

$$\dot{u}_i = \dot{u}(t_i) = \sum_{j=0}^{N} \dot{l}_j(t_i) u_j = \sum_{j=0}^{N} G_{ij} u_j \qquad (i = 0, 1, ..., N) \qquad (1.56)$$

where G_{ij} is the weighting coefficient of the first-order derivative with respect to time t. Note that different symbol is used to distinguish that the right end point is not used in determining the weighting coefficient of the first-order derivative with respect to time t. Since G_{ij} are the same as the A_{ij}, they can be computed explicitly by using Eq. (1.30), namely,

$$G_{ij} = \dot{l}_j(t_i) = \begin{cases} \dfrac{\omega'_N(t_i)}{(t_i - t_j)\omega'_N(t_j)} & (i \neq j) \\[3mm] \displaystyle\sum_{\substack{k=1 \\ k \neq i}}^{N} \dfrac{1}{(t_i - t_k)} & (i = j) \end{cases} \qquad (1.57)$$

In Eq. (1.57), $\omega'_N(t_i)$ and $\omega'_N(t_j)$ are computed, respectively, by

$$\omega'_N(t_i) = (t_i - t_0)(t_i - t_1) \cdots (t_i - t_{i-1})(t_i - t_{i+1}) \cdots (t_i - t_{N-1})(t_i - t_N) \qquad (1.58)$$

$$\omega'_N(t_j) = (t_j - t_0)(t_j - t_1) \cdots (t_j - t_{j-1})(t_j - t_{j+1}) \cdots (t_j - t_{N-1})(t_j - t_N) \tag{1.59}$$

In matrix form, Eq. (1.56) is given by

$$\begin{Bmatrix} \dot{u}_0 \\ \dot{u}_1 \\ \vdots \\ \dot{u}_N \end{Bmatrix} = \begin{pmatrix} G_{00} & G_{01} & \cdots & G_{0N} \\ G_{10} & G_{11} & \cdots & G_{1N} \\ \vdots & \vdots & \ddots & \vdots \\ G_{N0} & G_{N1} & & G_{NN} \end{pmatrix} \begin{Bmatrix} u_0 \\ u_1 \\ \vdots \\ u_N \end{Bmatrix} \tag{1.60}$$

Let $\{u\}^T = \begin{bmatrix} u_1 & \cdots & u_N \end{bmatrix}$ and $\{\dot{u}\}^T = \begin{bmatrix} \dot{u}_1 & \cdots & \dot{u}_N \end{bmatrix}$, where $\{u\}$ and $\{\dot{u}\}$ are the displacement vector and the velocity vector, respectively.

Taking out the first equation in Eq. (1.60), the remaining equations, that is, i taking the value of 1, 2,..., N, can be rewritten by Refs. [38,39]

$$\{\dot{u}\} = \{G_0\} u_0 + [G]\{u\} \tag{1.61}$$

where $\{G_0\}$ and $[G]$ are defined by

$$\{G_0\} = \begin{pmatrix} A_{10} \\ \vdots \\ A_{N0} \end{pmatrix}, \quad [G] = \begin{pmatrix} A_{11} & \cdots & A_{1n} \\ \vdots & \ddots & \vdots \\ A_{N1} & \cdots & A_{NN} \end{pmatrix} \tag{1.62}$$

The second-order derivative with respect to time t can be expressed in terms of u_0, v_0, and $\{u\}$ in a similar way. Replacing u_0, $\{u\}$, and $\{\dot{u}\}$ by $\dot{u}_0 = v_0$, $\{\dot{u}\}$, and $\{\ddot{u}\}$ in Eq. (1.61) and using Eq. (1.61) yield

$$\begin{aligned} \{\ddot{u}\} &= \{G_0\} \dot{u}_0 + [G]\{\dot{u}\} \\ &= \{G_0\} v_0 + [G]\{G_0\} u_0 + [G][G]\{u\} \\ &= \{G_0\} v_0 + \{GG_0\} u_0 + [GG]\{u\} \end{aligned} \tag{1.63}$$

It is shown [39] that if the abscissas of Gaussian quadrature together with $t_0 = 0$ is used as the grid points, the DQ-based time-integration scheme is unconditionally stable, higher-order accurate, and computationally efficient. Thus, t_j is computed by

$$t_j = \frac{\Delta t(1 + x_j)}{2} \qquad (j = 1, 2, ..., N) \tag{1.64}$$

where x_j is the abscissa of Gaussian quadrature in $[-1, 1]$. For example, x_j takes the values of $-\sqrt{0.6}, 0, \sqrt{0.6}$ if $N = 3$.

Remember that the right end point ($t = \Delta t$) is not used as a grid point in determining the weighting coefficients, thus, different symbol is used for the weighting coefficients of the first-order derivative with respect to time t.

For a given second-order initial-value problem, u_0 and $v_0 = \dot{u}_0$ are known. Solving the resultant algebraic equations in the time interval yields the vector $\{u\}$. Once $\{u\}$ is known, $\{\dot{u}\}$ and $\{\ddot{u}\}$ can be computed by using Eqs. (1.61) and (1.63), respectively.

In order to proceed to the next time interval, u and v at the end of time interval ($t = \Delta t$) should be found since they are the initial conditions for the next time interval. This can be done by using the method of extrapolation, namely, $u_{\Delta t}$ and $v_{\Delta t} = \dot{u}_{\Delta t}$ can be computed by [39]

$$\begin{cases} u_{\Delta t} = u(t = \Delta t) = \sum_{j=0}^{N} l_j (t = \Delta t) u_j \\ \\ v_{\Delta t} = \dfrac{du(t = \Delta t)}{dt} = \sum_{j=0}^{N} l_j (t = \Delta t) \dot{u}_j \end{cases} \tag{1.65}$$

where $\{\dot{u}\}$ is obtained by Eq. (1.61).

In this way, the time integration can be performed step by step.

1.9.2 APPLICATION AND DISCUSSION

An assessment is made on the DQ-based time-integration scheme [42,43]. To demonstrate the method, two examples cited from [42,43] are presented.

EXAMPLE 1.3

Consider a hardening elastic spring. The nonlinear dynamic equation in nondimensional form is given by

$$\ddot{u} + 100u\left(1 + 10u^2\right) = 0 \tag{1.66}$$

where the double over dots represent the second-order derivative with respect to time t.

The initial displacement u and velocity v are $u_0 = 1.5$ and $v_0 = 0.0$, respectively.

To assess the time-integration scheme, the percent error in terms of the energy is introduced [43], namely,

$$Er = \left| \frac{(E - E_0)}{E_0} \right| \times 100\% \tag{1.67}$$

where E_0 is the total energy at time $t = 0$ and the total energy E is computed by

$$E = \frac{1}{4}\left(2\dot{u}^2 + 200u^2 + 1000u^4\right) \tag{1.68}$$

Using Eq. (1.63), Eq. (1.66) becomes

$$\{G_0\}v_0 + \{GG_0\}u_0 + [GG]\{u\} + 100\{u\} + 1000\{u^3\} = 0 \tag{1.69}$$

or

$$([GG] + 100[I])\{u\} = -\{G_0\}v_0 - \{GG_0\}u_0 - 1000\{u^3\} \tag{1.70}$$

where $[I]$ is the unit matrix and $\{u^3\}^T = \lfloor u_1^3, u_2^3, ..., u_N^3 \rfloor$.

Eq. (1.70) can be further simplified as

$$\{u\} = \left[\overline{G}\right]^{-1} f(\{u\}) \tag{1.71}$$

where the definition of $[\overline{G}]$ and $f(\{u\})$ is apparent.

Since Eq. (1.71) is nonlinear algebraic equations, direct iteration is used to obtain the solutions, that is,

$$\{u\}_{n+1} = \left[\overline{G}\right]^{-1} f(\{u\}_n) \qquad (n = 0, 1, 2, ...) \tag{1.72}$$

The iteration stops if $error \leq eps = 10^{-10}$, where the *error* is defined by

$$error = \sqrt{\sum_{j=1}^{N} \left((u_j)_{n+1} - (u_j)_n\right)^2} \tag{1.73}$$

Since $[\overline{G}]$ is a constant matrix, its inverse is only performed once. To start the iteration, the initial guess could be $\{u\}_0 = \{0\}$.

For each time interval, if $error \leq eps = 10^{-10}$ is satisfied, the approximate solution is obtained. Use Eq. (1.61) to compute $\{\dot{u}\}$. Then use Eq. (1.65) to compute $u_{\Delta t}$ and $v_{\Delta t} = \dot{u}_{\Delta t}$. The total energy E_0 and the approximate total energy E at the end of each time step are given by

$$\left\{ \begin{aligned} E_0 &= \frac{1}{4}\left(2v_0^2 + 200u_0^2 + 1000u_0^4\right) \\[2em] E &= \frac{1}{4}\left(2\dot{u}_{\Delta t}^2 + 200u_{\Delta t}^2 + 1000u_{\Delta t}^4\right) \end{aligned} \right. \tag{1.74}$$

If it is necessary, then go to the next time step with $u_{\Delta t}$ and $v_{\Delta t} = \dot{u}_{\Delta t}$ as the initial conditions.

Figure 1.5 [43] shows the phase portraits of the exact solution and the ones obtained by various numerical integration methods. $\Delta t = T/20$ and time duration is $100T$ ($T = 0.15$).

In Fig. 1.5, the abscissa is displacement and the ordinate is velocity. Figure 1.5a–h represents the exact solution (Fig. 1.5a), solutions obtained by the DQ-based time integration with $N = 3$ (Fig. 1.5b), the precision integration method (Fig. 1.5c), Newmark method ($\beta = 0.25$, $\gamma = 0.5$) (Fig. 1.5d), Newmark method ($\beta = 0.3025$, $\gamma = 0.6$) (Fig. 1.5e), Wilson-θ method (Fig. 1.5f), the collocation method (Fig. 1.5g), and the HHT-α method ($\alpha = -0.1$, $\beta = 0.3025$, $\gamma = 0.6$) (Fig. 1.5h) [43].

It is seen that good agreement is achieved for the DQ-based time integration with $N = 3$, the precision integration method, and Newmark method ($\beta = 0.25$, $\gamma = 0.5$). With the $\Delta T = T/20$, however, the other four methods, that is, Newmark method ($\beta = 0.3025$, $\gamma = 0.6$), Wilson-θ method, the collocation method, and the HHT-α method ($\alpha = -0.1$, $\beta = 0.3025$, $\gamma = 0.6$) cannot give satisfied results. Smaller time step should be used although they are all unconditional stable integration schemes.

Table 1.3 summarizes the maximum percentage errors of the total energy over the time duration of $100T$ ($T = 0.15$) [43]. Different time intervals are tried.

From Table 1.3, it is seen that the DQ-based numerical-integration scheme is the best one among all methods presented in the table. Accurate results can be obtained by the DQ-based time-integration scheme with much larger time increment. Care should be taken in choosing the time step when using various unconditional stable time-integration schemes. If improper time step is used, the obtained dynamic response may not be correct for nonlinear systems.

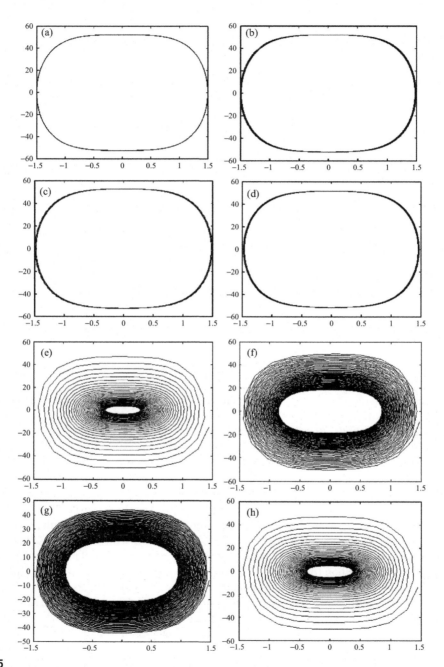

FIGURE 1.5

Phase Portraits of the Hardening Spring Obtained by Various Methods

Table 1.3 Maximum Percentage Errors of the Total Energy Over the Time Duration of 100T (T = 0.15)

Δt	T/10	T/15	T/20	T/25	T/50	T/100	T/200	T/1000
DQM ($N = 3$)	0.0	0.0	0.0	0.0	0.0	0.0	0.0	0.0
Precision int.	6.7812	3.455	1.9976	1.295	0.32889	0.082511	0.01054	0.0
Average accl.	18.596	9.6511	5.7821	3.8177	0.99718	0.25215	0.06322	0.0
Newmark	99.909	99.766	99.572	99.338	97.701	93.126	82.878	39.44
Wilson-θ	97.863	94.007	87.981	79.989	37.119	7.423	0.61655	0.0087
Collocation	95.793	90.687	83.83	75.646	36.542	6.8032	1.16215	0.0223
HHT-α	99.657	99.418	99.141	98.827	96.848	91.79	81.052	37.853

EXAMPLE 1.4

Consider a softening elastic spring. The nonlinear dynamic equation in nondimensional form is given by

$$\ddot{u} + 100\tanh(u) = 0 \tag{1.75}$$

where the double over dots represent the second-order derivative with respect to time t.

The initial displacement u and velocity v are $u_0 = 4.0$ and $v_0 = 0.0$, respectively. The total energy E is given by

$$E = \frac{1}{2}\dot{u}^2 + 100\ln\left[\cosh(u)\right] \tag{1.76}$$

Figure 1.6 [43] shows the phase portraits of the exact solution and the ones obtained by various methods. $\Delta T = T/20$ and time duration is $100T(T = 1.14)$. Similar to Fig. 1.5, the abscissa in Fig. 1.6 is displacement and the ordinate is velocity. Figure 1.6a–h represents the exact solution (Fig. 1.6a), solutions obtained by the DQ-based time-integration method with $N = 3$ (Fig. 1.6b), the precision integration method (Fig. 1.6c), Newmark method ($\beta = 0.25$, $\gamma = 0.5$) (Fig. 1.6d), Newmark method ($\beta = 0.3025$, $\gamma = 0.6$) (Fig. 1.6e), Wilson-θ method (Fig. 1.6f), the collocation method (Fig. 1.6g), and the HHT-α method ($\alpha = -0.1$, $\beta = 0.3025$, $\gamma = 0.6$) (Fig. 1.6f).

From Fig. 1.6, it is seen that good agreement is achieved for the DQ-based time integration scheme with $N = 3$ (Fig. 1.6b), the precision integration method (Fig. 1.6c), and Newmark method ($\beta = 0.25$, $\gamma = 0.5$) (Fig. 1.6d) with $\Delta T = T/20$. However, the solution accuracy for the softening spring is slightly lower than the case for the hardening spring. The other four methods, that is, Newmark method ($\beta = 0.3025$, $\gamma = 0.6$), Wilson-θ method, the collocation method, and the HHT-α method ($\alpha = -0.1$, $\beta = 0.3025$, $\gamma = 0.6$) cannot give satisfying results with $\Delta T = T/20$. Smaller time step should be used although the four methods are all unconditional stable time-integration methods.

Table 1.4 summarizes the maximum percentage errors of the total energy over the time duration of $100T$ ($T = 1.14$) [43]. Different time intervals are tried.

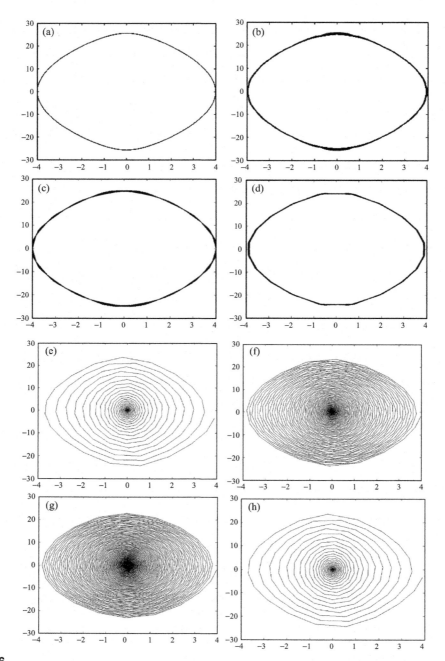

FIGURE 1.6

Phase Portraits of the Softening Spring Obtained by Various Methods

Table 1.4 Maximum Percentage Errors of the Total Energy Over the Time Duration of 100T (T = 1.14)

Δt	T/10	T/15	T/20	T/25	T/50	T/100	T/200	T/1000
DQM (N = 3)	2.5116	0.0816	0.0582	0.0	0.0	0.0	0.0	0.0
Precision int.	7.5957	9.7252	5.2867	3.2788	0.75535	0.01875	0.00619	0.0
Average accl.	19.143	8.8489	5.7309	3.8056	0.93205	0.22752	0.0184	0.0
Newmark	100	100	100	100	100	99.989	96.456	30.016
Wilson-θ	100	100	100	100	36.0	5.5	0.8	0.0223
Collocation	100	100	100	99.278	15.141	4.8713	0.85683	0.0035
HHT-α	100	100	100	100	100	99.989	96.456	30.016

In Table 1.4, the percent error in terms of the energy is computed by using Eq. (1.68) in which the total energy E_0 and the approximate total energy E at the end of each time step can, respectively, be computed by

$$\begin{cases} E_0 = \frac{1}{2}v_0^2 + 100\ln\left[\cosh\left(u_0\right)\right] \\ \\ E = \frac{1}{2}v_{\Delta t}^2 + 100\ln\left[\cosh\left(u_{\Delta t}\right)\right] \end{cases} \tag{1.77}$$

From Table 1.4, it is seen again that the DQ-based numerical integration scheme is the best one among all methods presented in the table. Accurate results can be obtained by the DQ-based time-integration scheme with much larger time increment. However, the accuracy is not as good as the one shown in Table 1.3. Similar results are observed for all other methods. Again care should be taken in choosing the time step when using various unconditional stable time-integration schemes. If improper time step is used, the dynamic response may be incorrect for nonlinear systems.

More information on the DQ-based time integration may be found in Refs. [38,39,42,43].

1.10 SUMMARY

The basic principle of the DQM or DQ method is presented in this chapter. For completeness considerations, the original DQ method to determine the weighting coefficients is included. Two types of explicit formulas to compute the weighting coefficients are given, one is based on the polynomial functions and the other is based on the harmonic functions. LaDQM, HDQM and the DQ-based time-integration scheme are also given. Various grid distributions are summarized and the error analysis of the DQM is briefly discussed.

On the basis of the results reported in literature, several conclusions may be drawn and some suggestions are given to help researchers and engineers to make a choice between a DQM and the DQ-based element method for solving practical problems.

1. For reliable and accurate considerations, nonuniform grid spacing should be used in the applications of the DQM. Among the six nonuniform grid distributions, Grid III is the most widely used grid spacing in literature perhaps due to its accuracy and convenience. However, Grid V is the most reliable grid spacing although its accuracy may not be as good as Grid III sometimes. For dynamic analysis by using the DQM, Grid V is recommended especially when the central FDM is to be used for time integration.

2. The common knowledge is that the discrete error of the DQM at the boundary point is the largest and larger discrete errors usually occur at points near the boundary. However, this is not true for the DQM with Grids II and V; the maximum discrete error occurs near the middle point. At the middle point, the discrete error of the DQM with Grid I is almost an order smaller than the one of the DQM with all other grid distributions.

3. If the number of grid points is large, the LaDQM is recommended to minimize the round-off errors as well as to raise the computational efficiency.

4. The solution accuracy between the DQM and the HDQM is similar if the number of grid points is large. This is the reason why the HDQM is not widely used by the researchers.

5. The DQ-based time-integration scheme can use a larger time step as compared with other existing time-integration schemes and is suitable for numerical integration over long time durations.

REFERENCES

[1] T.Y. Yang, Finite Element Structural Analysis, Prentice-Hall Inc., Englewood Cliffs, NJ 07632, 1986.

[2] D. Appel, N.A. Petersson, A stable finite difference method for the elastic wave equation on complex geometries with free surfaces, Commun. Comput. Phys. 5 (2009) 84–107.

[3] N. Fallah, A. Parayandeh-Shahrestany, A novel finite volume based formulation for the elasto-plastic analysis of plates, Thin Wall. Struct. 77 (2014) 153–164.

[4] C.D. Xu, The Weighted Residual Method in Solid Mechanics, Tong Ji University Press, Shanghai, China, 1987 (in Chinese).

[5] B.A. Finlayson, L.E. Scriven, The method of weighted residuals – a review, Appl. Mech. Rev. 19 (9) (1966) 735–748.

[6] S.K. Ha, C. Keilers, F.K. Chang, Finite element analysis of composite structures containing distributed piezoceramic sensors and actuators, AIAA J. 30 (1992) 772–780.

[7] C.W. Bert, M. Malik, Differential quadrature in computational mechanics: A review, Appl. Mech. Rev. 49 (1) (1996) 1–27.

[8] X. Zhang, Y. Liu, S. Ma, Meshfree methods and their applications, Adv. Mech. 39 (1) (2009) 1–36.

[9] B. Li, X. Chen, Wavelet-based numerical analysis: a review and classification, Finite Elem. Anal. Des. 81 (2014) 14–31.

[10] G.W. Wei, Discrete singular convolution for the solution of the Fokker–Planck equation, J. Chem. Phys. 110 (18) (1999) 8930–8942.

[11] C. Willberg, S. Duczek, J.M. Vivar Perez, D. Schmicker, U. Gabbert, Comparison of different higher order finite element schemes for the simulation of Lamb waves, Comput. Meth. Appl. Mech. Eng. 241–244 (2012) 246–261.

[12] D. Komatitsch, C.H. Barnes, J. Tromp, Simulation of anisotropic wave propagation based upon a spectral element method, Geophysics 4 (2000) 1251–1260.

[13] P. Kudela, M. Krawczuk, W. Ostachowicz, Wave propagation modeling in 1D structure using spectral finite elements, J. Sound Vib. 300 (2007) 88–100.

[14] L. Ge, X. Wang, C. Jin, Numerical modeling of PZT-induced Lamb wave-based crack detection in plate-like structures, Wave Motion 51 (2014) 867–885.

[15] C. Jin, X. Wang, L. Ge, Novel weak form quadrature element method with expanded Chebyshev nodes, Appl. Math. Lett. 34 (2014) 51–59.

[16] P.W. Randles, L.D. Libersky, Smoothed particle hydrodynamics: some recent improvements and applications, Comput. Meth. Appl. Mech. Eng. 139 (1996) 375–408.

[17] N. Moes, J. Dolbow, T. Belytschko, A finite element method for crack growth without remeshing, Int. J. Numer. Meth. Eng. 46 (1999) 131–150.

[18] S.L. Ho, S.Y. Yang, Wavelet-Galerkin method for solving parabolic equations in finite domains, Finite Elem. Anal. Des. 37 (2001) 1023–1037.

[19] S. Zhao, G.W. Wei, Y. Xiang, DSC analysis of free-edged beams by an iteratively matched boundary method, J. Sound Vib. 284 (2005) 487–493.

[20] W.X. Wu, C. Shu, C.M. Wang, Mesh-free least-squares-based finite difference method for large amplitude free vibration analysis of arbitrarily shaped thin plates, J. Sound Vib. 317 (2008) 955–974.

[21] J.W. Xiang, X.F. Chen, Z.J. He, Y.H. Zhang, A new wavelet-based thin plate element using B-spline wavelet on the interval, Comput. Math. 41 (2008) 243–255.

[22] L. Jameson, T. Adachi, O. Ukai, A. Yuasa, Wavelet-based numerical methods, Int. J. Comput. Fluid Dyn. 10 (1998) 267–280.

[23] R.E. Bellman, J. Casti, Differential quadrature and long-term integration, J. Math. Anal. Appl. 34 (1971) 235–238.

[24] R.E. Bellman, B.G. Kashef, J. Casti, Differential quadrature: a technique for the rapid solution of nonlinear partial differential equations, J. Comput. Phys. 10 (1972) 40–52.

[25] C.W. Bert, S.K. Jang, A.G. Striz, Two new approximate methods for analyzing free vibration of structural components, AIAA J. 26 (1988) 612–618.

[26] C.W. Bert, X. Wang, A.G. Striz, Differential quadrature for static and free vibration analyses of anisotropic plates, Int. J. Solids Struct. 30 (13) (1993) 1737–1744.

[27] Y. Wang, Differential quadrature method and differential quadrature element method – theory and applications. PhD Dissertation. Nanjing University of Aeronautics & Astronautics, China, 2001 (in Chinese).

[28] X. Wang, Differential quadrature in the analysis of structural components, Adv. Mech. 25 (2) (1995) 232–240.

[29] C. Shu, Differential Quadrature and Its Application in Engineering, Springer-Verlag, London, 2000.

[30] W. Chen, Differential quadrature method and its applications to structural engineering – applying special matrix product to nonlinear computations. PhD Dissertation. Shanghai Jiaotong University, Shanghai, China, 1997.

[31] C.H.W. Ng, Y.B. Zhao, G.W. Wei, Comparison of the DSC and GDQ methods for the vibration analysis of plates, Comput. Meth. Appl. Mech. Eng. 193 (2004) 2483–2506.

[32] A.N. Sherbourne, M.D. Pandey, Differential quadrature method in the buckling analysis of beams and composite plates, Comput. Struct. 40 (4) (1991) 903–913.

[33] B. He, X. Wang, Error analysis in differential quadrature method, T. NUAA 11 (2) (1994) 194–200.

[34] A.G. Striz, X. Wang, C.W. Bert, Harmonic differential quadrature method and applications to structural components, Acta Mech. 111 (1995) 85–94.

[35] P.J. Davis, P. Rabinowitz, Methods of Numerical Integration, Academic Press, New York, 1975.

[36] Y. Wang, Y.B. Zhao, G.W. Wei, A note on the numerical solution of high-order differential equations, J. Comput. Appl. Math. 159 (2003) 387–398.

[37] L. Zhang, Y. Xiang, G.W. Wei, Local adaptive differential quadrature for free vibration analysis of cylindrical shells with various boundary conditions, Int. J. Mech. Sci. 48 (2006) 1126–1138.

[38] T.C. Fung, Solving initial value problem by differential quadrature method – part 1: first-order equations, Int. J. Numer. Meth. Eng. 50 (2001) 1411–1427.

[39] T.C. Fung, Solving initial value problems by differential quadrature method – Part 2: second- and higher-order equations, Int. J. Numer. Meth. Eng. 50 (2001) 1429–1454.

[40] C. Shu, Q. Yao, K.S. Yeo, Block-marching in time with DQ discretization: an efficient method for time-dependent problems, Comput. Meth. Appl. Mech. Eng. 191 (2002) 4587–4597.

[41] W. Chen, M. Tanaka, A study on time schemes for DRBEM analysis of elastic impact wave, Comput. Mech. 28 (2002) 331–338.

[42] J. Liu, X. Wang, An assessment of the differential quadrature time integration scheme for non-linear dynamic equations, J. Sound Vib. 314 (2008) 246–253.

[43] J. Liu, On the differential quadrature method for analyzing dynamic problems. Master Thesis. Nanjing University of Aeronautics & Astronautics, China, 2007 (in Chinese).

DIFFERENTIAL QUADRATURE ELEMENT METHOD

2.1 INTRODUCTION

Although the DQM is an efficient method for analyzing problems in the area of structural mechanics, some critical shortcomings exist in the conventional DQM that have limited its application range. The DQM cannot be efficiently used to solve problems under discontinuously distributed loads or with sharp changes in geometry as well as with complex structures, such as uniform or stepped beams under general loadings, and frame structures. Therefore, the strong-form differential quadrature element method (DQEM), one kind of DQ-based element methods, is proposed by the author and his coworkers [1–6] to overcome these shortcomings. The DQEM is also called the new version of the differential quadrature method in [4–6]. It retains all advantages of the conventional DQM, such as the compactness and computational efficiency, and can be effectively used to solve structural mechanics problems with discontinuously distributed loads or changes in geometry. Besides, the method provides a new way to apply multiple boundary conditions if one element is used for the entire domain [7].

In this chapter, the basic principle of the DQEM and two different formulations are presented. One formulation is based on the Hermite interpolation and the other is based on the Lagrange interpolation. For the formulation of a DQ rectangular plate element, the mixed Hermite interpolation with Lagrange interpolation can also be used. Assemblage procedures are given and several examples are worked out in detail for illustrations.

2.2 DIFFERENTIAL QUADRATURE ELEMENT METHOD

In structural analysis, it is a common case for structural members under discontinuous loads or with discontinuous geometry. Fig. 2.1 shows a simply supported beam subjected to a concentrated load P and a partially distributed load q, Fig. 2.2 is an example of beam with multiple stepped changes in cross-section, and Fig. 2.3 is a fixed-plane portal frame. Although the DQM has advantages of compactness and computational efficiency, it cannot analyze these simple problems effectively.

The principle of the DQEM is similar to the well-known finite element method (FEM). In solving structural mechanics problems with discontinuous load or geometry by using the DQEM, the structure is divided into elements. For example, the beam shown in Fig. 2.1 is divided into three DQ beam elements at the locations where the concentrated load P is applied or where the discontinuity of the distributed load q occurs; the structure contains three beam elements, namely, element 1–2, element 2–3, and element 3–4. For the stepped beam shown in Fig. 2.2, the beam is divided into four DQ beam elements, namely, element 1–2, element 2–3, element 3–4, and element 4–5. For the clamped-plane portal frame shown in Fig. 2.3, the frame is divided into three DQ beam elements, namely, element 1–2, element 2–3, and element 3–4. Different from the conventional FEM, the end node of a DQ

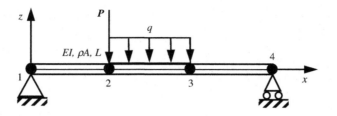

FIGURE 2.1

A Simply Supported Beam Under Discontinuous Loads

element is only placed at the locations where a concentrated load is applied or where the discontinuity of the distributed load q occurs (Fig. 2.1), or where stepped changes in cross-section occur (Fig. 2.2), or a connection point in a structure (Fig. 2.3) and certainly the boundary point.

For simplicity and without loss of the generality, the procedures to formulate a DQ beam element are presented. The idea can be readily extended to a DQ plate element or DQ shell element. Fig. 2.4 shows a seven-node DQ beam element. The two end nodes, that is, node 1 and node 7, have two degrees of freedom (DOFs) $\left(w_1, w_1^{[1]} = \theta_1 = (dw/dx)_{x=x_1}, \; w_7, w_7^{[1]} = \theta_7 = (dw/dx)_{x=x_7}\right)$; all inner nodes, that is, nodes 2–6, have only one DOF, namely, w_2, w_3, w_4, w_5, w_6. The positive direction of w is in the z-direction and the positive direction of θ, viewed as a vector, is in the negative y-direction. Here x–y–z is the right-hand Cartesian coordinate system and the y-axis is not shown in Fig. 2.4. For a DQ beam element, the number of DOFs at end points is equal to the number of boundary conditions. The rule can be extended to the DQ element based on higher-order differential equations. Three or four DOFs, namely, $w, w^{[1]}, w^{[2]}$ or $w, w^{[1]}, w^{[2]}, w^{[3]}$, should be placed on the end points for the sixth- or eighth-order governing equations. In such a way, the multiple boundary conditions can be easily applied by using the DQEM which will be discussed in Chapter 3.

For a uniform Bernoulli–Euler beam subjected to general loadings, the governing equation for static, buckling, and free vibration analysis is given by

$$EI \frac{d^4w}{dx^4} + P \frac{d^2w}{dx^2} - q(x) - \rho A \omega^2 w = 0 \quad x \in (0, L) \tag{2.1}$$

where $E, I, L, q, P, \rho, A, \omega$, and w are the elasticity modulus, principal moment of inertia about the y-axis, beam length, distributed load, axial compressive load, mass density, cross-sectional area, circular

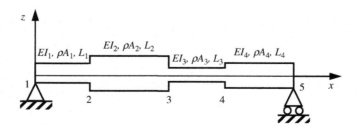

FIGURE 2.2

A Simply Supported Beam with Three-Stepped Changes in Cross-Section

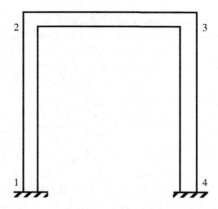

FIGURE 2.3

Sketch of a Clamped Plane Portal Frame

frequency, and deflection, respectively. The shear force Q and bending moment M at the beam ends are given by

$$EI\frac{d^3w}{dx^3} + P\frac{dw}{dx} = EIw^{[3]} + Pw^{[1]} = Q$$

$$EI\frac{d^2w}{dx^2} = EIw^{[2]} = M$$

(2.2)

Eqs. (2.1) and (2.2) are used to formulate the stiffness equation of a DQ beam element. Similar to the FEM, the DQ element stiffness equation can be symbolically written as

$$\left[K^e\right]\left\{u^e\right\} = \left\{F^e\right\}$$

(2.3)

where superscript e means element, $[K], \{u\}, \{F\}$ are element-stiffness matrix, displacement vector, and load vector, respectively.

Take the static analysis as an example, that is, $P = \omega = 0$. $\left\{u^e\right\}, \left\{F^e\right\}$ are given by

$$\left\{u^e\right\} = \left\{\begin{array}{ccccccccccc} w_1 & \theta_1 & w_2 & w_3 & w_4 & w_5 & w_6 & w_7 & \theta_7 \end{array}\right\}^T$$

$$\left\{F^e\right\} = \left\{\begin{array}{ccccccccccc} Q_1 & M_1 & q_2 & q_3 & q_4 & q_5 & q_6 & Q_7 & M_7 \end{array}\right\}^T$$

(2.4)

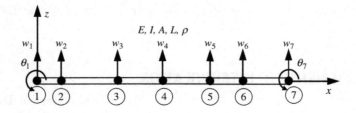

FIGURE 2.4

Sketch of a Seven-Node DQ-Beam Element

in which, superscript T denotes the transpose, and Q, M, and $q_i = q(x_i)$ are the shearing force, bending moment, and distributed load, respectively. It should be emphasized that the positive direction of the two shearing forces and distributed load is assumed along the positive z-direction, and the positive direction of the two bending moments is assumed along the negative y-direction. In other words, the positive direction of the generalized forces is the same as the direction of corresponding generalized displacements. The rule is exactly the same as the one used in the conventional FEM. The arrangement of displacement vector and force vector in the form of Eq. (2.4) is for convenience during the assemblage process. This is also similar to the FEM. Other arrangements can be used, especially if only one DQ beam element is used in the analysis.

The DQ beam element stiffness matrix $[K^e]$ is formulated by using the DQ rule, different from the conventional FEM, spectral element method (SEM), and weak-form quadrature element method (QEM). The elements in the stiffness matrix $[K^e]$ are actually the weighting coefficients obtained by using the differential quadrature rule. This is the main reason to call the beam element the DQ beam element. For the seven-node beam element, the stiffness matrix $[K^e]$ is given by

$$[K^e] = EI \begin{bmatrix} C_{11} & C_{12} & C_{13} & C_{14} & C_{15} & C_{16} & C_{17} & C_{18} & C_{19} \\ -B_{11} & -B_{12} & -B_{13} & -B_{14} & -B_{15} & -B_{16} & -B_{17} & -B_{18} & -B_{19} \\ D_{21} & D_{22} & D_{23} & D_{24} & D_{25} & D_{26} & D_{27} & D_{28} & D_{29} \\ D_{31} & D_{32} & D_{33} & D_{34} & D_{35} & D_{36} & D_{37} & D_{38} & D_{39} \\ D_{41} & D_{42} & D_{43} & D_{44} & D_{45} & D_{46} & D_{47} & D_{48} & D_{49} \\ D_{51} & D_{52} & D_{53} & D_{54} & D_{55} & D_{56} & D_{57} & D_{58} & D_{59} \\ D_{61} & D_{62} & D_{63} & D_{64} & D_{65} & D_{66} & D_{67} & D_{68} & D_{69} \\ -C_{71} & -C_{72} & -C_{73} & -C_{74} & -C_{75} & -C_{76} & -C_{77} & -C_{78} & -C_{79} \\ B_{71} & B_{72} & B_{73} & B_{74} & B_{75} & B_{76} & B_{77} & B_{78} & B_{79} \end{bmatrix} \tag{2.5}$$

where B_{ij}, C_{ij}, D_{ij} are weighting coefficients of the second-, third-, and fourth-order derivatives with respect to x at nodal point i; the negative sign is due to the sign convention of the shearing force and bending moment. Different from the FEM, the stiffness matrix of a DQ element is a nonsymmetric matrix.

Currently, there are two ways available in literature to formulate the weighting coefficients. One way uses the Lagrange interpolation and the method is called the new version of differential quadrature method [4–6] or the modified differential quadrature method. The other uses the Hermite interpolation and the method is called the DQEM [1–3] or generalized differential quadrature rule (GDQR) [8,9]. Explicit formulations to compute the weighting coefficients are available [3] to increase the solution accuracy as well as the efficiency of the DQEM. The details in derivations will be given in the following two sections.

For an N-node DQ beam element, the arrangement of $\{u^e\}, \{F^e\}$ is similar, that is, the first two DOFs are always the DOFs of node 1, and the last two DOFs are always the DOFs of node N. The formulation of the stiffness matrix is also similar to Eq. (2.5).

2.3 DQEM WITH HERMITE INTERPOLATION

Without loss of generality, consider an N-node DQ Euler–Bernoulli beam element. Deflection function $w(x)$ is continuous and differentiable with respect to x in the domain $[0, L]$. According to the differential quadrature rule, one has

$$L\{w(x)\}_{x=x_i} = \sum_{j=1}^{N+2} E_{ij}u_j \quad (i=1,2,...,N) \tag{2.6}$$

where L is the linear operator, $x_j(j=1,2,...,N)$ are N discrete points in the entire domain of the beam element including the two end points, that is, $0=x_1 < x_2 < \cdots < x_N = L$, E_{ij} are the weighting coefficients of the corresponding linear operator, and $u_j = w(x_j)(j=1,2,...,N)$, $u_{N+1}=\theta_1$, $u_{N+2}=\theta_N$. Note that the arrangement of u_j is slightly different from the one shown in Eq. (2.4). The obvious difference between Eq. (1.9) and Eq. (2.6) is that the range of summation changes from N to $N+2$.

It is well known that $w(x)$ can be expressed by using the Hermite interpolation functions once $x_j(j=1,2,...,N)$ and $u_j(j=1,2,...,N+2)$ are given, namely [3],

$$w(x) \approx \varphi_1(x)w(x_1) + \cdots + \varphi_N(x)w(x_N) + \psi_1(x)w^{[1]}(x_1) + \psi_N(x)w^{[1]}(x_N)$$
$$= \sum_{j=1}^{N} \varphi_j(x)w(x_j) + \psi_1(x)\theta(x_1) + \psi_N(x)\theta(x_N) \tag{2.7}$$

where $\varphi_j(x)$ and $\psi_j(x)$ are $(N+1)$th order polynomials. Function $w(x)$ should satisfy all conditions listed in Table 2.1 [3] and the interpolation is called the Hermite interpolation.

In order to derive the formula to compute explicitly the weighting coefficients of the DQEM, expressions of $\varphi_j(x)$ and $\psi_j(x)$ should be constructed. Define function $\psi_j(x)$ by

$$\psi_j(x) = \frac{1}{x_j - x_{N-j+1}} l_j(x)(x-x_j)(x-x_{N-j+1}) \quad (j=1,N) \tag{2.8}$$

where $l_j(x) = \prod_{\substack{k=1 \\ k \neq j}}^{N} \frac{x-x_k}{x_j-x_k}$ is the Lagrange interpolation function.

Differential Eq. (2.8) with respect to x results

$$\psi_j^{[1]}(x) = \frac{1}{x_j - x_{N-j+1}}\left\{ l_j^{[1]}(x)(x-x_j)(x-x_{N-j+1}) + l_j(x)\left[(x-x_j)+(x-x_{N-j+1})\right]\right\}(j=1,N) \tag{2.9}$$

Since $l_j(x_i) = \delta_{ij}$, where δ_{ij} is the well-known Kronecker symbol, substituting x_i into Eqs. (2.8) and (2.9) results

$$\begin{aligned}
\psi_j(x_i) &= 0 \quad (i=1,2,\cdots,N; \ j=1,N) \\
\psi_1^{[1]}(x_1) &= \psi_N^{[1]}(x_N) = 1 \\
\psi_1^{[1]}(x_N) &= \psi_N^{[1]}(x_1) = 0
\end{aligned} \tag{2.10}$$

Table 2.1 Conditions Satisfied by the Interpolation Function $w(x)$

x	x_1	x_2	$\cdots$	x_i	$\cdots$	x_N
$w(x)$	$w(x_1)$	$w(x_2)$	$\cdots$	$w(x_i)$	$\cdots$	$w(x_N)$
$w^{[1]}(x)$	$w^{[1]}(x_1)$	—	—	—	—	$w^{[1]}(x_N)$

To satisfy all conditions listed in Table 2.1, define function $\varphi_j(x)$ by

$$\varphi_j(x) = \begin{cases} \dfrac{1}{x_j - x_{N-j+1}} l_j(x)(x - x_{N-j+1}) - \left[l_j^{[1]}(x_j) + \dfrac{1}{x_j - x_{N-j+1}} \right] \psi_j(x) & (j = 1, N) \\[4mm] \dfrac{1}{(x_j - x_1)(x_j - x_N)} l_j(x)(x - x_1)(x - x_N) & (j = 2, 3, \cdots, N-1) \end{cases}$$

(2.11)

It is easy to show that the following equations are valid, namely,

$$\begin{aligned} \varphi_j(x_i) &= \delta_{ij} \\ \varphi_j^{[1]}(x_1) &= 0 \quad (i, j = 1, 2, \cdots N) \\ \varphi_j^{[1]}(x_N) &= 0 \end{aligned}$$

(2.12)

In other words, Eq. (2.7) satisfies all conditions listed in Table 2.1 if $\varphi_j(x)$ and $\psi_j(x)$ are defined by Eqs. (2.11) and (2.8).

By using Eq. (2.7), the k-th-order derivative of function $w(x)$ can be computed by

$$\begin{aligned} w^{[k]}(x_i) &\approx \sum_{j=1}^{N} \varphi_j^{[k]}(x_i) w(x_j) + \psi_1^{[k]}(x_i) w^{[1]}(x_1) + \psi_N^{[k]}(x_i) w^{[1]}(x_N) \\ &= \sum_{j=1}^{N+2} E_{ij} u_j \quad (i = 1, 2, \ldots, N) \end{aligned}$$

(2.13)

where,

$$\psi_j^{[k]}(x_i) = \begin{cases} \dfrac{1}{x_j - x_{N-j+1}} \left\{ l_j^{[1]}(x_i)(x_i - x_1)(x_i - x_N) + l_j(x_i)\left[(x_i - x_1) + (x_i - x_N)\right] \right\} \\[2mm] \hspace{8cm} k = 1 \\[3mm] \dfrac{1}{x_j - x_{N-j+1}} \left\{ l_j^{[k]}(x_i)(x_i - x_1)(x_i - x_N) + k l_j^{[k-1]}(x_i)\left[(x_i - x_1) + (x_i - x_N)\right] \right. \\[2mm] \left. \hspace{3cm} + k(k-1) l_j^{[k-2]}(x_i) \right\} \\[3mm] \hspace{8cm} k \geq 2 \\ \hspace{5cm} (j = 1, N; \ i = 1, 2, \cdots, N) \end{cases}$$

(2.14)

$$\varphi_j^{[k]}(x_i) = \dfrac{1}{x_j - x_{N-j+1}} \left[l_j^{[k]}(x_i)(x_i - x_{N-j+1}) + k l_j^{[k-1]}(x_i) \right] - \left[l_j'(x_j) + \dfrac{1}{x_j - x_{N-j+1}} \right] \psi_j^{[k]}(x_i)$$

$$(j = 1, N; i = 1, 2, \cdots, N)$$

(2.15)

and

$$\varphi_j^{[k]}(x_i) = \dfrac{1}{(x_j - x_1)(x_j - x_N)} \left\{ l_j^{[k]}(x_i)(x_i - x_1)(x_i - x_N) + k l_j^{[k-1]}(x_i)\left[(x_i - x_1) + (x_i - x_N)\right] \right.$$

$$\left. + k(k-1) l_j^{[k-2]}(x_i) \right\} \quad (j = 2, 3, \cdots, N-1; i = 1, 2, \cdots, N)$$

(2.16)

In Eqs. (2.14)–(2.16), $l_j^{[1]}(x_i)$ is given by

$$l_j^{[1]}(x_i) = \begin{cases} \dfrac{1}{x_j - x_i} \displaystyle\prod_{\substack{m=1 \\ m \neq i,j}}^{N} \dfrac{x_i - x_m}{x_j - x_m} & (i \neq j) \\[3mm] \displaystyle\sum_{\substack{m=1 \\ m \neq i}}^{N} \dfrac{1}{x_i - x_m} & (i = j) \end{cases} \qquad (2.17)$$

and $l_j^{[k]}(x_i)\,(k > 1)$ is given by

$$l_j^{[k]}(x_i) = \begin{cases} k\left[l_j'(x_i) l_i^{[k-1]}(x_i) - \dfrac{l_j^{[k-1]}(x_i)}{x_i - x_j} \right] & (i \neq j) \\[3mm] \displaystyle\sum_{m=1}^{N} l_j'(x_m) l_m^{[k-1]}(x_j) & (i = j) \end{cases} \qquad (2.18)$$

Detailed derivations can be found in reference [3]. A FORTRAN subroutine is given in Appendix II (SUBROUTINE DQEM) for readers' reference. Since the arrangement of $\{u^e\}$ in Eq. (2.3) and $\{u\}$ in Eq. (2.6) is slightly different, the rearrangement is made for $A_{ij}, B_{ij}, C_{ij}, D_{ij}$ in the subroutine according to the arrangement of $\{u^e\}$. The converted MATLAB function is also provided.

It is seen that Eqs. (2.14)–(2.18) for explicitly computing the weighting coefficients of the DQEM are far more complicated than Eqs. (1.30)–(1.31) for computing the weighting coefficients of the DQM. Therefore, other way, much simpler than the method based on the Hermite interpolation, has been proposed to compute the weighting coefficients explicitly and is presented in the following section.

2.4 DQEM WITH LAGRANGE INTERPOLATION

For the conventional DQM, the explicit formulations derived based on the Lagrange interpolation are given by Eqs. (1.30) and (1.31). For N grid points, the order of Lagrange interpolation functions is $(N\text{-}1)$th-order polynomials, different from the Hermite interpolation functions.

To solve the difficulty in applying the multiple boundary conditions by using the conventional DQM, various methods have been proposed [7] and will be discussed in detail in Chapter 3. Among them, two methods, called the method of modification of weighting coefficient-3 (MMWC-3) and the method of modification of weighting coefficient-4 (MMWC-4), can also be called the DQEM, since they can be used as an element and the form of the stiffness equation is exactly the same as Eq. (2.3). The only difference from the DQEM presented in the previous section is the way to determine the weighting coefficients.

Similar to the DQEM presented in Section 2.2, $w_1^{[1]}$ and $w_N^{[1]}$ are introduced as additional DOFs. However, the weighting coefficients of the first- and second-order derivatives are the same as the ones in conventional DQM, since Lagrange interpolation functions are used. The weighting coefficients of the second-order derivative at the end points are slightly modified in order to introduce the two additional DOFs. The weighting coefficients of the third- and the fourth-order derivatives are computed by using Eqs. (1.22) and (1.23) but based on the modified weighting coefficients of the second-order derivatives.

Denote the weighting coefficients of the first-, second-, third- and fourth-order derivatives by $A_{ij}, B_{ij}, C_{ij}, D_{ij}$ in the DQEM with Lagrange interpolation. Denote the weighting coefficients of the first- and second-order derivatives by $\bar{A}_{ij}$ and $\bar{B}_{ij}$ in the conventional DQM. The $\bar{A}_{ij}$ can be computed explicitly by using Eq. (1.30), namely,

$$\bar{A}_{ij} = \begin{cases} \displaystyle\prod_{\substack{k=1 \\ k\neq i,j}}^{N}(x_i - x_k) \Big/ \prod_{\substack{k=1 \\ k\neq j}}^{N}(x_j - x_k) & (i \neq j) \\ \displaystyle\sum_{\substack{k=1 \\ k\neq i}}^{N}\frac{1}{(x_i - x_k)} & (i = j) \end{cases} \tag{2.19}$$

and $\bar{B}_{ij}$ can be computed explicitly by Eq. (1.21) as follows:

$$\bar{B}_{ij} = \sum_{k=1}^{N}\bar{A}_{ik}\bar{A}_{kj} \quad (i,j = 1,2,\cdots,N) \tag{2.20}$$

For convenience, the introduced two additional DOFs, $\theta_1 = \omega_1^{[1]}, \theta_N = w_N^{[1]}$, are placed at the end of the displacement vector $\{u\}$, namely, $u_j = w(x_j)\,(j=1,2,\ldots,N)$, $u_{N+1} = \theta_1$ and $u_{N+2} = \theta_N$. Then A_{ij} are modified as follows:

$$A_{ij} = \begin{cases} \bar{A}_{ij} & (i,j=1,2,\ldots,N) \\ 0 & (i=1,2,\ldots,N,\ j=N+1,N+2) \end{cases} \tag{2.21}$$

For all inner grid points $(i=2,3,\ldots,N-1)$, B_{ij} are defined by

$$B_{ij} = \begin{cases} \bar{B}_{ij} & (j=1,2,\ldots,N) \\ 0 & (j=N+1,N+2) \end{cases} \tag{2.22}$$

B_{ij} at two end points $(i=1,N)$ are computed differently as follows,

$$B_{ij} = \sum_{k=2}^{N-1}\bar{A}_{ik}\bar{A}_{kj} \quad (j=1,2,\ldots,N),\ B_{i(N+1)} = \bar{A}_{i1},\ B_{i(N+2)} = \bar{A}_{iN} \tag{2.23}$$

Eq. (2.23) is derived based on the fact that the second-order derivative with respect to x at the two end points can be computed by using the DQ rule, namely,

$$\begin{aligned} w_i^{[2]} &= \sum_{k=1}^{N}\bar{A}_{ik}w_k^{[1]} = \sum_{k=2}^{N-1}\bar{A}_{ik}w_k^{[1]} + \bar{A}_{i1}w_1^{[1]} + \bar{A}_{iN}w_N^{[1]} \\ &= \sum_{k=2}^{N-1}\bar{A}_{ik}\sum_{j=1}^{N}\bar{A}_{kj}w_j + \bar{A}_{i1}w_1^{[1]} + \bar{A}_{iN}w_N^{[1]} = \sum_{j=1}^{N+2}B_{ij}u_j \quad (i=1,N) \end{aligned} \tag{2.24}$$

With the modified B_{ij}, the weighting coefficients of the third- and the fourth-order derivatives with respect to x, that is, C_{ij}, D_{ij}, can be computed by

$$C_{ij} = \sum_{k=1}^{N}\bar{A}_{ik}B_{kj} \quad (i=1,2,\ldots,N,\quad j=1,2,\ldots,N+2) \tag{2.25}$$

and

$$D_{ij} = \sum_{k=1}^{N} \bar{B}_{ik} B_{kj} \quad (i=1,2,...,N, \quad j=1,2,...,N+2) \tag{2.26}$$

respectively, since the third- and fourth-order derivatives with respect to x can be expressed by using the DQ rule as follows,

$$w_i^{[3]} = \sum_{k=1}^{N} \bar{A}_{ik} w_k^{[2]} \quad (i=1,2,...,N) \tag{2.27}$$

$$w_i^{[4]} = \sum_{k=1}^{N} \bar{B}_{ik} w_k^{[2]} \quad (i=1,2,...,N) \tag{2.28}$$

Because $\bar{A}_{ij}$ and $\bar{B}_{ij}$ are computed by using the conventional DQ rule, the weighting coefficients of the DQEM in this section are computed based on the Lagrange interpolations. It is seen that Eq. (2.24) is the key step to introduce $w_1^{[1]}$ and $w_N^{[1]}$ as the additional DOFs. By this simple way, the order of polynomials has been raised from $(N-1)$ to $(N+1)$. This fact is easy to be verified by directly computing the weighting coefficients of the fourth-order derivative, D_{ij}. For conventional DQM, $\bar{D}_{ij} \equiv 0$ if $N < 5$. For the DQEM, however, $D_{ij} \neq 0$ if $N > 2$.

The way to introduce the two additional DOFs is called MMWC-3 [7]. Actually, another simple way, called MMWC-4, can also be used to introduce the additional DOF. Instead of Eqs. (2.23) and (2.24), B_{ij} at two end points $(i=1,N)$ can be computed differently as

$$\begin{cases} B_{1j} = \sum_{k=2}^{N} \bar{A}_{1k} \bar{A}_{kj} \quad (j=1,2,...,N), B_{1(N+1)} = \bar{A}_{11}, B_{1(N+2)} = 0 \\ B_{Nj} = \sum_{k=1}^{N-1} \bar{A}_{Nk} \bar{A}_{kj} \quad (j=1,2,...,N), B_{N(N+1)} = 0, B_{N(N+2)} = \bar{A}_{NN} \end{cases} \tag{2.29}$$

since the second-order derivative with respect to x can be expressed by using the DQ rule as

$$\begin{cases} \begin{aligned} w_1^{[2]} &= \sum_{k=1}^{N} \bar{A}_{1k} w_k^{[1]} = \sum_{k=2}^{N} \bar{A}_{1k} w_k^{[1]} + \bar{A}_{11} w_1^{[1]} \\ &= \sum_{k=2}^{N} \bar{A}_{1k} \sum_{j=1}^{N} \bar{A}_{kj} w_j + \bar{A}_{11} w_1^{[1]} + 0 \times w_N^{[1]} = \sum_{j=1}^{N+2} B_{1j} u_j \\ w_N^{[2]} &= \sum_{k=1}^{N} \bar{A}_{Nk} w_k^{[1]} = \sum_{k=1}^{N-1} \bar{A}_{Nk} w_k^{[1]} + \bar{A}_{NN} w_N^{[1]} \\ &= \sum_{k=1}^{N-1} \bar{A}_{Nk} \sum_{j=1}^{N} \bar{A}_{kj} w_j + 0 \times w_1^{[1]} + \bar{A}_{NN} w_N^{[1]} = \sum_{j=1}^{N+2} B_{Nj} u_j \end{aligned} \end{cases} \tag{2.30}$$

The weighting coefficients of the first-order derivative (A_{ij}) and the remaining terms of the weighting coefficient of the second-order derivative (B_{ij}) are the same as defined by Eqs. (2.21) and (2.22). The weighting coefficients of the third- and the fourth-order derivatives, C_{ij}, D_{ij}, are still computed by using Eqs. (2.25) and (2.26), respectively.

2.5 ASSEMBLAGE PROCEDURES

For structural analysis by the DQEM, the assemblage procedures are similar to the conventional FEM if more than one DQ elements are used in the analysis. Physically, at each common nodal point, the generalized displacement components at the DQ element level, the end generalized displacement components in $\{u^e\}$, are the same as those after assemblage. If the generalized forces at end points in $\{F^e\}$ are regarded as internal forces, then the sum of such internal forces at each common nodal point should be equal to the corresponding external applied generalized loads. Since generalized forces and generalized displacements are vectors, the vectorial sum is usually replaced by the scale sum of their components in the common coordinate system at the connecting nodal point. Similar to the FEM, the coordinate transformation may be necessary. Remember that the coordinate transformation is only necessary for the quantities at the end points and not at inner points. Besides, the additional DOF is the first-order derivative with respect to x and not to the nondimensional coordinate $\xi\,(\xi = x/l_e)$, where l_e is the element length. This is extremely important. Otherwise incorrect results may be obtained if the DQ beam element lengths are different.

For illustrations of the assemblage procedures, consider a cantilever beam under a partially uniform distributed load and a concentrated load. The beam length is $2L$. E, A, and I are the elasticity modulus, cross-sectional area, and the principal moment of inertia about the y-axis, respectively. The uniformly distributed load is applied on the half of the beam and the concentrated load is applied at the middle point of the beam, as shown in Fig. 2.5.

The problem belongs to the type of discontinuous load. Two DQ beam elements shown in Fig. 2.5 are used in the analysis. Since no distributed load is applied at the portion of 1–2, thus a two-node DQ beam element is used in the region. The stiffness matrix is exactly the same as the one of a standard finite beam element if Eqs. (2.14)–(2.18) are used to compute the weighting coefficient. Since uniformly distributed load is applied at portion 2–3, a three-node DQ beam element is used in the region. DQ beam elements with higher number of nodes can be used too. However, the accuracy of results will remain the same, since exact solution will be obtained by using the DQEM.

For the two-node DQ beam element 1–2, Eq. (2.3) becomes

$$\frac{EI}{L^4}\begin{bmatrix} 12L & 6L^2 & -12L & 6L^2 \\ 6L^2 & 4L^3 & -6L^2 & 2L^3 \\ -12L & -6L^2 & 12L & -6L^2 \\ 6L^2 & 2L^3 & -6L^2 & 4L^3 \end{bmatrix}\begin{Bmatrix} w_1 \\ \theta_1 \\ w_2 \\ \theta_2 \end{Bmatrix}=\begin{Bmatrix} Q_1 \\ M_1 \\ Q_2 \\ M_2 \end{Bmatrix} \tag{2.31}$$

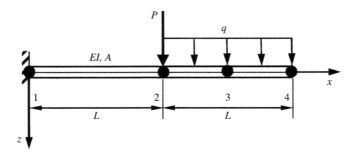

FIGURE 2.5

A Cantilever Beam Under Combined Concentrated and Partially Uniform Distributed Loads

For the three-node DQ beam element 2–4 with uniformly distributed nodal points, Eq. (2.3) becomes

$$
\frac{EI}{L^4}
\begin{bmatrix}
108L & 30L^3 & 192L & 84L & -18L^2 \\
22L^2 & 8L^3 & -32L^2 & 10L^2 & -2L^3 \\
-192 & -48L & 384 & -192 & -48L \\
84L & 18L^2 & -192L & 108L & -30L^2 \\
-10L^2 & -2L^3 & 32L^2 & -22L^2 & 8L^3
\end{bmatrix}
\begin{Bmatrix}
w_2 \\ \theta_2 \\ w_3 \\ w_4 \\ \theta_4
\end{Bmatrix}
=
\begin{Bmatrix}
Q_2 \\ M_2 \\ q_3 \\ Q_4 \\ M_4
\end{Bmatrix}
\tag{2.32}
$$

In Eqs. (2.31) and (2.32), the number of the subscript denotes the nodal number shown in Fig. 2.5. It should be emphasized that similar to the conventional FEM, the direction of all generalized forces and displacements is the same for all nodes in writing Eqs. (2.31) and (2.32).

For the problem considered, there is only one common nodal point, namely, node 2. Besides, coordinate transformation is not necessary. At the common nodal point 2, a concentrated force P is applied and no bending moment is applied externally. According the rule of the assemblage, one has

$$
\begin{cases}
w_2^{1-2} = w_2^{2-4} = w_2, & \theta_2^{1-2} = \theta_2^{2-4} = \theta_2 \\
Q_2^{1-2} + Q_2^{2-4} = P, & M_2^{1-2} + M_2^{2-4} = 0
\end{cases}
\tag{2.33}
$$

where the superscript denotes the two end nodal numbers of each DQ beam element, and the internal shear force and bending moment can be computed by Eq. (2.31) for DQ beam element 1–2 and by Eq. (2.32) for DQ beam element 2–4, respectively.

After assemblage, the stiffness equation for the entire beam structure is given by

$$
\frac{EI}{L^4}
\begin{bmatrix}
12L & 6L^2 & -12L & 6L^2 & 0 & 0 & 0 \\
6L^2 & 4L^3 & -6L^2 & 2L^3 & 0 & 0 & 0 \\
-12L & -6L^2 & 120L & 24L^2 & -192L & 84L & -18L^2 \\
6L^2 & 2L^3 & 16L^2 & 12L^3 & -32L^2 & 10L^2 & -2L^3 \\
0 & 0 & -192 & -48L & 384 & -192 & 48L \\
0 & 0 & 84L & 18L^2 & -192L & 108L & -30L^2 \\
0 & 0 & -10L^2 & -2L^3 & 32L^2 & -22L^2 & 8L^3
\end{bmatrix}
\begin{Bmatrix}
w_1 \\ \theta_1 \\ w_2 \\ \theta_2 \\ w_3 \\ w_4 \\ \theta_4
\end{Bmatrix}
=
\begin{Bmatrix}
Q_1 \\ M_1 \\ P \\ 0 \\ q \\ 0 \\ 0
\end{Bmatrix}
\tag{2.34}
$$

In writing Eq. (2.34), the sum of all internal generalized forces at each common node equals to the external applied load, and generalized forces at all other unconstrained nodes equal to the external applied loads, that is, $Q_2 = P$, $M_2 = 0$, $q_3 = q$. For the free end, $Q_4 = M_4 = 0$. For the fixed end, $w_1 = \theta_1 = 0$ and the generalized forces, Q_1 and M_1 are unknown reaction shearing force and bending moment.

The first two rows and two columns of the structural stiffness matrix in Eq. (2.34) are eliminated by applying the displacement boundary conditions of $w_1 = \theta_1 = 0$. Then the unknown displacements can be solved by matrix inversion. The results are given by

$$
\begin{Bmatrix}
w_2 \\ \theta_2 \\ w_3 \\ w_4 \\ \theta_4
\end{Bmatrix}
= \frac{L}{EI}
\begin{bmatrix}
\dfrac{L^2}{3} & \dfrac{L}{2} & \dfrac{7L^3}{12} & \dfrac{5L^2}{6} & \dfrac{L}{2} \\[2mm]
\dfrac{L}{2} & 1 & L^2 & \dfrac{3L}{2} & 1 \\[2mm]
\dfrac{7L^2}{12} & L & \dfrac{433L^3}{384} & \dfrac{27L^2}{16} & \dfrac{9L}{8} \\[2mm]
\dfrac{5L^2}{6} & \dfrac{3L}{2} & \dfrac{41L^3}{24} & \dfrac{8L^2}{3} & 2L \\[2mm]
\dfrac{L}{2} & 1 & \dfrac{7L^2}{6} & 2L & 2
\end{bmatrix}
\begin{Bmatrix}
P \\ 0 \\ q \\ 0 \\ 0
\end{Bmatrix}
= \frac{PL^3}{EI}
\begin{Bmatrix}
\dfrac{1}{3} \\[2mm] \dfrac{1}{2L} \\[2mm] \dfrac{7}{12} \\[2mm] \dfrac{5}{6} \\[2mm] \dfrac{1}{2L}
\end{Bmatrix}
+ \frac{qL^4}{EI}
\begin{Bmatrix}
\dfrac{7}{12} \\[2mm] \dfrac{1}{L} \\[2mm] \dfrac{433}{384} \\[2mm] \dfrac{41}{24} \\[2mm] \dfrac{7}{6L}
\end{Bmatrix}
\tag{2.35}
$$

As is expected, exact solution is obtained by using the DQEM with only two elements. All other types of discontinuous loads can be treated in a similar way without any difficulty.

The reaction shearing force and bending moment at node 1 can be obtained by substituting the known displacements into Eq. (2.34), namely,

$$\left\{ \begin{matrix} Q_1 \\ M_1 \end{matrix} \right\} = \frac{EI}{L^3} \left[\begin{matrix} -12 & 6L \\ -6L & 2L^2 \end{matrix} \right] \left(\frac{PL^3}{EI} \left\{ \begin{matrix} \frac{1}{3} \\ \frac{1}{2L} \end{matrix} \right\} + \frac{qL^4}{EI} \left\{ \begin{matrix} \frac{7}{12} \\ \frac{1}{L} \end{matrix} \right\} \right) = - \left\{ \begin{matrix} P+qL \\ PL + \frac{3qL^2}{2} \end{matrix} \right\} \tag{2.36}$$

It is easy to verify that both Q_1 and M_1 are exact. The negative sign means that the direction is opposite to the assumed direction in deriving the element stiffness equation, that is, the direction of Q_1 is opposite to the positive z-axis and the direction of M_1 is along with the positive y-axis.

To illustrate the assemblage procedures further, consider the free vibration of a simply supported stepped beam shown in Fig. 2.6. For simplicity, take a seven-node uniform DQ beam element shown in Fig. 2.4 as an example. Three DQ beam elements shown in Fig. 2.6, that is, element 1–7, 7–13, and 13–19, are required for the analysis. Totally 19 nodes are involved.

For free vibration analysis, $P = q = 0$, the stiffness equation for a seven-node uniform DQ Euler–Bernoulli beam element is symbolically the same as Eq. (2.3), namely,

$$\left[K^e \right] \left\{ u^e \right\} = \left\{ F^e \right\} \tag{2.37}$$

where $[K^e]$ is given by Eq. (2.5), $\{u^e\}$ and $\{F^e\}$ are given by Eq. (2.4). In Eq. (2.4), q_i $(i = 2,3,4,5,6)$ are now replaced by

$$q_i = \rho A \omega^2 w_i \quad (i = 2,3,...,6) \tag{2.38}$$

where ω is the circular frequency. $L, E, I, A,$ and ρ in Eq. (2.37) are the length, modulus of elasticity, principal moment of inertia about the y-axis, cross-sectional area, and the mass density of the beam element, respectively. For uniform Euler–Bernoulli beam, the bending rigidity EI and mass per unit length ρA are constant.

To formulate the stiffness equation of a beam structure shown in Fig. 2.6, the assemblage procedures are exactly the same as the ones in conventional FEM. In details, (1) the displacement and the slope are the same at common nodes, that is,

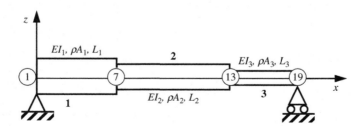

FIGURE 2.6

Sketch of a Two-Stepped S–S Beam

$$\begin{cases} w_7^{1-7} = w_7^{7-13} = w_7, & \theta_7^{1-7} = \theta_7^{7-13} = \theta_7 \\ w_{13}^{7-13} = w_{13}^{13-19} = w_{13}, & \theta_{13}^{7-13} = \theta_{13}^{13-19} = \theta_{13} \end{cases}$$

(2.39)

and (2) the resultants of the generalized internal forces at common nodes are equal to the corresponding generalized external forces applied at the common nodes, namely,

$$\begin{cases} Q_7^{1-7} + Q_7^{7-13} = Q_7, & M_7^{1-7} + M_7^{7-13} = M_7 \\ Q_{13}^{7-13} + Q_{13}^{13-19} = Q_{13}, & M_{13}^{7-13} + M_{13}^{13-19} = M_{13} \end{cases}$$

(2.40)

where the superscript denotes the two end nodal numbers of each DQ beam element, and subscript denotes the nodal number. For free vibration of the stepped beam shown in Fig. 2.6, $Q_7 = Q_{13} = 0$, $M_7 = M_{13} = 0$. The boundary conditions are $M_1 = M_{19} = 0$ and $w_1 = w_{19} = 0$.

For the stepped beam structure shown in Fig. 2.6, coordinate transformation is not necessary to formulate the structure stiffness matrix equation. After assemblage, the stiffness equation for the entire beam structure is given by

$$\begin{bmatrix} C_{11}^1 & C_{12}^1 & C_{1i}^1 & C_{18}^1 & C_{19}^1 & 0 & 0 & 0 & 0 & 0 & 0 \\ -B_{11}^1 & -B_{12}^1 & -B_{1i}^1 & -B_{18}^1 & -B_{19}^1 & 0 & 0 & 0 & 0 & 0 & 0 \\ D_{i1}^1 & D_{i2}^1 & D_{ii}^1 & D_{i8}^1 & D_{i9}^1 & 0 & 0 & 0 & 0 & 0 & 0 \\ -C_{71}^1 & -C_{72}^1 & -C_{7i}^1 & k_{88} & k_{89} & C_{1j}^2 & C_{18}^2 & C_{19}^2 & 0 & 0 & 0 \\ B_{71}^1 & B_{72}^1 & B_{7i}^1 & k_{98} & k_{99} & -B_{1j}^2 & -B_{18}^2 & -B_{19}^2 & 0 & 0 & 0 \\ 0 & 0 & 0 & D_{j1}^2 & D_{j2}^2 & D_{jj}^2 & D_{j8}^2 & D_{j9}^2 & 0 & 0 & 0 \\ 0 & 0 & 0 & -C_{71}^2 & -C_{72}^2 & -C_{7j}^2 & k_{15,15} & k_{15,16} & C_{1k}^3 & C_{18}^3 & C_{19}^3 \\ 0 & 0 & 0 & B_{71}^2 & B_{72}^2 & B_{7j}^2 & k_{16,15} & k_{16,16} & -B_{1i}^3 & -B_{18}^3 & -B_{19}^3 \\ 0 & 0 & 0 & 0 & 0 & 0 & D_{k1}^3 & D_{k2}^3 & D_{kk}^3 & D_{k8}^3 & D_{k9}^3 \\ 0 & 0 & 0 & 0 & 0 & 0 & -C_{71}^3 & -C_{72}^3 & -C_{7k}^3 & -C_{78}^3 & -C_{79}^3 \\ 0 & 0 & 0 & 0 & 0 & 0 & B_{71}^3 & B_{72}^3 & B_{7k}^3 & B_{78}^3 & B_{79}^3 \end{bmatrix} \begin{Bmatrix} w_1 \\ \theta_1 \\ w_{i-1} \\ w_7 \\ \theta_7 \\ w_{j-1} \\ w_{13} \\ \theta_{13} \\ w_{k-1} \\ w_{19} \\ \theta_{19} \end{Bmatrix} = \begin{Bmatrix} Q_1 \\ 0 \\ \rho A_1 \omega^2 w_{i-1} \\ 0 \\ 0 \\ \rho A_2 \omega^2 w_{j-1} \\ 0 \\ 0 \\ \rho A_3 \omega^2 w_{k-1} \\ Q_{19} \\ 0 \end{Bmatrix}$$

(2.41)

where superscript is the element number, element 1 is element 1–7, element 2 is element 7–13, and element 3 is element 13–19, as shown in Fig. 2.6, $i = 3, 4, 5, 6, 7$ for element 1–7, $j = 3, 4, 5, 6, 7$ for element 7–13, and $k = 3, 4, 5, 6, 7$ for element 13–19. For the beam structure, $i = 3, 4, 5, 6, 7, j = 8, 9, 10, 11, 12$, and $k = 14, 15, 16, 17, 18$. Elements $k_{88}, k_{89},$ and so on. are given by

$$\begin{aligned} k_{88} &= k_{88}^1 + k_{11}^2 = -C_{78}^1 + C_{11}^2, & k_{89} &= k_{89}^1 + k_{12}^2 = -C_{79}^1 + C_{12}^2 \\ k_{98} &= k_{98}^1 + k_{21}^2 = B_{78}^1 - B_{11}^2, & k_{99} &= k_{99}^1 + k_{22}^2 = B_{79}^1 + B_{12}^2 \\ k_{15,15} &= k_{88}^2 + k_{11}^3 = -C_{78}^2 + C_{11}^3, & k_{15,16} &= k_{89}^2 + k_{12}^3 = -C_{79}^2 + C_{12}^3 \\ k_{16,15} &= k_{98}^2 + k_{21}^3 = B_{78}^2 - B_{11}^3, & k_{16,16} &= k_{99}^2 + k_{22}^3 = B_{79}^2 + B_{12}^3 \end{aligned}$$

(2.42)

The global stiffness equation of a beam structure, Eq. (2.41), can be symbolically written by

$$\{F\} = [K]\{\Delta\}$$

(2.43)

Denote the subscripts e and r for the quantity to be eliminated and retained, respectively. After some rearrangements, Eq. (2.43) can be partitioned as

$$\begin{bmatrix} K_{ee} & K_{er} \\ K_{re} & K_{rr} \end{bmatrix} \begin{Bmatrix} \Delta_e \\ \Delta_r \end{Bmatrix} = \begin{bmatrix} I & 0 \\ 0 & M_{rr} \end{bmatrix} \begin{Bmatrix} F_e \\ \Delta_r \end{Bmatrix} \tag{2.44}$$

where $[M_{rr}]$ is a diagonal mass matrix and given by

$$[M_{rr}] = \omega^2 [M] \tag{2.45}$$

From the first matrix equation in Eq. (2.44), one has

$$\{\Delta_e\} = [K_{ee}]^{-1} \{F_e\} - [K_{ee}]^{-1} [K_{er}] \{\Delta_r\} \tag{2.46}$$

For free vibration analysis, $\{F_e\}$ is zero if both ends of the beam are free. For other boundary conditions, either the displacement and/or the slope are zero if the corresponding generalized forces are not zero. For the simply supported beam shown in Fig. 2.6, Q_1 and Q_{19} are not zero since w_1 and w_{19} are zero. In such cases, simply replace the corresponding diagonal term in $[K_{ee}]$ by a big number, for example, 10^{30}, to enforce the zero-generalized displacement. Therefore, Eq. (2.46) can be further simplified as

$$\{\Delta_e\} = -[K_{ee}]^{-1} [K_{er}] \{\Delta_r\} \tag{2.47}$$

Substituting Eq. (2.47) into the second part of Eq. (2.44) yields

$$\left([K_{rr}] - [K_{re}][K_{ee}]^{-1} [K_{er}] \right) \{\Delta_r\} = \omega^2 [M] \{\Delta_r\} \tag{2.48}$$

or

$$[\bar{K}]\{\Delta_r\} = \omega^2 [M]\{\Delta_r\} \tag{2.49}$$

If the terms ρA_i $(i = 1, 2, 3)$ are moved to the left-hand side of Eq. (2.49), then Eq. (2.49) can be changed to a standard eigenvalue equation, namely,

$$[K^*]\{\Delta_r\} = \omega^2 [I]\{\Delta_r\} = \omega^2 \{\Delta_r\} \tag{2.50}$$

Solving Eq. (2.50) by using a standard eigenvalue solver yields the natural frequencies and corresponding mode shapes.

If an N-node DQ beam element is used, the assemblage procedures are similar to the seven-node DQ beam element.

2.6 DISCUSSION

For static analysis of an Euler–Bernoulli beam subjected to distributed loading, it has been pointed out [2] that the stiffness matrix of the DQ beam element is exactly the same as the stiffness matrix of the two-node-four-DOF finite beam element and that the load vector is exactly the same as the work equivalent load vector, if all inner nodal DOFs are eliminated at the element level. Eqs. (2.14)–(2.18) are used to compute the weighting coefficients.

For illustration, consider the three-node DQ element 2–4 shown in Fig. 2.5. The element stiffness equation is given by

$$\frac{EI}{L^4}\begin{bmatrix} 108L & 30L^2 & -192L & 84L & -18L^2 \\ 22L^2 & 8L^3 & -32L^2 & 10L^2 & -2L^3 \\ -192 & -48L & 384 & -192 & 48L \\ 84L & 18L^2 & -192L & 108L & -30L^2 \\ -10L^2 & -2L^3 & 32L^2 & -22L^2 & 8L^3 \end{bmatrix}\begin{Bmatrix} w_2 \\ \theta_2 \\ w_3 \\ w_4 \\ \theta_4 \end{Bmatrix} = \begin{Bmatrix} Q_2 \\ M_2 \\ q \\ Q_4 \\ M_4 \end{Bmatrix} \tag{2.51}$$

To eliminate the inner nodal DOF w_3, the third equation in Eq. (2.51) is used. One has

$$w_3 = \frac{qL^4}{384EI} + \frac{1}{384}\lfloor 192 \quad 48L \quad 192 \quad -48L \rfloor\begin{Bmatrix} w_2 \\ \theta_2 \\ w_4 \\ \theta_4 \end{Bmatrix} \tag{2.52}$$

Substituting Eq. (2.52) into Eq. (2.51) yields

$$\begin{Bmatrix} Q_2 \\ M_2 \\ Q_4 \\ M_4 \end{Bmatrix} = \frac{EI}{L^4}\begin{bmatrix} 108L & 30L^2 & 84L & -18L \\ 22L^2 & 8L^3 & 10L^2 & -2L^3 \\ 84L & 18L^2 & 108L & -30L^2 \\ -10L^2 & -2L^3 & -22L^2 & 8L^3 \end{bmatrix}\begin{Bmatrix} w_2 \\ \theta_2 \\ w_4 \\ \theta_4 \end{Bmatrix} +$$

$$+\frac{EI}{L^4}\begin{Bmatrix} -192L \\ -32L^2 \\ -192L \\ 32L^2 \end{Bmatrix}\left(\lfloor \frac{192}{384} \quad \frac{48L}{384} \quad \frac{192}{384} \quad \frac{-48L}{384} \rfloor\begin{Bmatrix} w_2 \\ \theta_2 \\ w_4 \\ \theta_4 \end{Bmatrix} + \frac{qL^4}{384EI}\right) \tag{2.53}$$

$$= \frac{EI}{L^4}\begin{bmatrix} 12L & 6L^2 & -12L & 6L \\ 6L^2 & 4L^3 & -6L^2 & 2L^3 \\ -12L & -6L^2 & 12L & -6L^2 \\ 6L^2 & 2L^3 & -6L^2 & 4L^3 \end{bmatrix}\begin{Bmatrix} w_2 \\ \theta_2 \\ w_4 \\ \theta_4 \end{Bmatrix} + \frac{qL}{12}\begin{Bmatrix} -6 \\ -L \\ -6 \\ L \end{Bmatrix}$$

It is seen that the first part in Eq. (2.53) is exactly the same as the stiffness matrix in Eq. (2.31), and the second part is exactly the same as the work equivalent load vector. If the DQ beam element has more inner nodes, whether uniformly or nonuniformly distributed nodes, the conclusion remains the same. For the DQ plate element in bending, however, no such equivalence exists. For buckling and dynamic analysis by using the DQEM, the condensation at the element level cannot be done.

It has been demonstrated that if Eq. (2.7) is used to introduce the additional DOFs $w_1^{[1]}$ and $w_N^{[1]}$, then the stiffness matrix of a two-node DQ beam element is exactly the same as one of the four-DOF finite beam elements. To formulate the stiffness matrix, only Eq. (2.2) is used and Eq. (2.1) is not involved, since inner nodal point does not exist. In other words, assumption is made that the governing equation is satisfied automatically.

A two-node DQ beam element can also be formulated by using Eq. (2.23) or Eq. (2.29). Since the way to introduce the additional DOFs $w_1^{[1]}$ and $w_N^{[1]}$ is different, the stiffness matrix of a two-node DQ beam element is different. The stiffness matrix equation now becomes

$$\frac{EI}{L^4}\begin{bmatrix} 2L & L^2 & -2L & L^2 \\ L^2 & L^3 & -L^2 & 0 \\ -2L & -L^2 & 2L & -L^2 \\ L^2 & 0 & -L^2 & L^3 \end{bmatrix}\begin{Bmatrix} w_1 \\ \theta_1 \\ w_2 \\ \theta_2 \end{Bmatrix} = \begin{Bmatrix} Q_1 \\ M_1 \\ Q_2 \\ M_2 \end{Bmatrix} \tag{2.54}$$

It is obvious that Eq. (2.54) is quite different from Eq. (2.31). Since Eq. (2.31) is accurate for static analysis, the weighting coefficients computed by Lagrange interpolation are not as accurate as the ones obtained by using Hermite interpolation if the number of grid points is small.

Numerical experience shows that the two methods do not give much difference in solution accuracy if the number of grid points is large enough, say, $N>13$. However, the computation of the weighting coefficients by the method of Lagrange interpolation is much simpler.

For two-dimensional problems such as plate in bending, the DQ plate element formulated by using the DQ rule is different if the weighting coefficients are computed by using different interpolation functions. The corner nodal points has four DOFs, that is, w, w_x, w_y, w_{xy}, if the weighting coefficients are determined by using Hermite interpolation functions, and has only three DOFs, that is, w, w_x, w_y, if the weighting coefficients are determined by using Lagrange interpolation functions. The method of Lagrange interpolation is preferred and recommended in practical applications, since the method is simple and without any difficulties in applying multiple boundary conditions at the corner point.

2.7 SUMMARY

One type of the DQ-based element methods, the strong-form DQEM or simply called the DQEM, is presented in this chapter. The basic principle of the DQEM is given. Two different formulations, one based on the Hermite interpolation and the other based on the Lagrange interpolation, are presented. For the formulation of a DQ rectangular plate element, the third formulation, the mixed Hermite interpolation with Lagrange interpolation, can also be used. Assemblage procedures are given and several examples are worked out in detail for illustrations.

Numerical results show that the DQEM can yield accurate results for beams and rectangular plates under discontinuously distributed loads and beams with step changes in cross-sections. Numerically exact results as to the analytical solutions are obtained. The DQEM overcomes the difficulty existing in the original DQM, such as the difficulties to deal with the loading and geometric discontinuities. Besides, the DQEM can also be used for analysis of frame structures. More applications by using the DQEM can be found in Refs. [1–7,10–20].

It is interesting to see that the stiffness matrix and work equivalent load vector of the two-node-four-DOF finite beam element can be deduced by eliminating DOFs at all inner points of the proposed DQ beam element. If one element is used, the DQEM provides a way to apply the multiple boundary conditions and will be discussed in Chapter 3.

REFERENCES

[1] X. Wang, H. Gu, B. Liu, On buckling analysis of beams and frame structures by the differential quadrature element method, Proc. Eng. Mech. 1 (1996) 382–385.
[2] X. Wang, H. Gu, Static analysis of frame structures by the differential quadrature element method, Int. J. Numer. Meth. Eng. 40 (1997) 759–772.

[3] Y. Wang, Differential quadrature method and differential quadrature element method-theory and application. PhD Thesis. Nanjing University of Aeronautics and Astronautics, China, 2001 (in Chinese).

[4] X. Wang, M. Tan, Y. Zhou, Buckling analyses of anisotropic plates and isotropic skew plates by the new version differential quadrature method, Thin Wall. Struct. 41 (2003) 15–29.

[5] Y. Wang, X. Wang, Y. Zhou, Static and free vibration analyses of rectangular plates by the new version of differential quadrature element method, Int. J. Numer. Meth. Eng. 59 (9) (2004) 1207–1226.

[6] X. Wang, Y. Wang, Free vibration analyses of thin sector plates by the new version of differential quadrature method, Comput. Meth. Appl. Mech. Eng. 193 (2004) 3957–3971.

[7] X. Wang, F. Liu, X. Wang, L. Gan, New approaches in application of differential quadrature method for fourth-order differential equations, Commun. Numer. Meth. Eng. 21 (2) (2005) 61–71.

[8] T.Y. Wu, G.R. Liu, The generalized differential quadrature rule for initial value differential equations, J. Sound. Vib. 233 (2) (2000) 195–213.

[9] T.Y. Wu, G.R. Liu, The generalized differential quadrature rule for fourth-order differential equations, Int. J. Numer. Meth. Eng. 50 (2001) 1907–1929.

[10] X. Wang, Y. Wang, R. Chen, Static and free vibrational analysis of rectangular plates by the differential quadrature element method, Commun. Numer. Meth. Eng. 14 (1998) 1133–1141.

[11] Y. Wang, R. Liu, X. Wang, On free vibration analysis of nonlinear piezoelectric circular shallow spherical shells by the differential quadrature element method, J. Sound Vib. 245 (2001) 179–185.

[12] Y. Wang, X. Wang, On a high-accuracy curved differential quadrature beam element and its applications, J. NUAA 33 (6) (2001) 516–520 (in Chinese).

[13] Y. Wang, X. Wang, Analysis of nonlinear piezoelectric circular shallow spherical shells by the differential quadrature element method, T. NUAA 18 (2) (2001) 130–136.

[14] X. Wang, Y. Wang, On nonlinear behavior of spherical shallow shells bonded with piezoelectric actuators by the differential quadrature element method (DQEM), Int. J. Numer. Meth. Eng. 53 (2002) 1477–1490.

[15] X. Wang, Y. Wang, Y. Zhou, Application of a new differential quadrature element method for free vibrational analysis of beams and frame structures, J. Sound Vib. 269 (2004) 1133–1141.

[16] C.W. Dai, Buckling analysis of stiffened plates with differential quadrature element method. Master Thesis. Nanjing University of Aeronautics and Astronautics, China, 2006 (in Chinese).

[17] L.H. Jiang, Buckling analysis of stiffened circular cylindrical panels with differential quadrature element method. Master Thesis. Nanjing University of Aeronautics and Astronautics, China, 2007 (in Chinese).

[18] L. Jiang, Y. Wang, X. Wang, Buckling analysis of stiffened circular cylindrical panels using differential quadrature element method, Thin Wall. Struct. 46 (4) (2008) 390–398.

[19] Z.B. Zhou, Application of differential quadrature element method to fracture analysis. Master Thesis. Nanjing University of Aeronautics and Astronautics, China, 2008 (in Chinese).

[20] X. Wang, Y. Wang, Free vibration analysis of multiple-stepped beams by the differential quadrature element method, Appl. Math. Comput. 219 (2013) 5802–5810.

METHODS OF APPLYING BOUNDARY CONDITIONS

3.1 INTRODUCTION

For success in applications of the differential quadrature method (DQM), one of the key steps is to use an accurate way to apply the boundary conditions. For second-order ordinary differential equations, only one boundary condition at each boundary point is to be satisfied. This is easy to do by the DQM to satisfy the boundary condition by placing the discrete boundary condition at the boundary point. For higher-order differential equations, however, multiple boundary conditions should be satisfied at each boundary point. The way of satisfying the boundary conditions is important to get reliable and accurate solutions by using the DQM.

In this chapter, various methods existing in literature for applying multiple boundary conditions are summarized and discussed. Numerical examples are given to demonstrate the importance of choosing an appropriate way to apply the multiple boundary conditions. Finally, some recommendations are made.

3.2 BASIC EQUATIONS OF A BERNOULLI–EULER BEAM

For simplicity and illustrations, a Bernoulli–Euler beam is considered first. The governing equation is given by

$$EI\frac{d^4w}{dx^4} + P\frac{d^2w}{dx^2} - q(x) - \rho A\omega^2 w = 0, x \in (0, L) \tag{3.1}$$

where $E, I, L, q, P, \rho, A, \omega$, and w are the elasticity modulus, principal moment of inertia about the y-axis, beam length, distributed load, axial compressive load, mass density, cross-sectional area, circular frequency, and deflection, respectively. The shear force Q and bending moment M are given by

$$\begin{cases} EI\dfrac{d^3w}{dx^3} + P\dfrac{dw}{dx} = EIw^{[3]} + Pw^{[1]} = Q \\ EI\dfrac{d^2w}{dx^2} = EIw^{[2]} = M \end{cases} \tag{3.2}$$

Four classical boundary conditions are,

1. Simply supported (S):

$$w = M = 0 \tag{3.3}$$

Differential Quadrature and Differential Quadrature Based Element Methods. 978-0-12-803081-3

2. Clamped (C):

$$w = w^{[1]} = 0 \tag{3.4}$$

3. Free (F):

$$Q = M = 0 \tag{3.5}$$

4. Guided (G):

$$w^{[1]} = Q = 0 \tag{3.6}$$

$A_{ij}, B_{ij}, C_{ij}, D_{ij}$ denote the weighting coefficients of the first-, second-, third-, and fourth-order derivatives with respect to x. In terms of conventional differential quadrature, Eq. (3.1) can be written as

$$\sum_{j=1}^{N}\left[EID_{ij}w_j + PB_{ij}w_j\right] - q(x_i) - \rho A\omega^2 w_i = 0, \quad (i = M, M+1, ..., N-M+1) \tag{3.7}$$

where N is the total number of grid points and M is the starting number of inner grid points and varies with the method for applying multiple boundary conditions.

3.3 METHODS FOR APPLYING MULTIPLE BOUNDARY CONDITIONS

For successful applications by using the DQM, various methods for applying multiple boundary conditions have been proposed. In what follows, these methods are illustrated and briefly discussed. Among them, two approaches proposed by the author and his research associates are recommended for practical applications, since they are the most accurate ones, especially for solving plate problems with free corners.

3.3.1 δ APPROACH

The δ approach [1,2] is the earliest method used for applying the multiple boundary conditions by using the DQM. In the δ method, grids 2 and $N-1$ are also viewed as the boundary points besides grids 1 and N. Grids 1 and 2 (or grids $N-1$ and N) are separated by a very small distance δL as compared to the beam length L. The two boundary conditions are applied at points 1 and 2, or at points $N-1$ and N, respectively. Take the S-C beam as an example, Eq. (3.3) is applied at points 1 and 2, and Eq. (3.4) is applied at points $N-1$ and N; thus, one has

$$w_1 = 0, \quad \sum_{j=1}^{N} B_{2j}w_j = 0; \quad w_N = 0, \quad \sum_{j=1}^{N} A_{(N-1)j}w_j = 0 \tag{3.8}$$

Later, Eq. (3.8) is further modified as [3]

$$w_1 = 0, \quad \sum_{j=1}^{N} B_{1j}w_j = 0; \quad w_N = 0, \quad \sum_{j=1}^{N} A_{Nj}w_j = 0 \tag{3.9}$$

Numerical experience shows that there is not much difference between the differential quadrature (DQ) results obtained by using Eq. (3.8) or Eq. (3.9) in the analysis, since δL is usually small as compared to the beam length L and can be adjusted for differential boundary condition. If the δ approach is employed for applying multiple boundary conditions, M in Eq. (3.7) is 3.

The δ method is simple in programming and can be used for all combinations of boundary conditions. Its drawback is, however, that the method is not very accurate and sometimes δ is problem dependent. In literature, δ is usually set to 0.001.

3.3.2 EQUATION REPLACED APPROACH

The equation replaced approach [3] or CBCGE (directly coupling the boundary conditions with the governing equations) [4] is similar to the δ approach. In the method, one of the two boundary conditions in terms of differential quadrature is usually put at the inner point nearby the boundary point, i.e., at points 2 and $N - 1$. In other words, the governing equation in terms of the differential quadrature at points 2 and $N - 1$ are not used but replaced by the boundary conditions, thus, M in Eq. (3.7) is also 3. For analyzing the S-C beam by using the differential quadrature method, Eq. (3.9) is used for applying the boundary conditions.

Since the method is similar to the δ approach, the programming is also simple and the method can be used for all combinations of boundary conditions. The drawback of the method is, however, that the method is not very accurate and sensitive to grid spacing for certain problems. If the multiple boundary conditions are applied by using the equation replaced approach, the DQM is equivalent to the mixed collocation methods [5]. Thus, one of the two boundary equations can be placed at any inner point. Numerical experience shows that the solution accuracy is the highest if the boundary equations are placed at the inner point nearby the boundary point.

The difference of the equation replaced approach from the δ method is that point 2 (or $N - 1$) is not necessary δL apart from point 1 (or N). Published results indicate that the accuracy of this method is not as good as the δ method, especially for plate problems with free corners.

3.3.3 METHOD OF MODIFICATION OF WEIGHTING COEFFICIENT-1

Since one can only apply one boundary condition at the boundary point directly, one of the two boundary conditions is built in when computing the weighting coefficients of higher-order derivatives [6]. The method is called the method of modification of weighting coefficient-1 or simply (MMWC-1). Take an S-C beam as an example. The weighting coefficients for the second-, third-, and fourth-order derivatives are modified by

$$\bar{B}_{ij} = \sum_{k=1}^{N-1} A_{ik} A_{kj}; \quad \bar{C}_{ij} = \sum_{k=2}^{N} A_{ik} \bar{B}_{kj}; \quad \bar{D}_{ij} = \sum_{k=2}^{N} B_{ik} \bar{B}_{kj} \quad (i, j = 1, 2, ..., N) \tag{3.10}$$

It is seen that parts of the boundary conditions have been built in by simply changing the summation range during formulations of the weighting coefficients. In detail, the upper summation index in the first equation of Eq. (3.10) is changed from N to $N - 1$ for building in the condition of $w_N^{[1]} = 0$. The lower summation index in the second and third equations of Eq. (3.10) is changed from 1 to 2 for building in the condition of $w_1^{[2]} = 0$. Therefore, only one boundary condition at each end is left. This can be applied easily, i.e., $w_1 = 0$ and $w_N = 0$ for the case of an S-C beam.

When MMWC-1 is employed for applying multiple boundary conditions, M in Eq. (3.7) is 2. Numerical experience shows that the accuracy is higher than the previous two methods if the total number of grid points N is small. MMWC-1 is also simple in programming for both one-dimensional and two-dimensional problems, and is more accurate than the δ method and the equation replaced approach. Its drawback is, however, that the method cannot be used for C-C beam and beams with free boundaries.

3.3.4 DQEM OR GDQR

In order to overcome the difficulty in applying the multiple boundary conditions for thin beam and plate problems as well as to extend the application range of the conventional DQM, multiple degrees of freedom (DOFs) at the boundary points are used. The number of the DOF equals to the number of the boundary conditions. The method is called the differential quadrature element method (DQEM) [7,8] or the generalized differential quadrature rule (GDQR) [9,10]. Since the method can be used to analyze structural problems with more elements, as was presented in Chapter 2, the method is called the DQEM in this book.

For the Bernoulli–Euler beam, two DOFs at each end point, i.e., $w_1, w_1^{[1]}$, $w_N, w_N^{[1]}$, are used. The major difference of the DQEM or the GDQR from the conventional DQM is that Hermite-type polynomials are used to determine the weighting coefficients. For the conventional DQM, Lagrange polynomials are used to determine the weighting coefficients. Let $A_{ij}, B_{ij}, C_{ij}, D_{ij}$ be the weighting coefficients of the first-, second-, third-, and fourth-order derivatives of the DQEM or the GDQR, and N the number of grid points; in terms of the DQEM, the generalized boundary conditions at points 1 and N can be written as

$$EI \sum_{j=1}^{N+2} C_{1j} u_j + P \sum_{j=1}^{N+2} A_{1j} u_j = Q_1$$

$$EI \sum_{j=1}^{N+2} B_{1j} u_j = -M_1$$

$$EI \sum_{j=1}^{N+2} C_{Nj} u_j + P \sum_{j=1}^{N+2} A_{Nj} u_j = -Q_N$$

$$EI \sum_{j=1}^{N+2} B_{Nj} u_j = M_N$$

(3.11)

And Eq. (3.1) at all inner grid points can be written by

$$\sum_{j=1}^{N+2} \{EI\ D_{ij} u_j + P\ B_{ij} u_j\} - q(x_i) - \rho A \omega^2 w_i = 0, \quad (i = 2,3,...,N-1)$$

(3.12)

where $\{u\}^T = \lfloor w_1, w_1^{[1]}, w_2, ..., w_{N-1}, w_N, w_N^{[1]} \rfloor$.

Equations (3.11) and (3.12) have $N + 2$ algebraic equations. After applying the boundary conditions, they can be solved for $N + 2$ unknowns. Comparing Eq. (3.12) with Eq. (3.7) reveals that the upper summation index of the DQEM is slightly different from that of the conventional DQM. Usually four equations of boundary conditions, i.e., Eq. (3.11), are put at the position where the corresponding DOF is, namely, at the rows of 1, 2, $N + 1$, and $N + 2$, since the corresponding DOF are $w_1, w_1^{[1]}, w_N, w_N^{[1]}$.

It is seen that the multiple boundary conditions can be applied without any difficulty for all combinations of boundary conditions. Take the S-C beam as an example; one has $u_1 = u_{N+1} = u_{N+2} = 0, M_1 = 0$. Perhaps, its drawback is not as convenient in programming as the other three methods described previously. Besides, the explicit formulation of computing the weighting coefficients is far more complicated as compared to the conventional DQM. It should be mentioned that the DQEM extends the application range of the DQM, since it can be used for analyzing beams with discontinuous cross-sectional area, under discontinuous distributed loads, and even for beam structural problems. Details can be found in Chapter 2.

3.3.5 METHOD OF MODIFICATION OF WEIGHTING COEFFICIENT-2

Although MMWC-1 is simple and accurate, it cannot be used for all combinations of boundary conditions. On the other hand, although the method of multiple degrees of freedom (the DQEM or the GDQR) can be used for all combinations of boundary conditions, the DQEM is different from the conventional DQM, since Hermite-type polynomials are used in the determination of the weighting coefficients. Combining these two ideas, a new way to apply the boundary condition is proposed by Karami and Malekzadeh [11]. The method is called MMWC-2. The essence of MMWC-2 is that $w_1^{[2]}$ and $w_N^{[2]}$, the second-order derivative at two end points, are viewed as independent variables in computing the weighting coefficients of the third- and the fourth-order derivatives with respect to x by using Eqs. (1.22) and (1.23), namely,

$$w_i^{[3]} = \sum_{k=2}^{N-1} A_{ik} \sum_{l=1}^{N} B_{kl} w_l + A_{i1} w_1^{[2]} + A_{iN} w_N^{[2]} = \sum_{j=1}^{N+2} C_{ij} u_j \quad (i=1,2,...,N) \tag{3.13}$$

$$w_i^{[4]} = \sum_{k=2}^{N-1} B_{ik} \sum_{l=1}^{N} B_{kl} w_l + B_{i1} w_1^{[2]} + B_{iN} w_N^{[2]} = \sum_{j=1}^{N+2} D_{ij} u_j \quad (i=1,2,...,N) \tag{3.14}$$

where $\{u\}^T = \lfloor w_1, w_1^{[2]}, w_2,..., w_{N-1}, w_N, w_N^{[2]} \rfloor$, and A_{ij}, B_{ij} are the weighting coefficients of the first- and second-order derivatives of the conventional DQM. The first- and second-order derivatives at all grid points can also be expressed in terms of $u_j (j=1,2,...,N+2)$, namely,

$$w_i^{[1]} = \sum_{j=1}^{N+2} \bar{A}_{ij} u_j, \quad w_i^{[2]} = \sum_{j=1}^{N+2} \bar{B}_{ij} u_j \quad (i=1,2,...,N) \tag{3.15}$$

where the rearranged $\bar{A}_{ij} = \bar{B}_{ij} = 0$ $(i=1,2,...,N; j=2,N+2)$.

The DQ equations at all inner grids are symbolically the same as the ones of the DQEM, i.e., similar to Eq. (3.12). The major difference from the DQEM is that Lagrange polynomials are used in determination of the weighting coefficients of A_{ij} and B_{ij}. For S-S beams, both $w_1^{[2]}$ and $w_N^{[2]}$ are zero and MMWC-2 is reduced to MMWC-1. Note that Eq. (3.15) should be slightly modified to apply the zero slopes at boundary points. For example, to apply $w_1^{[1]} = 0$, $w_1^{[2]}$ is modified by

$$w_1^{[2]} = \sum_{k=2}^{N} A_{1k} w_k^{[1]} = \sum_{k=2}^{N} \sum_{l=1}^{N} A_{1k} A_{kl} w_l = \sum_{j=1}^{N+2} B_{1j} u_j \tag{3.16}$$

All combinations of boundary conditions can be applied without any difficulties. The programming may require a little more effort, especially for two-dimensional problems, since the $w_1^{[2]}$ or $w_N^{[2]}$ in the expression of the bending moment should be modified to apply the zero slope condition if an edge is clamped.

3.3.6 METHOD OF MODIFICATION OF WEIGHTING COEFFICIENT-3

MMWC-3, called the new version of DQM in references [12–14] or the modified differential quadrature method, is also an extension of MMWC-1. The method is similar to MMWC-2, since additional DOFs are introduced during formulations of weighting coefficients of higher-order derivatives. The essential difference of MMWC-3 from MMWC-2 is on the additional DOF. Instead of $w_1^{[2]}$ and $w_N^{[2]}$, $w_1^{[1]}$ and $w_N^{[1]}$ are used as additional DOF in computing the weighting coefficients of the third- and fourth-order derivatives. The weighting coefficients of the second-order derivative at inner grid points remain the same as the ones in the conventional DQM, but the weighting coefficients of the second-order derivative at two end points are formulated slightly different from the conventional DQM, namely,

$$w_i^{[2]} = \sum_{k=2}^{N-1} A_{ik} \sum_{j=1}^{N} A_{kj} w_j + A_{i1} w_1^{[1]} + A_{iN} w_N^{[1]} = \sum_{j=1}^{N+2} \bar{B}_{ij} u_j \quad (i = 1, N)$$

$$w_i^{[2]} = \sum_{k=1}^{N} A_{ik} \sum_{j=1}^{N} A_{kj} w_j = \sum_{j=1}^{N+2} B_{ij} w_j = \sum_{j=1}^{N+2} \bar{B}_{ij} u_j \quad (i = 2, 3, ..., N-1)$$

(3.17)

where $\{u\}^T = \left\lfloor w_1, w_1^{[1]}, w_2, ..., w_{N-1}, w_N, w_N^{[1]} \right\rfloor$, A_{ij}, B_{ij} are the weighting coefficients of the first- and second-order derivatives of the conventional DQM, and $\bar{B}_{ij} = 0 \quad (i = 2, 3, ..., N-1; \ j = 2, N+2)$.

The weighting coefficients of the third- and the fourth-order derivatives are formulated by

$$w_i^{[3]} = \sum_{k=1}^{N} A_{ik} w_k^{[2]} = \sum_{k=1}^{N} A_{ik} \sum_{j=1}^{N+2} \bar{B}_{kj} u_j = \sum_{j=1}^{N+2} C_{ij} u_j \quad (i = 1, 2, ..., N)$$

(3.18)

$$w_i^{[4]} = \sum_{k=1}^{N} B_{ik} w_k^{[2]} = \sum_{k=1}^{N} B_{ik} \sum_{j=1}^{N+2} \bar{B}_{kj} u_j = \sum_{j=1}^{N+2} D_{ij} u_j \quad (i = 1, 2, ..., N)$$

(3.19)

The first-order derivative can also be expressed in terms of $u_j (j = 1, 2, ..., N+2)$ as

$$w_i^{[1]} = \sum_{j=1}^{N+2} \bar{A}_{ij} u_j \quad (i = 1, 2, ..., N)$$

(3.20)

where the rearranged $\bar{A}_{ij} = 0 \quad (i = 1, 2, ..., N; \ j = 2, N+2)$.

The DQ equations at all inner grids are symbolically similar to Eq. (3.12). Slightly different from MMWC-2, all combinations of boundary conditions can be directly applied. Note that Eq. (3.17), not the B_{ij} in the conventional DQM, should be used if the boundary condition contains the second-order derivative.

MMWC-3 can be readily extended to the DQEM, since $w_1^{[1]}$ and $w_N^{[1]}$ are used as the additional DOFs. The major difference of MMWC-3 from the DQEM or the GDQR is that different polynomials are used to determine the weighting coefficients of various derivatives. Details can be found in Chapter 2.

3.3.7 METHOD OF MODIFICATION OF WEIGHTING COEFFICIENT-4

The MMWC-4 is similar to MMWC-3. Again $w_1^{[1]}$ and $w_N^{[1]}$ are introduced as additional DOFs in computing the weighting coefficients of the third- and the fourth-order derivatives. The weighting coefficients of the second-order derivative at inner grid points remain the same as the ones in the conventional DQM, but the weighting coefficients of the second-order derivative at two end points are formulated slightly different from MMWC-3, namely,

$$w_1^{[2]} = \sum_{k=2}^{N} A_{1k} \sum_{j=1}^{N} A_{kj} w_j + A_{11} w_1^{[1]} = \sum_{j=1}^{N+2} \bar{B}_{1j} u_j$$

$$w_N^{[2]} = \sum_{k=1}^{N-1} A_{Nk} \sum_{j=1}^{N} A_{kj} w_j + A_{NN} w_N^{[1]} = \sum_{j=1}^{N+2} \bar{B}_{Nj} u_j \qquad (3.21)$$

$$w_i^{[2]} = \sum_{k=1}^{N} A_{ik} \sum_{j=1}^{N} A_{kj} w_j = \sum_{j=1}^{N} B_{ij} w_j = \sum_{j=1}^{N+2} \bar{B}_{ij} u_j \, (i = 2, 3, ..., N-1)$$

where $\{u\}^T = \lfloor w_1, w_1^{[1]}, w_2, ..., w_{N-1}, w_N, w_N^{[1]} \rfloor$, A_{ij}, B_{ij} are the weighting coefficients of the first- and second-order derivatives in the conventional DQM, $\bar{B}_{1(N+2)} = 0$, $\bar{B}_{N2} = 0$, and $\bar{B}_{ij} = 0$ $(i = 2, 3, ..., N-1; \, j = 2, N+2)$.

The weighting coefficients of the third- and fourth-order derivatives are formulated by

$$w_i^{[3]} = \sum_{k=1}^{N} A_{ik} w_k^{[2]} = \sum_{k=1}^{N} A_{ik} \sum_{j=1}^{N+2} \bar{B}_{kj} u_j = \sum_{j=1}^{N+2} C_{ij} u_j \, (i = 1, 2, ..., N) \qquad (3.22)$$

$$w_i^{[4]} = \sum_{k=1}^{N} B_{ik} w_k^{[2]} = \sum_{k=1}^{N} B_{ik} \sum_{j=1}^{N+2} \bar{B}_{kj} u_j = \sum_{j=1}^{N+2} D_{ij} u_j \, (i = 1, 2, ..., N) \qquad (3.23)$$

The first-order derivative can also be expressed in terms of $u_j (j = 1, 2, ..., N+2)$, i.e.,

$$w_i^{[1]} = \sum_{j=1}^{N+2} \bar{A}_{ij} u_j \ \ (i = 1, 2, ..., N) \qquad (3.24)$$

where the rearranged $\bar{A}_{ij} = 0$ $(i = 1, 2, ..., N; \, j = 2, N+2)$.

Comparing Eq. (3.21) with Eq. (3.17) reveals that only the weighting coefficients of the second-order derivative at two end points, i.e., $w_i^{[2]}(i = 1, N)$, are slightly different from each other. In other words, $w_1^{[1]}$ and $w_N^{[1]}$ are introduced as the additional DOF differently.

The DQ equations at all inner grids are symbolically similar to Eq. (3.12). Again, all combinations of boundary conditions can be conveniently applied and the effort for programming is the same as MMWC-3 for two-dimensional problems.

Similar to MMWC-3, MMWC-4 can be readily extended to the DQEM, since $w_1^{[1]}$ and $w_N^{[1]}$ are used as the additional DOF. The major difference of MMWC-4 from the DQEM is that different polynomials are used to determine the weighting coefficients of various derivatives. Besides, MMWC-4 can also be used if local adaptive differential quadrature method (La-DQM) is to be employed in the analysis.

It is seen that there are three different ways to introduce the additional DOF $w_1^{[1]}$ and $w_N^{[1]}$, i.e., the DQEM, MMWC-3, and MMWC-4. Numerical experience shows that there is not much difference in solution accuracy by using the three different ways if the total number of grid point is large enough, say, $N > 13$.

3.3.8 VIRTUAL BOUNDARY POINT METHOD OR LA-DQM

Instead of introducing additional DOF at each end point, virtual boundary point method [15] or La-DQM [16] introduces two additional grid points outside the beam, one on each side. The added point is not necessary δL apart from the corresponding boundary point. Usually Chebyshev–Gauss–Lobatto grid distribution (Grid III) is used as the grid points in the DQM to ensure the solution accuracy and numerical stability. In such a case, the nonuniform grid points can be computed by

$$\begin{cases} \bar{x}_i = \left(1 - \cos\left[(i-1)\pi / (N-1)\right]\right) / 2 \\ x_i = L\left(\bar{x}_i - \bar{x}_2\right) / \left(\bar{x}_{N-1} - \bar{x}_2\right) \quad (i = 1, 2, ..., N) \end{cases} \tag{3.25}$$

or

$$\bar{x}_i = L\left(1 - \cos\left[(i-1)\pi / (N-3)\right]\right) / 2, \quad (i = 1, 2, ..., N-2)$$
$$x_1 = -\Delta, \ x_{i+1} = \bar{x}_i \ (i = 1, 2, ..., N-2), \ x_N = L + \Delta \tag{3.26}$$

where L is the beam length.

Note that boundary points are not 1 and N but 2 and $N-1$, and points 1 and N are located outside the beam. Besides, Δ is not necessarily very small. Take an F-C beam as an example. The boundary conditions in terms of differential quadrature become

$$\sum_{j=1}^{N} B_{2j} w_j = 0, \ EI\sum_{j=1}^{N} C_{2j} w_j + P\sum_{j=1}^{N} A_{2j} w_j = 0, \ w_{N-1} = 0, \ \sum_{j=1}^{N} A_{(N-1)j} w_j = 0 \tag{3.27}$$

The four boundary conditions shown in Eq. (3.27) are located on the points 1, 2, $N-1$, and N, respectively. The DQ equations at all remaining points are

$$\sum_{j=1}^{N} \{EI \ D_{ij} w_j + P \ B_{ij} w_j\} - q(x_i) - \rho A \ \omega^2 w_i = 0 \quad (i = 3, 4, ..., N-2) \tag{3.28}$$

The method is simple and similar to the equation replaced approach or CBCGE. The La-DQM introduces two additional grid points outside the free end of a beam [17]. Besides the two boundary conditions, the governing equation is also used at the free end point. La-DQM with Grid I can also yield accurate solutions. The good feature of the method is that spurious eigenvalues can be avoided due to using exterior grid points to apply the boundary conditions.

3.3.9 METHOD OF MODIFICATION OF WEIGHTING COEFFICIENT-5

The MMWC-5 is similar to MMWC-1. One of the two boundary conditions is built in when computing the weighting coefficients of higher-order derivatives [15]. Thus, employing the multidegree of freedoms at the end points is not needed. The major difference between MMWC-5 and MMWC-1 is that the weighting coefficients are modified differently. Take the C-F beam as an example to describe the modification procedures. The weighting coefficients are modified by

$$\bar{B}_{ij} = \sum_{k=2}^{N} A_{ik} A_{kj}; \ \bar{C}_{ij} = \sum_{k=1}^{N} A_{ik} \bar{B}_{kj}; \ \bar{D}_{ij} = \sum_{k=1}^{N} B_{ik} \bar{B}_{kj} (i = 1, 2, ..., N_h) \tag{3.29}$$

$$\bar{C}_{ij} = \sum_{k=1}^{N-1} A_{ik} B_{kj}; \ \bar{D}_{ij} = \sum_{k=1}^{N-1} B_{ik} B_{kj} (i = N - N_h + 1, N - N_h + 2, ..., N) \tag{3.30}$$

where A_{ij} and B_{ij} are original weighting coefficients of the first- and second-order derivatives with respect to x; $w_1^{[1]} = 0$ and $w_N^{[2]} = 0$ have been built in.

Note that N_h may vary and its maximum value is $N \div 2$. The governing equation at points 2 to N_h and $(N - N_h + 1)$ to $(N - 1)$ are expressed in term of modified weighting coefficients calculated by Eq. (3.29) and Eq. (3.30), respectively, and the governing equation at remaining inner points is expressed in term of unmodified weighting coefficients. The remaining boundary conditions are

$$w_1 = 0, \quad \sum_{j=1}^{N} \bar{C}_{Nj} w_j = 0 \tag{3.31}$$

The two boundary conditions in Eq. (3.31) are applied at points 1 and N. In this simply way, the limitation that existed in MMWC-1 has been completely removed without employing either the multidegree of freedom at end points or δ approach. Accurate frequency parameters are obtained by the DQM with MMWC-5 for beams with 10 combinations of boundary conditions and clamped square plate [15].

3.4 DISCUSSION

Nine methods in applying the multiple boundary conditions are presented in Section 3.3. Although all methods have no difficulty for one-dimensional problems and can be extended to two-dimensional problems, such as problems of thin plates in bending; however, limitations or difficulties may exist for some methods.

Figure 3.1 schematically shows a mesh used for analyzing problem of thin rectangular plates in bending by using the DQM with the δ approach. Uniform grid spacing 5×5 is shown.

The δ approach or the modified δ approach is to be used for applying multiple boundary conditions. Thus, the nine unfilled circles in Fig. 3.1 are inner grid points, and all filled circles are regarded as

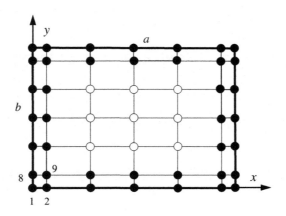

FIGURE 3.1

Schematic Diagram of Uniform Mesh for a Thin Rectangular Plate (δ Approach)

boundary points separated by a small distance δ. One boundary condition is applied at each boundary point. Difficulty arises, however, at the corner point, especially when the corner is free. Seven boundary conditions are available, since there are actually three boundary points at the free corner. Only four of them can be applied, since there are only four points at the corner. Take the free corner (0,0) shown in Fig. 3.1 as an example. Nodes 1, 2, and 8 are actual boundary points and thus have seven boundary conditions, i.e., $R(0,0) = 0$, $M_y(0,0) = 0$, $M_y(0,0) = 0$, $Q_y(\delta,0) = 0$, $M_y(\delta,0) = 0$, $Q_x(0,\delta) = 0$, and $M_x(0,\delta) = 0$. But only four grid points, i.e., points 1, 2, 8, and 9, are available to apply the seven boundary conditions.

In the modified δ approach, the concentrated force $R(0,0) = 0$ is usually applied at node 1, $Q_x(0,\delta) = 0$ is applied at node 8, $Q_y(\delta,0) = 0$ is applied at node 2, and any one of the remaining four conditions, i.e., $M_y(0,0) = 0$, $M_y(\delta,0) = 0$, $M_x(0,0) = 0$, and $M_x(0,\delta) = 0$, can be applied at node 9. Since δ is small, the four moment conditions have similar effect on the solution accuracy.

Numerical experience shows that reliable and accurate results can be obtained for isotropic and orthotropic rectangular plates if $\delta = 0.001L$ (L is the length in the x and y directions, i.e., either a or b). The major disadvantage of the δ approach is that δ may be problem dependent. Besides, δ cannot be too small and improper δ will degrade the solution accuracy, since δ is actually used in computing the weighting coefficients. For the mesh shown in Fig. 3.1, actually a 7×7 nonuniform grid spacing but not a 5×5 uniform grid spacing is used to compute the weighting coefficients. Occasionally numerical instability may be encountered if improper δ is used. In other words, too small δ may degrade the behavior of the weighting coefficients. For anisotropic rectangular plate or isotropic skew plate with large skew angles, the DQ solutions may be very sensitive to δ and reliable solutions cannot be even obtained, since boundary conditions have not been appropriately applied at the free corners. Due to the stress singularity at the obtuse angles of the skew plates, many approximate and numerical methods have encountered serious convergence problems when the skew angle is large. Thus, the DQM with the modified δ approach to apply the multiple boundary conditions may not yield reliable frequency for skew plates with free corners and large skew angles [17].

Figure 3.2 schematically shows the mesh used for analyzing problems of thin rectangular plates in bending by the DQM with the equation replaced approach or CBCGE for applying the multiple boundary conditions. The nonuniform grid spacing (7×7) is shown in Fig. 3.2. The 9 unfilled circles and 16 half-filled circles shown in Fig. 3.2 are inner grid points and only the filled circles are boundary points, different from the δ approach, although the figure is similar to Fig. 3.1.

One boundary condition, either zero deflection or zero shear force, is applied at each boundary point. The other boundary condition, i.e., either the zero slope or the zero bending moment, is applied at the half-filled circles.

Similar to the modified δ approach, difficulty arises at the corner point, since only four out of seven boundary conditions can be applied. Theoretically speaking, the remaining three boundary conditions can be applied at any other inner points, i.e., the DQ equations at any three inner points are replaced by the boundary conditions. However, numerical experience shows that it affects the solution accuracy and may even cause numerical instability. Therefore, the remaining boundary conditions are usually not applied in practice.

From Figs. 3.1 and 3.2, it seems that no difference exists between the modified δ approach and the equation replaced approach if the same mesh is employed, since the weighting coefficients are all determined by using 7×7 nonuniform grid spacing and the way to apply the boundary conditions is the same. The only difference from the modified δ approach is that the filled circles and corresponding

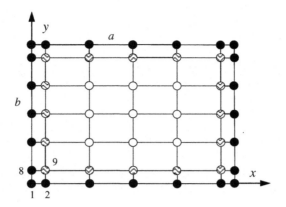

FIGURE 3.2

Schematic Diagram of Nonuniform Mesh (7 × 7) for a Thin Rectangular Plate

half-filled circles are not necessarily δ apart. However, there may be problems of thin plate with free edges and free corners, since not all boundary conditions around corners have been appropriately applied. Therefore, the method is no better than the modified δ approach if free corners are involved. For example, for the case of free vibration of thin isotropic rectangular plates, the frequencies obtained by the DQM with CBCGE are sensitive to grid spacing [4]. If the boundary conditions at the free corner are appropriately applied, the DQ solution should be insensitive to grid spacing and the DQM with even uniform grid points can yield accurate frequencies.

The method of modification of weighting coefficient-1 is very convenient to be used for applying the multiple boundary conditions for isotropic or orthotropic thin rectangular plates and very accurate results can be obtained by the DQM. Since one of the two boundary conditions has been built in during formulation of the weighting coefficients of higher-order derivatives and the displacement at boundary is zero, only equations at inner points, the unfilled and half-filled circles shown in Fig. 3.2, need to be formulated. The corner points are not involved. In other words, the three boundary conditions at a plate corner are satisfied automatically. Therefore, the programming is very simple. The major drawback is that MMWC-1 cannot be used for applying general boundary conditions of isotropic and orthotropic rectangular thin plates or for applying boundary conditions of anisotropic rectangular thin plates. Therefore, its application range is limited.

One of the purposes for development of the DQEM is to overcome the difficulty in applying the multiple boundary conditions existing in the conventional DQM. For thin rectangular plates, all inner points have one DOF, the boundary points have two DOFs, and the four corner points have four DOFs. The major disadvantages of the DQEM are that the implementation is more complicated than the DQM and one additional condition at each corner is needed, since only three boundary conditions are available. To simplify the implementation for analyzing plate problems, a simple way has been proposed by the author and will be discussed at the end of this section. Figure 3.3 shows schematically a uniform mesh (5 × 5) of a rectangular plate to be used for implementation of the multiple boundary conditions by using the DQEM.

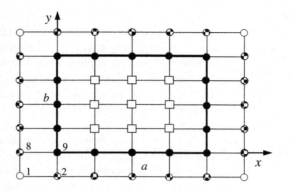

FIGURE 3.3

Schematic Diagram of Uniform Mesh (5 × 5) for a Rectangular Plate

Different from Fig. 3.1 or Fig. 3.2, the nine unfilled rectangles in Fig. 3.3 are inner grid points, the filled circles are boundary points, and the half-filled and unfilled circles are virtual points outside the plate. The DOF at points denoted by the half-filled circles is the first-order derivative of the displacement with respect to x or y at boundary points, namely, either $\partial w/\partial x = w_x$ or $\partial w/\partial y = w_y$. The DOF at the points denoted by unfilled circles is the mixed second-order derivative of the displacement with respect to x and y at corner points, namely, $\partial^2 w/\partial x\partial y = w_{xy}$. The DOF at the points located on the rectangular plate is w. The shear force condition is applied at points denoted by the filled circles except the four corner points, the bending moment condition is applied at points denoted by the half-filled circles, and the concentrated force R condition is applied at corner points. Similar to finite element method, additional effort is needed to deal with the fourth DOF, i.e., w_{xy}. The missing condition should be carefully sought in order to get accurate results. The governing equation may be served as the missing condition. However, other conditions can also be used. For example, the condition $w_{xy} = 0$ could be used if the corner is free for an isotropic thin rectangular plate. Therefore, the difficulty is to find an appropriate boundary condition applied at the corner points.

For MMWC-2, some differences from the DQEM exist. The DOF at points denoted by the half-filled circles is now the second-order derivative of the displacement with respect to x or y at the boundary points, namely, either $\partial^2 w/\partial x^2 = w_{xx}$ or $\partial^2 w/\partial y^2 = w_{yy}$. The DOF at the points located on the rectangular plate is w. The four points denoted by unfilled circles in Fig. 3.3 are not involved. Thus, the difficulty to find the missing boundary condition does not exist.

Similar to the DQEM, the shear force condition is applied at points denoted by the filled circles except the four corner points, the bending moment (or the slope) condition is applied at points denoted by the half-filled circles, and the concentrated force condition is applied at corner points. It is seen that no difficulty arises for rectangular thin plates with isotropic, orthotropic, or anisotropic materials; since two boundary conditions can be applied at all boundary nodes and three boundary conditions can be applied at each corner point. It should be emphasized that the weighting coefficient of the first-order derivative is based on the 5 × 5 meshes. The additional points outside the rectangular plate are introduced for the purpose of convenience in programming. This is similar to the DQEM.

For MMWC-3 and MMWC-4, the nine unfilled rectangles in Fig. 3.3 are inner grid points, the filled circles are boundary points, and the half-filled and unfilled circles are virtual points outside the plate.

The DOF at points denoted by the half-filled circles is the first-order derivative of the displacement with respect to x or y at boundary points, namely, either $\partial w/\partial x = w_x$ or $\partial w/\partial y = w_y$. The DOF at the points located on the rectangular plate is w. No difficulty arises for plates with isotropic, orthotropic, or anisotropic materials, since two boundary conditions applied at each boundary points and three boundary conditions are applied at each corner point. This is the major difference between MMWC-3 or MMWC-4 and the DQEM, although the DOFs in one dimension are exactly the same. Similar to MMWC-2, the four points denoted by unfilled circles in Fig. 3.3 are not involved. In other words, the DOF $\partial^2 w/\partial x \partial y = w_{xy}$ at corner points is not involved.

It is worth noting that for isotropic or orthotropic plates with some boundaries simply supported, the mixed MMWC-1 with MMWC-3 or MMWC-4 can be used to applying the general boundary conditions. MMWC-1 is used to apply the simply supported boundary conditions, and MMWC-3 or MMWC-4 is used to apply the clamped or free boundary conditions. If no free edges are involved, the programming is also very simple, since only DQ equations at inner grid points are needed.

Figure 3.4 schematically shows the meshes of a rectangular plate by the virtual boundary point method or La-DQM to apply multiple boundary conditions. The unfilled circles are inner grid points, and the filled circles are boundary points. All half-filled circles outside the rectangular plate are virtual points, used to apply one of the two boundary conditions. Although the figure seems similar to Fig. 3.3, differences exist between two figures. The DOF at all grid points is the same, namely, the displacement w at the corresponding point. Besides, the formulations of weighting coefficients of various derivatives are based on all points shown in Fig. 3.4. In other words, the mesh size is 7×7 in Fig. 3.4, but only 5×5 in Fig. 3.3.

Similar to the DQEM, difficulty arises in applying the boundary conditions at the corner points, since four boundary conditions are required. Take the corner point 9 shown in Fig. 3.4 as an example. In order to eliminate the DOFs at points 1, 2, 8, and 9, one additional boundary condition is needed. The governing equation may be served as the missing condition. However, other conditions can also be used. Therefore the missing boundary condition should be carefully sought in order to get the correct and accurate results.

For free boundary edges, La-DQM introduces two fictitious points outside the free edge [17]. Although the governing equation and one boundary condition can be used to eliminate the two fictitious

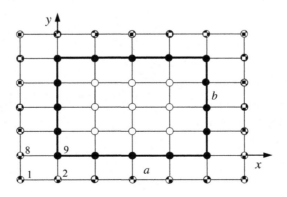

FIGURE 3.4

Schematic Diagram of Uniform Mesh for a Rectangular Plate (Virtual Boundary Point Method)

points in beam problems, it is difficult for the problems of thin plate with free corners. Total nine points related to one corner and only three boundary conditions and one governing equation can be used. Five more equations are needed. The problem may be solved by the iteratively matched boundary method [18], but is not very convenient in implementation. Alternatively, MMWC-4 can be used in the La-DQM. In this way the difficulty can be overcome completely.

The advantage and limitation of MMWC-5 are similar to mixed MMWC-1 with MMWC-3. The method is very convenient to be used for applying the multiple boundary conditions for isotropic or orthotropic thin rectangular plates without a free edge. Since one of the two boundary conditions has been built in during formulation of the weighting coefficients of higher-order derivatives and the displacement at boundary is zero, only equations at inner points, the unfilled and half-filled circles shown in Fig. 3.2, need to be formulated. Therefore, the programming is very simple. The major drawback is that MMWC-5 cannot be used for applying free boundary conditions of isotropic and orthotropic rectangular thin plates or for applying simply supported and free boundary conditions of anisotropic rectangular thin plates. Therefore, its application range is limited.

Based on the previous discussions, some recommendations are made. For beams, isotropic or orthotropic plates, combinations of MMWC-1 with MMWC-3 or MMWC-4, or with the DQEM, are recommended. In other words, MMWC-1 is used for applying simply supported boundary conditions, and MMWC-3, MMWC-4, or the DQEM is used for applying the clamped, free, and guided boundary conditions. This will make the programming much simpler and easier.

For general cases, MMWC-3 or MMWC-4 is recommended for applying the multiple boundary conditions in the application of DQM to analyze plate problems with or without discontinuous loads or geometries. Mixed MMWC-3 or MMWC-4 with the DQEM, i.e., MMWC-3 or MMWC-4 used in one direction and the DQEM used in the other direction, can also be used for general cases, especially when one direction has clamped edges. In this way, the DOF w_{xy} at corners is not involved; thus, the difficulty to find an appropriate missing boundary condition is circumvented.

In literature, the importance of using the right way to apply the multiple boundary conditions has not been emphasized, especially for solving thin plate problems. For simplicity consideration of the implementation, the modified δ method has been widely used with success. For anisotropic rectangular plates or isotropic skew plates with free corners, the advantage or disadvantage of various methods discussed in this chapter would be revealed completely. Four examples, free vibration of SSSS and SFFS isotropic rectangular plate and skew plate with very large skew angle (75°), will be given in the following section to demonstrate the importance of using a correct way to apply the multiple boundary conditions as well as the importance of choosing right grid distributions.

It is mentioned previously that a simple way has been proposed by the author to simplify the implementation for analyzing thin plate problems. The way will be briefly described. Various subroutines and programs are supplied in Appendices for readers' reference. Note that in the FORTRAN programs and converted MATLAB functions for general boundary conditions, the arrangement of DOFs is based on Fig. 3.5, not on Fig. 3.3. Details are also given in Appendix IV.

The major difference between Fig. 3.3 and Fig. 3.5 is the way to place the virtual points outside the rectangular plate. Similar to Fig. 3.3, the nine unfilled rectangles shown in Fig. 3.5 are inner grid points, the filled circles are boundary points, and the half-filled and unfilled circles are virtual points outside the plate. The DOF at points denoted by the half-filled circles is the first-order derivative of the displacement with respect to x or y at boundary points. In details, the DOF at points located on the line of $N_\xi = 6$ is $\partial w / \partial \xi = w_\xi$ at the corresponding boundary points of $\xi = 0$. The DOF

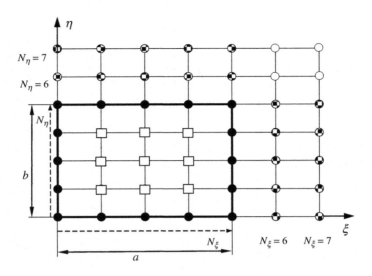

FIGURE 3.5

Sketch of a Rectangular Plate with Grid Spacing of 5×5 ($N = N_\xi = N_\eta = 5$)

at points located on the line of $N_\xi = 7$ is $\partial w / \partial \xi = w_\xi$ at the corresponding boundary points of $\xi = a$. Similarly, the DOF at points located on the line of $N_\eta = 6$ is $\partial w / \partial \eta = w_\eta$ at the corresponding boundary points of $\eta = 0$. The DOF at points located on the line of $N_\eta = 7$ is $\partial w / \partial \eta = w_\eta$ at the corresponding boundary points of $\eta = b$. The DOF at the points located on the rectangular plate, i.e., all boundary points and inner points, is w. The four unfilled circles are not used for MMWC-3 or MMWC-4.

Again, the shear force condition is applied at points denoted by the filled circles except the four corner points, the bending moment condition is applied at points denoted by the half-filled circles, and the concentrated force R condition is applied at corner points. Since all points have only one DOF, the program is as simple as the DQM with the modified δ approach.

It is also worthy of noting that the nine methods and their combinations can be readily extended to apply the boundary conditions more than two.

3.5 NUMERICAL EXAMPLES

To demonstrate the importance of applying the multiple boundary conditions appropriately, several examples are given in this section.

Consider first the free vibration of an SSSS square thin plate and a rhombic thin plate schematically shown in Fig. 3.6. Two skew angles, $\theta = 0°$ and $75°$ corresponding to a rectangular plate and a skew plate with a large skew angle, are analyzed. The plate material is isotropic and Poisson's ratio is 0.3. Nondimensional frequency parameters ($\omega a^2 \sqrt{\rho h / D}$) are presented, where a, h, and D are the plate side length shown in Fig. 3.6, plate thickness, and bending rigidity of the plate, and ρ and ω are mass density and circular frequency, respectively.

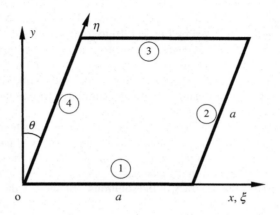

FIGURE 3.6

Sketch of a Rhombic Plate

According to the edge number shown in Fig. 3.6, the symbol SSSS stands for all four edges simply supported. Grid III, widely used in literatures, is used.

Seven ways to apply the multiple boundary conditions in using DQM are investigated for the square plate. They are the modified δ approach, MMWC-1, MMWC-2, MMWC-3, the DQEM, the equation replaced approach or CBCGE, and the mixed method of MMWC-3 with the DQEM. For the modified δ approach, three values of δ, i.e., 0.0003, 0.001, and 0.003, are tried. For the DQEM, four more constraints at corners, namely, $w_{xy} = 0$, are applied.

Table 3.1 lists the first ten mode frequency parameters ($\omega a^2 \sqrt{\rho h/D}$) for the isotropic square thin plate with all edges simply supported. Grid spacing is 21×21.

Table 3.1 Comparison of Frequency Parameters $\omega a^2 \sqrt{\rho h/D}$ for SSSS Square Plates by the DQM with Various Methods to Apply the Multiple Boundary Conditions ($N = 21, \theta = 0^\circ, \mu = 0.3$, Grid III)

	Mode Number					
Methods	**1**	**2 and 3**	**4**	**5 and 6**	**7 and 8**	**9 and 10**
Mixed	19.7392	49.3480	78.9568	98.6960	128.3049	167.7833
MMWC-1	19.7392	49.3480	78.9568	98.6960	128.3049	167.7833
MMWC-2 or MMWC-3	19.7392	49.3480	78.9568	98.6960	128.3049	167.7833
DQEM	19.7392	49.3480	78.9567	98.6959	128.3046	167.7832
δ Method (0.0003a)	19.7392	49.3480	78.9568	98.6960	128.3049	167.7833
δ Method (0.001a)	19.7392	49.3480	78.9568	98.6960	128.3049	167.7833
δ Method (0.003a)	19.7392	49.3480	78.9568	98.6960	128.3049	167.7833
CBCGE	19.7392	49.3480	78.9568	98.6960	128.3049	167.7833
Upper bound solutions [20]	19.7392	49.3480	78.9568	98.6960	128.3049	167.7833

From Table 3.1 it can be seen that except for the DQEM, all other methods yield exactly the same solutions as the analytical solutions given by Leissa [20] to four places of decimal. The DQEM yields slightly lower results for some mode frequencies; however, the results are also very accurate. For the modified δ method, the DQ solutions are insensitive to the value of δ and the same accurate results are obtained for the three chosen values of δ. It is seen that all methods work very well for the isotropic rectangular plate without free edges. Besides, the DQM with Grid III can yield reliable and accurate solutions for the case investigated.

Different from the case of the isotropic rectangular plate, only six ways to apply the multiple boundary conditions are investigated for the SSSS rhombic plate, since MMWC-1 cannot be used in this case. Table 3.2 lists the first five mode frequency parameters ($\omega a^2 \sqrt{\rho h/D}$) for the rhombic plate with a very large skew angle. Again Grid III is used in the analysis.

From Table 3.2, it is seen that the accuracy of the solutions, obtained by the DQM with various ways to apply the multiple boundary conditions, varies and is not as good as the ones in the isotropic square plate. Overall speaking, the mixed method and MMWC-3 yield similar accurate solutions as compared to the existing accurate results reported in Refs. [21,22]. The solution accuracy of the modified δ method varies with the value of δ, different from the case of the isotropic square plate. The method of CBCGE yields the least accurate solutions. It is seen that the way of applying multiple boundary conditions affects the solution accuracy.

Consider next the free vibration of an SFFS square thin plate and a rhombic thin plate schematically shown in Fig. 3.6. Two skew angles, $\theta = 0°$ and $75°$ corresponding to the square plate and rhombic plate with a large skew angle, are investigated. According to the edge number shown in Fig. 3.6, the symbol SFFS stands for edge 1($\eta = 0$) and edge 4 ($\xi = 0$) of the plate simply supported, and edge 2 ($\xi = a$) and edge 3 ($\eta = a$) are free. Numerical experience shows that except for Grid V, all other grid points cannot yield reliable solutions for the case of rhombic plates with free edges and large skew angles. Therefore, Grid V is used to ensure accurate and reliable results by using the DQM.

Table 3.2 Comparison of Frequency Parameters $\omega a^2 \sqrt{\rho h/D}$ for SSSS Rhombic Plates by the DQM with Various Methods to Apply the Multiple Boundary Conditions ($N = 21, \theta = 75°, \mu = 0.3$, Grid III)

Methods	Mode Number				
	1	2	3	4	5
Mixed	204.66	283.09	359.72	438.84	522.28
MMWC-2 or MMWC-3 [19]	196.64	283.05	357.66	438.78	521.03
DQEM	206.63	283.36	360.54	439.15	522.68
δ Method (0.0003a)	206.92	283.57	360.77	439.48	523.22
δ Method (0.001a)	206.10	283.57	360.53	439.48	523.10
δ Method (0.003a)	202.28	283.57	359.48	439.48	522.58
CBCGE	224.61	285.71	378.16	441.73	537.96
MLS-Ritz solutions [21]	197.02	283.06	357.76	438.77	520.83
Upper bound solutions [22]	199.28	283.06	358.45	438.83	524.41

Table 3.3 Comparison of Frequency Parameters $\omega a^2 \sqrt{\rho h/D}$ for SFFS Square Plates by DQM with Various Methods to Apply the Boundary Conditions ($N = 21, \theta = 0°, \mu = 0.3$, Grid V)

Methods	Mode Number					
	1	2	3	4	5	6
Mixed	3.3670	17.316	19.293	38.211	51.035	53.487
MMWC-2 or MMWC-3	3.3670	17.316	19.293	38.211	51.035	53.487
DQEM	3.3670	17.316	19.293	38.211	51.035	53.487
δ Method (0.0003a)	3.3671	17.316	19.293	38.211	51.035	53.487
δ Method (0.001a)	3.3670	17.316	19.293	38.211	51.035	53.487
δ Method (0.003a)	3.3670	17.316	19.293	38.211	51.035	53.487
CBCGE	3.0476	17.316	18.529	37.381	51.036	52.737
Upper bound solutions [20]	3.3687	17.407	19.367	38.291	51.324	53.738

Six ways to apply the multiple boundary conditions in using DQM are investigated. They are the modified δ approach, MMWC-2, MMWC-3, the DQEM, the equation replaced approach or CBCGE, and the mixed method of MMWC-3 with the DQEM. For modified δ approach, three values of δ, i.e., 0.0003, 0.001, and 0.003, are tried. For the DQEM, four more constrains at corners are applied, namely, $Q_\xi(a,a) = 0$ at the free corner ($\xi = a$, $\eta = a$) and $w_{\xi\eta} = 0$ at other three corners.

Table 3.3 lists the first six mode frequency parameters ($\omega a^2 \sqrt{\rho h/D}$) for the isotropic square thin plate. It can be seen that all solutions are slightly lower than the upper bound solutions given by Leissa [20]. Except CBCGE, all other methods yield the same accurate solutions. The accuracy of the CBCGE is slightly lower than that of all other methods and different from the square plate without free edges. For the modified δ method, the DQ solutions are insensitive to the value of δ and almost the same accurate results are obtained for the three different δ values.

Table 3.4 lists the first five mode frequency parameters ($\omega a^2 \sqrt{\rho h/D}$) for the isotropic rhombic thin plate with a very large skew angle. It can be seen that both MMWC-3 (or MMWC-2) and the mixed method can yield reliable and accurate solutions. The modified δ method is very sensitive to the value of δ and the fundamental frequency cannot be even caught when δ takes certain values. This is quite different from the case of isotropic square thin plate, shown in Table 3.3.

For CBCGE, reliable solutions cannot be obtained. It is obvious that the first five mode frequencies have not been caught. The problem is extremely sensitive to grid spacing even with MMWC-3 or the mixed method to apply the multiple boundary conditions. The DQM with only Grid V can yield reliable solutions. With CBCGE, the DQ solution is very sensitive to grid spacing even for isotropic rectangular thin plate with free edges, and the widely used Grid III cannot yield convergence solutions [4]. Therefore, the method cannot be successfully used for solving this problem. The problem demonstrates the importance of the way to apply the multiple boundary conditions and of choice the appropriate grid points. For accuracy and convenience considerations, either MMWC-3 (or MMWC-4) or the mixed method should be used. Besides, Grid V should be adopted to obtain reliable solutions by the DQM.

Table 3.4 Comparison of Frequency Parameters $\omega a^2 \sqrt{\rho h/D}$ for SFFS Rhombic Plates by DQM with Various Methods to Apply the Boundary Conditions ($N = 21, \theta = 75°, \mu = 0.3$, Grid V)

Methods	Mode Number				
	1	2	3	4	5
Mixed [19]	6.1705	26.926	65.535	83.451	119.98
MMWC-2 or MMWC-3 [19]	6.3145	27.321	66.075	83.456	120.15
DQEM(1) [19]	6.6780	29.329	70.583	88.814	126.16
DQEM(2)	6.4267	28.033	67.712	85.745	122.65
δ Method (0.0003a) [19]	-	25.038	69.055	86.574	124.81
δ Method (0.001a) [19]	9.4606	29.450	68.997	86.845	123.85
δ Method (0.003a) [19]	-	30.063	65.988	87.283	123.57
CBCGE	270.40	351.36	468.20	570.64	693.81
MLS-Ritz solutions [21]	6.2578	27.096	65.546	82.705	119.29
Upper bound solutions [23]	6.2683	27.137	65.635	82.837	119.46

The DQEM yields slightly higher frequencies as compared to the accurate upper bound solutions, since more constrains are applied. DQEM(1) applies $Q_\xi(a,a) = 0$ at the free corner ($\xi = a, \eta = a$) and $w_{\xi\eta} = 0$ at other three corners, DQEM(2) applies $Q_\xi(a,a) = 0$ ($\xi = a, \eta = a$) and the governing equation at corners; thus, $w_{\xi\eta} = 0$ at other three corners has not been applied. It is obvious that DQEM(2) yields more accurate results than DQEM(1), since the condition, $w_{\xi\eta} = 0$, at other three corner points is only valid approximately. Using appropriate missing boundary condition is important in using DQEM. For isotropic square plates, DQEM(1) and DQEM(2) yield exactly the same results; thus, only the results obtained by one of them are given in Table 3.3.

3.6 SUMMARY

Various approaches to apply the multiple boundary conditions are summarized and discussed. Although all methods work equally well for one-dimensional problems, some of them have difficulty in applying the multiple boundary conditions at plate corners for two-dimensional problems. A way to overcome the difficulty in implementation of methods of modifying the weighting coefficients is proposed. For demonstration, four examples are given. Numerical results show that using an appropriate way to apply the multiple boundary conditions is very important for success by using the DQM.

The methods of modifying the weighting coefficients, such as MMWC-1, MMWC-3, and MMWC-4, are recommended for solving practical problems with general boundary conditions. Numerical experience shows that spurious eigenvalues may be eliminated with using proper grid spacing and an appropriate method to apply the multiple boundary conditions. This is extremely important if the DQM is to be used for dynamic analysis. Existing spurious eigenvalues will cause numerical instability during time integration by using the central finite difference method. The idea of MMWC-3 and MMWC-4 can also be extended to the cases of a boundary point containing three or four boundary conditions.

REFERENCES

[1] C.W. Bert, S.K. Jang, A.G. Striz, Two new approximate methods for analyzing free vibration of structural components, AIAA J. 26 (1988) 612–618.

[2] S.K. Jang, C.W. Bert, A.G. Striz, Application of differential quadrature to deflection and buckling of structural components, Int. J. Numer. Meth. Eng. 28 (1989) 561–577.

[3] Y. Wang, Differential quadrature method and differential quadrature element method – theory and practice. Ph.D. Dissertation. Nanjing University of Aeronautics and Astronautics, China, 2001 (in Chinese).

[4] C. Shu, H. Du, A generalized approach for implementing general boundary conditions in the GDQ free vibration analysis of plates, Int. J. Solids Struct. 34 (1997) 837–846.

[5] C.W. Bert, X. Wang, A.G. Striz, Differential quadrature for static and free vibration analyses of anisotropic plates, Int. J. Solids Struct. 30 (1993) 1737–1744.

[6] X. Wang, C.W. Bert, A new approach in applying differential quadrature to static and free vibration analyses of beams and plates, J. Sound Vib. 162 (3) (1993) 566–572.

[7] X. Wang, H. Gu, B. Liu, On buckling analysis of beams and frame structures by the differential quadrature element method, Proc. Eng. Mech. 1 (1996) 382–385.

[8] X. Wang, H. Gu, Static analysis of frame structures by the differential quadrature element method, Int. J. Numer. Meth. Eng. 40 (1997) 759–772.

[9] T.Y. Wu, G.R. Liu, The generalized differential quadrature rule for initial value differential equations, J. Sound Vib. 233 (2) (2000) 195–213.

[10] T.Y. Wu, G.R. Liu, The generalized differential quadrature rule for fourth-order differential equations, Int. J. Numer. Meth. Eng. 50 (2001) 1907–1929.

[11] G. Karami, P. Malekzadeh, A new differential quadrature methodology for beam analysis and the associated differential quadrature element method, Comput. Meth. Appl. Mech. Eng. 191 (2002) 3509–3526.

[12] X. Wang, M. Tan, Y. Zhou, Buckling analyses of anisotropic plates and isotropic skew plates by the new version differential quadrature method, Thin Wall Struct. 41 (2003) 15–29.

[13] Y. Wang, X. Wang, Y. Zhou, Static and free vibration analyses of rectangular plates by the new version of differential quadrature element method, Int. J. Numer. Meth. Eng. 59 (9) (2004) 1207–1226.

[14] X. Wang, Y. Wang, Free vibration analyses of thin sector plates by the new version of differential quadrature method, Comput. Meth. Appl. Mech. Eng. 193 (2004) 3957–3971.

[15] X. Wang, F. Liu, X. Wang, L. Gan, New approaches in application of differential quadrature method for fourth-order differential equations, Commun. Numer. Meth. Eng. 21 (2) (2005) 61–71.

[16] Y. Wang, Y.B. Zhao, G.W. Wei, A note on the numerical solution of high-order differential equations, J. Comput. Appl. Math. 159 (2003) 387–398.

[17] L. Zhang, Y. Xiang, G.W. Wei, Local adaptive differential quadrature for free vibration analysis of cylindrical shells with various boundary conditions, Int. J. Mech. Sci. 48 (2006) 1126–1138.

[18] S. Zhao, G.W. Wei, Y. Xiang, DSC analysis of free-edged beams by an iteratively matched boundary method, J. Sound Vib. 284 (2005) 487–493.

[19] X. Wang, Z. Wu, Differential quadrature analysis of free vibration of rhombic plates with free edges, Appl. Math. Comput. 225 (2013) 171–183.

[20] A.W. Leissa, The free vibration of rectangular plates, J. Sound Vib. 31 (3) (1973) 257–293.

[21] L. Zhou, W.X. Zheng, Vibration of skew plates by the MLS-Ritz method, Int. J. Mech. Sci. 50 (2008) 1133–1141.

[22] C.S. Huang, O.G. McGee, A.W. Leissa, J.W. Kim, Accurate vibration analysis of simply supported rhombic plates by considering stress singularities, J. Vib. Acoust. 117 (1995) 245–251.

[23] O.G. McGee, J.W. Kim, A.W. Leissa, The influence of corner stress singularities on the vibration characteristics of rhombic plates with combinations of simply supported and free edges, Int. J. Mech. Sci. 41 (1999) 17–41.

4

QUADRATURE ELEMENT METHOD

4.1 INTRODUCTION

Although the strong-form DQEM is an efficient method for analyzing structural mechanics problems, the method has not been widely used thus far. As compared to the DQM, much fewer applications by using the DQEM have been reported. Besides, the applications to two-dimensional thin-plate problems by either the DQM or the DQEM have mostly limited to regular domains, such as rectangular, skew, circular, and sectorial plates. With using the natural-to-Cartesian geometric mapping technique, the DQM can be applied to problems with any irregular shapes theoretically [1]; however, the accuracy and efficiency of the DQM may be lost. Similar to the DQM, the stiffness matrix of the DQEM is unsymmetrical, and thus may yield complex eigenvalues sometimes. This would cause numerical instability when the DQEM is used for dynamic analysis with an explicit time integration method.

The weak-form quadrature element method (QEM) [2–4], similar to the finite element method (FEM) and time-domain spectral element method (SPE) in principle, may be an alternative way to overcome the weaknesses existing in the ordinary DQM or the strong-form DQEM. With using the natural-to-Cartesian geometric mapping technique, the QEM can solve plate problems with any irregular shapes accurately. Besides, both stiffness and mass matrices are symmetric; thus spurious eigenvalues can be avoided. Since the QEM is essentially the FEM, it is more suitable than the DQEM to analyze complex structure problems.

In this chapter, the weak-form QEM or QEM is presented. Although the principle of the QEM is essentially the same as the FEM, the main feature of the QEM is that the differential quadrature (DQ) rule is used to explicitly compute the weighting coefficients at the integration points, thus the explicit expressions of the derivatives of shape functions are not needed. This feature greatly simplifies the formulations and implementations. In order to do so, a simple way is proposed to explicitly compute the weighting coefficients at the integration points for quadrature elements with nodes other than Gauss–Lobatto–Legendre (GLL) nodes.

The first part of this chapter is focused on the quadrature bar element, thick beam element, plane stress (strain) plate element, and thick plate element. The common characteristic of these elements is that the degrees of freedoms (DOFs) contain the generalized displacement components only. Thus, extension of plate elements from regular shape such as rectangular to irregular shape has no difficulties. Theoretically speaking, these elements are the same as the corresponding time-domain spectral elements [5] if the element nodes are the same and the GLL quadrature is used in the formulations. However, the formulation of the quadrature element is simpler and more flexible than the one of the SPE, especially for the element with nodes other than the GLL nodes. Besides, the derivatives at integration points can be expressed in terms of the DQ rule, the harmonic differential quadrature (HDQ) rule, or their combinations.

The second part of this chapter deals with the quadrature thin beam element and thin plate element in bending. These quadrature elements contain the first-order derivative DOFs, and thus are quite

Differential Quadrature and Differential Quadrature Based Element Methods. 978-0-12-803081-3

different from the ones without derivative DOFs. Although the principle is similar, the extension of the plate element from regular shape such as rectangular to irregular shape is far more complicated as compared to the ones without derivative DOFs. Finally some important things are emphasized to cause readers' attention.

4.2 QUADRATURE ELEMENT METHOD

Different from the strong-form differential QEM, the principle of the weak-form QEM is similar to the FEM or the time-domain SPE. For simplicity and illustration, GLL points are used as the nodes of the element and Lagrange interpolation functions are used as the shape function of the element. The stiffness matrix and mass matrix are approximately obtained by using the GLL quadrature. Thus, the strain fields at the GLL points can be expressed in terms of the DQ rule, namely, in terms of the weighted sum of the corresponding generalized displacement at all nodes in the entire element. This is the main reason why the method is called the QEM. The major difference between QEM and SEM will be more clearly seen when quadrature thin plate element in bending is presented or the nodes are not GLL points.

It is well known that the time-domain SPE combines the accuracy of the global pseudospectral method with the flexibility of the local FEM, thus it can accurately simulate the wave propagation in structures in terms of both phase and amplitude. Therefore, the QEM that possesses the similar properties as SEM is a more promising method for analyzing complex problems in the area of structure mechanics.

4.3 QUADRATURE BAR ELEMENT

The strain energy for a bar element with uniform cross-section is given by

$$U = \frac{1}{2} \int_0^L EA \left(\frac{\partial u(x,t)}{\partial x} \right)^2 dx \tag{4.1}$$

And the kinetic energy is given by

$$T = \frac{1}{2} \int_0^L \rho A \left(\frac{\partial u(x,t)}{\partial t} \right)^2 dx \tag{4.2}$$

where $u(x,t)$ is the axial displacement, t is the time, L and A are the length and cross-sectional area of the bar element, and E and ρ are the elasticity modulus and mass density, respectively. Eqs. (4.1) and (4.2) are to be used to derive the stiffness matrix and mass matrix of the quadrature bar element. Thus the method belongs to the QEM and not to the DQEM.

Consider an N-node quadrature bar element. Each node of the bar element has one nodal DOF, $u_i (i = 1, 2, ..., N)$. The distribution of nodes is the GLL points, namely, the extreme points of the $(N - 1)$th Legendre polynomial together with the two end points. Other distributions of nodes, such as the Chebyshev points or the approximate Lebesgue-optimal grid points, that is, the expanded-Chebyshev grid points [5,6], can be used. The advantage of the expanded-Chebyshev grid points is that the critical

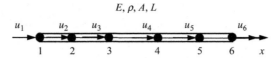

FIGURE 4.1

Sketch of a Six-Node Quadrature Bar Element

time step is the largest among various kinds of nodes for ensuring stable time integration by using the central finite difference method [6].

A six-node quadrature bar element is schematically shown in Fig. 4.1. The axial displacement function for an N-node quadrature bar element is given by

$$u(x,t) = \sum_{i=1}^{N} l_i(x)u(x_i,t) = \sum_{i=1}^{N} l_i(x)u_i(t) = \sum_{i=1}^{N} l_i(\xi)u_i(t) \tag{4.3}$$

where $u_i(t)$ are the nodal displacement, and the shape function $l_i(x) = l_i(\xi)$ is the Lagrange interpolation function, defined by Eq. (1.3). Note that $\xi = 2x/L - 1$ and $\xi \in [-1,1]$ in this chapter.

Substituting the displacement, Eq. (4.3), into the strain energy and kinematic energy expressions (4.1) and (4.2) results

$$U = \frac{1}{2}\int_0^L EA\left(\sum_{i=1}^N \frac{dl_i(x)}{dx}u_i(t)\right)^2 dx = \frac{1}{2}\int_{-1}^1 \frac{2EA}{L}\left(\sum_{i=1}^N \frac{dl_i(\xi)}{d\xi}u_i(t)\right)^2 d\xi = \frac{1}{2}\{u\}^T[k]\{u\} \tag{4.4}$$

$$T = \frac{1}{2}\int_0^L \rho A\left(\sum_{i=1}^N l_i(x)\,\dot{u}_i(t)\right)^2 dx$$
$$= \frac{1}{2}\int_0^L \frac{\rho AL}{2}\left(\sum_{i=1}^N l_i(\xi)\,\dot{u}_i(t)\right)^2 d\xi \tag{4.5}$$
$$= \frac{1}{2}\{\dot{u}\}^T[m]\{\dot{u}\}$$

where $[k]$ is the stiffness matrix, $[m]$ is the mass matrix, $\{u\}$ and $\{\dot{u}\}$ are nodal displacement vector and nodal velocity vector, and the over dot denotes the first-order derivative with respect to time t.

Since the number of DOFs is usually large, $[k]$ and $[m]$ are obtained by numerical integration. Assume that GLL nodes are used. Thus GLL quadrature is used to formulate the stiffness matrix and mass matrix. The elements k_{ij} in the stiffness matrix $[k]$ are

$$k_{ij} = \frac{2EA}{L}\int_{-1}^1 \frac{dl_i(\xi)}{d\xi}\frac{dl_j(\xi)}{d\xi}d\xi$$
$$= \frac{2EA}{L}\sum_{k=1}^N H_k \frac{dl_i(\xi_k)}{d\xi}\frac{dl_j(\xi_k)}{d\xi} = \frac{2EA}{L}\sum_{k=1}^N H_k A_{ki}A_{kj} \tag{4.6}$$

where ξ_i and H_i are abscissas and weights of the N-point GLL quadrature, which can be found in Table I.1 in Appendix I ($N < 22$), and A_{ki} or A_{kj} are the weighting coefficients of the first-order derivative with respect to ξ in the ordinary DQM, computed by Eq. (1.30) explicitly. This is the main reason why the element is called as the quadrature bar element and not the spectral bar element, although they are the same if GLL points are adopted as nodes and the GLL quadrature rule is used in formulations of stiffness matrix and mass matrix. Since the GLL quadrature is accurate up to a polynomial of degree of $(2N - 3)$, the stiffness matrix is fully integrated.

The elements m_{ij} in mass matrix $[m]$ are

$$
\begin{aligned}
m_{ij} &= \frac{\rho AL}{2} \int_{-1}^{1} l_i(\xi) l_j(\xi) \, d\xi \\
&= \frac{\rho AL}{2} \sum_{k=1}^{N} H_k l_i(\xi_k) l_j(\xi_k) = \frac{\rho AL}{2} \sum_{k=1}^{N} H_k \delta_{ki} \delta_{kj} = \frac{\rho AL}{2} H_i \delta_{ij}
\end{aligned}
\tag{4.7}
$$

where

$$
\delta_{ij} = \begin{cases} 1 & (i = j) \\ 0 & (i \ne j) \end{cases}
\tag{4.8}
$$

It is clearly seen that the mass matrix is a diagonal matrix. This is the major advantage of using the GLL quadrature although the mass matrix is not fully integrated. Since the integrand in Eq. (4.7) contains a polynomial of order of $(2N - 2)$ and the GLL quadrature is only accurate up to a polynomial of degree of $(2N - 3)$, the mass matrix is only obtained approximately and not exactly. This is different from the stiffness matrix. It is easy to show that the mass matrix is equivalent to the row or column summed consistent mass matrix that is fully integrated.

If other distributions of nodes, such as the Chebyshev points or the approximate Lebesgue- optimal grid points, that is, the expanded-Chebyshev grid points [5,6], are used, either GLL quadrature or Gauss quadrature can be used for numerical integration. Then the newly proposed method by the author and his research associates [7] should be used.

$\xi (k = 1, 2, ..., N)$ denote the node points and $\varsigma_k \ (k = 1,2,...,N)$ the Gauss quadrature points; the corresponding weights are $\bar{H}_k \ (k = 1,2,...,N)$. For the shape function $l_j(\xi)$ in Eq. (4.3), the function values at Gauss quadrature points can be calculated by

$$
l_j(\xi)_{\xi=\varsigma_k} = l_j(\varsigma_k) = l_{kj}
\tag{4.9}
$$

Since the shape function of the N-node bar element is an $(N - 1)$th-order polynomial, it can be interpolated by [7]

$$
l_j(\xi) = \sum_{k=1}^{N} \tilde{l}_k(\xi) l_j(\varsigma_k) = \sum_{k=1}^{N} \tilde{l}_k(\xi) l_{kj}
\tag{4.10}
$$

where $\tilde{l}_k(\xi)$ are Lagrange interpolation functions based on $\varsigma_k \ (k = 1,2,...,N)$, which can be calculated by

$$
\tilde{l}_j(\xi) = \frac{(\xi - \varsigma_1)(\xi - \varsigma_2)...(\xi - \varsigma_{j-1})(\xi - \varsigma_{j+1})...(\xi - \varsigma_N)}{(\varsigma_j - \varsigma_1)(\varsigma_j - \varsigma_2)...(\varsigma_j - \varsigma_{j-1})(\varsigma_j - \varsigma_{j+1})...(\varsigma_j - \varsigma_N)} = \prod_{\substack{k=1 \\ k \ne j}}^{N} \frac{\xi - \varsigma_k}{\varsigma_j - \varsigma_k}
\tag{4.11}
$$

Although the form of $\tilde{l}_k(\xi)$ is similar to $l_k(\xi)$, different points are used. $\tilde{l}_k(\xi)$ uses $\varsigma_k \ (k = 1,2,...,N)$ and $l_k(\xi)$ uses $\xi_k \ (k = 1,2,...,N)$. It is easy to show that Eq. (4.10) is exact since both sides of Eq. (4.10) are $(N - 1)$th-order polynomials.

The weighting coefficients of the first-order derivative of the shape function at the integration points can be computed by using the DQ rule, namely [7],

$$\bar{A}_{ij} = l'_j(\varsigma_i) = \sum_{k=1}^{N} \tilde{l}'_k(\varsigma_i) l_j(\varsigma_k) = \sum_{k=1}^{N} \tilde{l}'_k(\varsigma_i) l_{kj} = \sum_{k=1}^{N} \tilde{A}_{ik} l_{kj} \tag{4.12}$$

where the over bar means that the weighting coefficient of the first-order derivative with respect to ξ is computed on the basis of the integration points and not the element nodes, and $\tilde{A}_{ij}$ can be explicitly computed by

$$\tilde{A}_{ij} = \tilde{l}'_j(\varsigma_i) = \begin{cases} \displaystyle\prod_{\substack{k=1 \\ k \neq i,j}}^{N} (\varsigma_i - \varsigma_k) \Big/ \prod_{\substack{k=1 \\ k \neq j}}^{N} (\varsigma_j - \varsigma_k) & (i \neq j) \\ \displaystyle\sum_{\substack{k=1 \\ k \neq i}}^{N} \frac{1}{(\varsigma_i - \varsigma_k)} & (i = j) \end{cases} \tag{4.13}$$

Once $\bar{A}_{ij}$ are known, the elements of stiffness matrix can be computed by

$$k_{ij} = \frac{2EA}{L} \sum_{k=1}^{N} \bar{H}_k l'_i(\varsigma_k) l'_j(\varsigma_k) = \frac{2EA}{L} \sum_{k=1}^{N} \bar{H}_k \bar{A}_{ki} \bar{A}_{kj} \quad (i,j = 1,2,...,N) \tag{4.14}$$

The technique of row summation should be used to obtain a diagonal mass matrix. Thus, the diagonal term in the mass matrix can be computed by

$$m_{jj} = \frac{\rho AL}{2} \sum_{k=1}^{N} \bar{H}_k l_j(\varsigma_k) = \frac{\rho AL}{2} \sum_{k=1}^{N} \bar{H}_k l_{kj} \quad (j = 1,2,...,N) \tag{4.15}$$

It is easy to show that both stiffness matrix and mass matrix are fully integrated. Due to using the DQ rule, the explicit expression of $du(x,t)/dx$ is not needed; thus, the derivation and implementation of an N-node bar element become very simple. The proposed method is simple and can be used for any kinds of nodes and quadrature rules [7].

4.4 QUADRATURE TIMOSHENKO BEAM ELEMENT

The strain energy for a Timoshenko beam element with uniform cross-section is given by

$$U = \frac{1}{2} \int_0^L \left\{ EA \left(\frac{\partial u(x,t)}{\partial x} \right)^2 + EI \left(\frac{\partial \varphi(x,t)}{\partial x} \right)^2 + \kappa GA \left(\varphi(x,t) - \frac{\partial w(x,t)}{\partial x} \right)^2 \right\} dx \tag{4.16}$$

and the kinetic energy is given by

$$T = \frac{1}{2} \int_0^L \left\{ \rho A \left(\frac{\partial u(x,t)}{\partial t} \right)^2 + \rho A \left(\frac{\partial w(x,t)}{\partial t} \right)^2 + \rho I \left(\frac{\partial \varphi(x,t)}{\partial t} \right)^2 \right\} dx \tag{4.17}$$

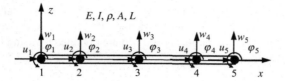

FIGURE 4.2

Sketch of a Five-Node Quadrature Timoshenko Beam Element

where $u(x, t)$, $w(x, t)$, $\varphi(x, t)$ are the axial displacement, flexural displacement, and the rotation, respectively; t is the time; L, A, and I are the length, cross-sectional area, and the moment of inertia of the beam element, respectively; E, G, ρ are the elasticity modulus, shear modulus, and mass density, respectively; and κ is the shear correction factor. Equations (4.16) and (4.17) are to be used to derive the stiffness matrix and mass matrix of the quadrature Timoshenko beam element, respectively.

Consider an N-node quadrature Timoshenko beam element. Each node of the beam element has three nodal DOFs, that is, u_i, w_i, φ_i, $(i = 1, 2, \ldots, N)$, the axial displacement, the deflection, and the rotation about the y-axis at the i-th node, respectively. For illustration, the distribution of nodes is the GLL points. Thus GLL quadrature rule is used for numerical integration. A five-node quadrature Timoshenko beam element is schematically shown in Fig. 4.2.

The generalized displacement functions for the N-node beam element are given by

$$\left\{ \begin{aligned} u(x,t) &= \sum_{i=1}^{N} l_i(x)u(x_i,t) = \sum_{i=1}^{N} l_i(x)u_i(t) = \sum_{i=1}^{N} l_i(\xi)u_i(t) \\ w(x,t) &= \sum_{i=1}^{N} l_i(x)w(x_i,t) = \sum_{i=1}^{N} l_i(x)w_i(t) = \sum_{i=1}^{N} l_i(\xi)w_i(t) \\ \varphi(x,t) &= \sum_{i=1}^{N} l_i(x)\varphi(x_i,t) = \sum_{i=1}^{N} l_i(x)\varphi_i(t) = \sum_{i=1}^{N} l_i(\xi)\varphi_i(t) \end{aligned} \right. \tag{4.18}$$

where $u_i(t)$, $w_i(t)$, and $\varphi_i(t)$ are the nodal generalized displacement components, and the shape function $l_i(x)$ or $l_i(\xi)$ is the Lagrange interpolation function, defined by Eq. (1.3).

Substituting the assumed displacements, Eq. (4.18), into the strain energy and kinematic energy expressions (4.16) and (4.17) results

$$U = \frac{1}{2} \int_0^L \left\{ EA \left(\sum_{i=1}^{N} \frac{dl_i(x)}{dx} u_i(t) \right)^2 + EI \left(\sum_{i=1}^{N} \frac{dl_i(x)}{dx} \varphi_i(t) \right)^2 \right.$$
$$\left. + \kappa GA \left(\sum_{i=1}^{N} \left[l_i(x)\varphi_i(t) - \frac{dl_i(x)}{dx} w_i(t) \right] \right)^2 \right\} dx = \frac{1}{2} \{\bar{u}\}^T [k]\{\bar{u}\} \tag{4.19}$$

$$T = \frac{1}{2} \int_0^L \left\{ \rho A \left(\sum_{i=1}^{N} l_i(x)\, \dot{u}_i(t) \right)^2 + \rho A \left(\sum_{i=1}^{N} l_i(x)\, \dot{w}_i(t) \right)^2 \right.$$
$$\left. + \rho I \left(\sum_{i=1}^{N} l_i(x)\, \dot{\varphi}_i(t) \right)^2 \right\} dx = \frac{1}{2} \{\bar{\dot{u}}\}^T [m]\{\bar{\dot{u}}\} \tag{4.20}$$

where $[k]$ is the stiffness matrix, $[m]$ is the mass matrix, $\{\bar{u}\}$ and $\{\bar{\dot{u}}\}$ are nodal displacement vector and nodal velocity vector, respectively, and the over dot denotes the first-order derivative with respect to time t. Since the number of DOFs is usually large, $[k]$ and $[m]$ are obtained by using the GLL quadrature rule.

Let $\{\bar{u}\}$ and $\{\bar{\dot{u}}\}$ be partitioned as

$$\begin{cases} \{\bar{u}\} = \left\lfloor \lfloor u \rfloor \ \lfloor w \rfloor \ \lfloor \varphi \rfloor \right\rfloor^T \\ \{\bar{\dot{u}}\} = \left\lfloor \lfloor \dot{u} \rfloor \ \lfloor \dot{w} \rfloor \ \lfloor \dot{\varphi} \rfloor \right\rfloor^T \end{cases} \tag{4.21}$$

and $[k]$, $[m]$ are partitioned as

$$[k] = \begin{bmatrix} [k^{uu}] & [k^{uw}] & [k^{u\varphi}] \\ [k^{wu}] & [k^{ww}] & [k^{w\varphi}] \\ [k^{\varphi u}] & [k^{\varphi w}] & [k^{\varphi\varphi}] \end{bmatrix}; \quad [m] = \begin{bmatrix} [m^{uu}] & [m^{uw}] & [m^{u\varphi}] \\ [m^{wu}] & [m^{ww}] & [m^{w\varphi}] \\ [m^{\varphi u}] & [m^{\varphi w}] & [m^{\varphi\varphi}] \end{bmatrix} \tag{4.22}$$

Then the elements k_{ij} in stiffness matrix are

$$\begin{aligned}
k_{ij}^{uu} &= EA \int_0^L \frac{dl_i(x)}{dx}\frac{dl_j(x)}{dx}dx = \frac{2EA}{L}\sum_{k=1}^N H_k A_{ki} A_{kj} \\
k_{ij}^{ww} &= \kappa GA \int_0^L \frac{dl_i(x)}{dx}\frac{dl_j(x)}{dx}dx = \frac{2\kappa GA}{L}\sum_{k=1}^N H_k A_{ki} A_{kj} \\
k_{ij}^{\varphi\varphi} &= EI \int_0^L \frac{dl_i(x)}{dx}\frac{dl_j(x)}{dx}dx + \kappa GA \int_0^L l_i(x)l_j(x)dx \\
&= \frac{2EI}{L}\sum_{k=1}^N H_k A_{ki} A_{kj} + \frac{\kappa GAL}{2}H_i \delta_{ij} \\
k_{ij}^{\varphi w} &= -\kappa GA \int_0^L l_i(x)\frac{dl_j(x)}{dx}dx = -\frac{\kappa GAL}{2}\sum_{k=1}^N H_k \delta_{ki} A_{kj} = -\frac{\kappa GAL}{2}H_i A_{ij} \\
k_{ij}^{w\varphi} &= -\kappa GA \int_0^L \frac{dl_i(x)}{dx}l_j(x)dx = -\frac{\kappa GAL}{2}\sum_{k=1}^N H_k A_{ki}\delta_{kj} = -\frac{\kappa GAL}{2}H_j A_{ji} \\
k_{ij}^{uw} &= k_{ij}^{u\varphi} = k_{ij}^{wu} = k_{ij}^{\varphi u} = 0
\end{aligned} \tag{4.23}$$

where $x_i = L(1+\xi_i)/2$; thus, the determinant of the Jacobian matrix $|J(x)| = |\partial x / \partial \xi| = L/2$, and ξ_i and H_i are abscissas and weights of the N-point GLL quadrature, respectively. A_{ki} and A_{kj} are the weighting coefficients of the first-order derivative with respect to ξ in the ordinary DQM, computed explicitly by Eq. (1.30).

The elements m_{ij} in mass matrix are

$$\begin{aligned}
m_{ij}^{uu} &= m_{ij}^{ww} = \rho A \sum_{k=1}^N H_k l_i(x_k) l_j(x_k)\frac{L}{2} = \frac{\rho AL}{2}\sum_{k=1}^N H_k \delta_{ki}\delta_{kj} = \frac{\rho AL}{2}H_i \delta_{ij} \\
m_{ij}^{\varphi\varphi} &= \rho I \sum_{k=1}^N H_k l_i(x_k) l_j(x_k)\frac{L}{2} = \frac{\rho IL}{2}\sum_{k=1}^N H_k \delta_{ki}\delta_{kj} = \frac{\rho IL}{2}H_i \delta_{ij} \\
m_{ij}^{uw} &= m_{ij}^{u\varphi} = m_{ij}^{wu} = m_{ij}^{w\varphi} = m_{ij}^{\varphi u} = m_{ij}^{\varphi w} = 0
\end{aligned} \tag{4.24}$$

where

$$\delta_{ij} = \begin{cases} 1 & (i = j) \\ 0 & (i \neq j) \end{cases} \tag{4.25}$$

Different from the quadrature bar element, the stiffness matrix is also not fully integrated. The N-point GLL integration is only accurate to the polynomial order of $(2N - 3)$, but the second part of $k_{ij}^{\varphi\varphi}$ in Eq. (4.23) contains the polynomial order of $(2N - 2)$ and thus is not exactly integrated. For the same reason, the diagonal mass matrix is not fully integrated but is equivalent to the row or column summation of a fully integrated consistent mass matrix.

Since explicit formulations are available for obtaining the weighting coefficients of various derivatives, the computation of the quadrature Timoshenko beam element and its implementation are very simple.

4.5 QUADRATURE PLANE STRESS (STRAIN) ELEMENT

For simplicity and illustration, a thin isotropic rectangular plane stress (strain) plate element with length a, width b, and uniform thickness h is considered.

An $N \times N$-node quadrature rectangular plate element is presented. GLL points that belong to $[-1,1]$ are used as the element nodes. Each node has two DOFs, that is, u, v, the displacement components in the x- and y-directions. A 5×5-node quadrature rectangular plate element is schematically shown in Fig. 4.3.

The strain energy for a rectangular thin plate element with uniform thickness is given by

$$U = \frac{1}{2} \int_0^a \int_0^b \{\varepsilon\}^T [C]\{\varepsilon\}\, dy\, dx \tag{4.26}$$

where $[C]$ is a 3×3 stiffness matrix of the material integrated through the thickness, and $\{\varepsilon\}$ is the strain vector defined by

$$\{\varepsilon\}^T = \left\lfloor\ \varepsilon_x\quad \varepsilon_y\quad \gamma_{xy}\ \right\rfloor = \left\lfloor\ \frac{\partial u}{\partial x}\quad \frac{\partial v}{\partial y}\quad \frac{\partial u}{\partial y}+\frac{\partial v}{\partial x}\ \right\rfloor \tag{4.27}$$

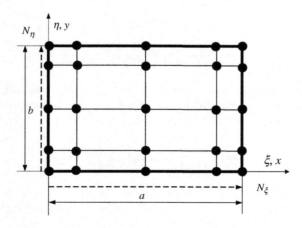

FIGURE 4.3

Sketch of a 5×5-Node Quadrature Rectangular Plate Element

For isotropic material and plane stress, the stiffness matrix $[C]$ is given by

$$[C] = \frac{Eh}{1-\mu^2} \begin{bmatrix} 1 & \mu & 0 \\ \mu & 1 & 0 \\ 0 & 0 & \frac{1-\mu}{2} \end{bmatrix} \tag{4.28}$$

where E and μ are the modulus of elasticity and Poisson's ratio.

For isotropic material and plane strain, the stiffness matrix $[C]$ is given by

$$[C] = \frac{Eh}{(1+\mu)(1-2\mu)} \begin{bmatrix} (1-\mu) & \mu & 0 \\ \mu & (1-\mu) & 0 \\ 0 & 0 & \frac{1-2\mu}{2} \end{bmatrix} \tag{4.29}$$

The kinetic energy expression of the rectangular plate element is given by

$$T = \frac{1}{2} \int_0^a \int_0^b \rho h \left[\left(\frac{\partial u(x,y,t)}{\partial t} \right)^2 + \left(\frac{\partial v(x,y,t)}{\partial t} \right)^2 \right] dy\, dx \tag{4.30}$$

where t is the time, and ρ the mass density of the plate material.

The displacement functions for an $N \times N$-node quadrature rectangular plate element are

$$u(x,y,t) = \sum_{i=1}^N \sum_{j=1}^N l_i(x) l_j(y) u(x_i, y_j, t) = \sum_{i=1}^N \sum_{j=1}^N l_i(\xi) l_j(\eta) u_{ij}(t) \tag{4.31}$$

$$v(x,y,t) = \sum_{i=1}^N \sum_{j=1}^N l_i(x) l_j(y) v(x_i, y_j, t) = \sum_{i=1}^N \sum_{j=1}^N l_i(\xi) l_j(\eta) v_{ij}(t) \tag{4.32}$$

where $u_{ij}(t)$ and $v_{ij}(t)$ are the nodal displacements, and the shape function $l_i(\xi)$ or $l_i(\eta)$ is the Lagrange interpolation functions, defined by Eq. (1.3).

Substituting Eqs. (4.31) and (4.32) into Eqs. (4.26) and (4.30) results

$$U = \frac{1}{2} \{\bar{u}\}^T [k]\{\bar{u}\} \tag{4.33}$$

$$T = \frac{1}{2} \{\ddot{\bar{u}}\}^T [m]\{\ddot{\bar{u}}\} \tag{4.34}$$

where $[k]$ is the stiffness matrix, $[m]$ is the mass matrix, $\{\bar{u}\}$ and $\{\dot{\bar{u}}\}$ are nodal displacement vector and nodal velocity vector, respectively, and the over dot denotes the first-order derivative with respect to time t. $\{\bar{u}\}$ and $\{\dot{\bar{u}}\}$ are partitioned by

$$\{\bar{u}\} = \left\{ \begin{array}{c} \{u\} \\ \{v\} \end{array} \right\}, \quad \{\dot{\bar{u}}\} = \left\{ \begin{array}{c} \{\dot{u}\} \\ \{\dot{v}\} \end{array} \right\} \tag{4.35}$$

Since the number of DOFs is usually large, $[k]$ and $[m]$ are obtained by numerical integration. GLL quadrature is used for the element with GLL nodes. The stiffness matrix $[k]$ is given by

$$[k] = \frac{ab}{4} \sum_{i=1}^N \sum_{j=1}^N H_i H_j [B(\xi_i, \eta_j)]^T [C][B(\xi_i, \eta_j)] \tag{4.36}$$

where (ξ_i, η_j) and H_i, H_j are abscissas and weights, respectively, of the $N \times N$-point GLL quadrature rule, and $[B(\xi_i, \eta_j)]$ is the strain matrix at point (ξ_i, η_j), containing the weighting coefficients of the first-order derivatives with respect to ξ and η in the ordinary DQM, computed explicitly by Eq. (1.30). In detail, $[B(\xi_i, \eta_j)]$ are defined by

$$[B(\xi_i, \eta_j)]\{\bar{u}\} = \begin{bmatrix} \dfrac{2}{a} \displaystyle\sum_{k=1}^{N} A_{ik}^{\xi} u_{kj} \\[2ex] \dfrac{2}{b} \displaystyle\sum_{k=1}^{N} A_{jk}^{\eta} v_{ik} \\[2ex] \dfrac{2}{b} \displaystyle\sum_{k=1}^{N} A_{jk}^{\eta} u_{ik} + \dfrac{2}{a} \displaystyle\sum_{k=1}^{N} A_{ik}^{\xi} v_{kj} \end{bmatrix} \tag{4.37}$$
$$(i, j = 1, 2, ..., N)$$

where A_{ik} are the weighting coefficients of the first-order derivatives with respect to ξ or η in the DQM; the superscripts ξ and η imply that the weighting coefficient of the corresponding derivative is taken with respect to ξ or η.

A diagonal mass matrix is obtained by using the GLL quadrature. The diagonal terms in the mass matrix are

$$m_{II} = m_{JJ} = \frac{\rho h a b}{4} H_i H_j \quad (i, j = 1, 2, ..., N) \tag{4.38}$$

where

$$I = N \times (i - 1) + j \quad (i, j = 1, 2, ..., N); \ J = I + N \times N \tag{4.39}$$

By using the idea of the isoparametric or subparametric element in the conventional FEM, the rectangular quadrature element can be easily extended to quadrilateral element even with curved edges.

Let x–y be Cartesian coordinate system and ξ–η the nondimensional curved coordinate system. Geometric functions to describe the geometry of the distorted element are introduced, which are defined by

$$\begin{cases} x = x(\xi, \eta) = \displaystyle\sum_{i=1}^{M} f_i(\xi, \eta) x_i \\[2ex] y = y(\xi, \eta) = \displaystyle\sum_{i=1}^{M} f_i(\xi, \eta) y_i \end{cases} \quad (-1 \le \xi, \eta \le 1) \tag{4.40}$$

where x_i, y_i are the coordinates of the i-th node, and $f_i(\xi, \eta)$ can be either Lagrange interpolation functions or the shape functions used for serendipity finite element.

Using the chain rule of partial differentiation gives

$$\begin{Bmatrix} \dfrac{\partial f_i}{\partial \xi} \\[2ex] \dfrac{\partial f_i}{\partial \eta} \end{Bmatrix} = \begin{bmatrix} \dfrac{\partial x}{\partial \xi} & \dfrac{\partial y}{\partial \xi} \\[2ex] \dfrac{\partial x}{\partial \eta} & \dfrac{\partial y}{\partial \eta} \end{bmatrix} \begin{Bmatrix} \dfrac{\partial f_i}{\partial x} \\[2ex] \dfrac{\partial f_i}{\partial y} \end{Bmatrix} = \begin{bmatrix} x_\xi & y_\xi \\[2ex] x_\eta & y_\eta \end{bmatrix} \begin{Bmatrix} \dfrac{\partial f_i}{\partial x} \\[2ex] \dfrac{\partial f_i}{\partial y} \end{Bmatrix} = [J] \begin{Bmatrix} \dfrac{\partial f_i}{\partial x} \\[2ex] \dfrac{\partial f_i}{\partial y} \end{Bmatrix} \tag{4.41}$$

where [J] is Jacobian matrix, which serves to transform the derivatives from ξ–η coordinates to x–y coordinates and can be computed from the given nodal coordinates for each element by using Eqs. (4.40) and (4.41).

From Eq. (4.41), the following equation results

$$\left\{ \begin{array}{c} \dfrac{\partial f_i}{\partial x} \\[2mm] \dfrac{\partial f_i}{\partial y} \end{array} \right\} = [J]^{-1} \left\{ \begin{array}{c} \dfrac{\partial f_i}{\partial \xi} \\[2mm] \dfrac{\partial f_i}{\partial \eta} \end{array} \right\} = \frac{1}{|J|} \left[\begin{array}{cc} y_\eta & -y_\xi \\ -x_\eta & x_\xi \end{array} \right] \left\{ \begin{array}{c} \dfrac{\partial f_i}{\partial \xi} \\[2mm] \dfrac{\partial f_i}{\partial \eta} \end{array} \right\} \tag{4.42}$$

where $|J|$ is the determinant of the Jacobian matrix and equals to $(x_\xi y_\eta - x_\eta y_\xi)$.

The differential area of dA is now given by

$$dA = dxdy = |J| \, d\xi d\eta \tag{4.43}$$

Therefore, the determinant of the Jacobian matrix should not be zero and its sign should not be changed in the domain of entire quadrature element. This requirement is the same as the one for the isoparametric finite element.

For the quadrilateral quadrature plate element, Eq. (4.36) is changed to

$$[k] = \sum_{i=1}^{N} \sum_{j=1}^{N} H_i H_j [B(\xi_i, \eta_j)]^T [C][B(\xi_i, \eta_j)] |J(\xi_i, \eta_j)| \tag{4.44}$$

A diagonal mass matrix is also resulted by using the GLL quadrature. The diagonal terms in the mass matrix are

$$m_{II} = m_{JJ} = \rho h H_i H_j |J(\xi_i, \eta_j)| \quad (i, j = 1, 2, ..., N) \tag{4.45}$$

where

$$I = N \times (i - 1) + j \quad (i, j = 1, 2, ..., N); \ J = I + N \times N \tag{4.46}$$

Note that coordinate transformation is not necessary during assemblage process, exactly the same as the assemblage of isoparametric finite elements.

4.6 QUADRATURE THICK PLATE ELEMENT
4.6.1 DISPLACEMENT AND STRAIN FIELDS

Consider a rectangular quadrature thick plate element. The coordinate system (x, y, z) is located on the mid-plane of the plate. The first-order shear deformation theory (FSDT) is adopted. In addition, the effect of the lateral contraction in the z-direction is also taken into account [5,8]. The displacement fields are

$$\left\{ \begin{array}{l} u(x, y, z, t) = u_0(x, y, t) + z\theta(x, y, t) \\ v(x, y, z, t) = v_0(x, y, t) + z\phi(x, y, t) \\ w(x, y, z, t) = w_0(x, y, t) + z\psi(x, y, t) \end{array} \right. \tag{4.47}$$

where $u_0(x, y, t)$, $v_0(x, y, t)$, and $w_0(x, y, t)$ are the in-plane and out-of-plane displacements on the mid-plane of the plate, $\theta(x, y, t)$ and $\phi(x, y, t)$ are the rotations of the cross-section normal to the mid-plane of the plate, respectively, $\psi(x, y, t)$ is the lateral contraction, and t is time.

Under the assumption of small deformation, the strain components in the thick plate are given by

$$\begin{cases} \varepsilon_x = \dfrac{\partial u_0}{\partial x} + z\dfrac{\partial \theta}{\partial x}, & \gamma_{yz} = \phi + \dfrac{\partial w_0}{\partial y} + z\dfrac{\partial \psi}{\partial y} \\[2mm] \varepsilon_y = \dfrac{\partial v_0}{\partial y} + z\dfrac{\partial \phi}{\partial y}, & \gamma_{zx} = \theta + \dfrac{\partial w_0}{\partial x} + z\dfrac{\partial \psi}{\partial x} \\[2mm] \varepsilon_z = \psi, & \gamma_{xy} = \dfrac{\partial u_0}{\partial y} + \dfrac{\partial v_0}{\partial x} + z\left(\dfrac{\partial \theta}{\partial y} + \dfrac{\partial \phi}{\partial x}\right) \end{cases} \quad (4.48)$$

The displacement fields, Eq. (4.47), consist of the antisymmetric and symmetric modes, namely,

$$\begin{cases} u(x, y, z, t) = u_A(x, y, z, t) + u_S(x, y, z, t) \\ v(x, y, z, t) = v_A(x, y, z, t) + v_S(x, y, z, t) \\ w(x, y, z, t) = w_A(x, y, z, t) + w_S(x, y, z, t) \end{cases} \quad (4.49)$$

where the subscripts A and S represent the antisymmetric and symmetric modes, and

$$\begin{cases} u_A(x, y, z, t) = z\theta(x, y, t) \\ v_A(x, y, z, t) = z\phi(x, y, t) \quad \text{and} \\ w_A(x, y, z, t) = w_0(x, y, t) \end{cases} \begin{cases} u_S(x, y, z, t) = u_0(x, y, t) \\ v_S(x, y, z, t) = v_0(x, y, t) \\ w_S(x, y, z, t) = z\psi(x, y, t) \end{cases} \quad (4.50)$$

Hence, Eq. (4.48) can be rewritten as

$$\begin{cases} \varepsilon_x = \varepsilon_x^A + \varepsilon_x^S, & \gamma_{yz} = \gamma_{yz}^A + \gamma_{yz}^S \\ \varepsilon_y = \varepsilon_y^A + \varepsilon_y^S, & \gamma_{zx} = \gamma_{zx}^A + \gamma_{zx}^S \\ \varepsilon_z = \varepsilon_z^A + \varepsilon_z^S, & \gamma_{xy} = \gamma_{xy}^A + \gamma_{xy}^S \end{cases} \quad (4.51)$$

where

$$\begin{cases} \varepsilon_x^A = z\dfrac{\partial \theta}{\partial x}, & \gamma_{yz}^A = \phi + \dfrac{\partial w_0}{\partial y} \\[2mm] \varepsilon_y^A = z\dfrac{\partial \phi}{\partial y}, & \gamma_{zx}^A = \theta + \dfrac{\partial w_0}{\partial x} \quad \text{and} \\[2mm] \varepsilon_z^A = 0, & \gamma_{xy}^A = z\left(\dfrac{\partial \theta}{\partial y} + \dfrac{\partial \phi}{\partial x}\right) \end{cases} \begin{cases} \varepsilon_x^S = \dfrac{\partial u_0}{\partial x}, & \gamma_{yz}^S = z\dfrac{\partial \psi}{\partial y} \\[2mm] \varepsilon_y^S = \dfrac{\partial v_0}{\partial y}, & \gamma_{zx}^S = z\dfrac{\partial \psi}{\partial x} \\[2mm] \varepsilon_z^S = \psi, & \gamma_{xy}^S = \dfrac{\partial u_0}{\partial y} + \dfrac{\partial v_0}{\partial x} \end{cases} \quad (4.52)$$

Again the superscripts A and S represent the antisymmetric and symmetric modes, respectively.

4.6.2 CONSTITUTIVE EQUATION

Let $\{\sigma\}$ and $\{\varepsilon\}$ denote the three-dimensional (3D) stress and strain vector, respectively. Since Eq. (4.48) or Eq. (4.51) is 3D strain field, the 3D constitutive equations could be directly used, namely,

$$
\begin{Bmatrix} \sigma_x \\ \sigma_y \\ \sigma_z \\ \tau_{yz} \\ \tau_{zx} \\ \tau_{xy} \end{Bmatrix} = \begin{bmatrix} C_{11} & C_{12} & C_{13} & 0 & 0 & 0 \\ C_{12} & C_{22} & C_{23} & 0 & 0 & 0 \\ C_{13} & C_{23} & C_{33} & 0 & 0 & 0 \\ 0 & 0 & 0 & C_{44} & 0 & 0 \\ 0 & 0 & 0 & 0 & C_{55} & 0 \\ 0 & 0 & 0 & 0 & 0 & C_{66} \end{bmatrix} \left(\begin{Bmatrix} \varepsilon_x^A \\ \varepsilon_y^A \\ \varepsilon_z^A \\ \gamma_{yz}^A \\ \gamma_{zx}^A \\ \gamma_{xy}^A \end{Bmatrix} + \begin{Bmatrix} \varepsilon_x^S \\ \varepsilon_y^S \\ \varepsilon_z^S \\ \gamma_{yz}^S \\ \gamma_{zx}^S \\ \gamma_{xy}^S \end{Bmatrix} \right)
\tag{4.53}
$$

or

$$
\{\sigma\} = [C]\{\varepsilon\} = [C]\left(\{\varepsilon^A\} + \{\varepsilon^S\}\right)
\tag{4.54}
$$

For isotropic materials with E the modulus of elasticity and μ the Poisson's ratio, one has

$$
C_{11} = C_{22} = C_{33} = \frac{E(1-\mu)}{(1+\mu)(1-2\mu)}
$$
$$
C_{12} = C_{13} = C_{23} = \frac{E\mu}{(1+\mu)(1-2\mu)}
\tag{4.55}
$$
$$
C_{44} = C_{55} = C_{66} = \frac{E}{2(1+\mu)}
$$

For thick plate problems, the thickness of the plate is usually much smaller than its other dimensions. Note that $\varepsilon_z^A = 0$ in Eq. (4.52); however, it is not independent due to Poisson's effect. Therefore, slightly different $[C]$ should be used for the antisymmetric mode to avoid the phenomenon of thickness locking. Equation (4.54) is modified as

$$
\{\sigma\} = [\bar{C}]\{\varepsilon^A\} + [C]\{\varepsilon^S\}
\tag{4.56}
$$

where $[C]$ remains the same as the ones given by Eqs. (4.53) and (4.55), but $[\bar{C}]$ is slightly modified to avoid the thickness locking. In view of the fact that the antisymmetric strain in Eq. (4.52) is exactly the same as the one in Mindlin plate theory, thus $\sigma_{zz} = 0$ is assumed for the antisymmetric mode [8]. Therefore, the modified $[\bar{C}]$ is given by

$$
[\bar{C}] = \begin{bmatrix} \bar{C}_{11} & \bar{C}_{12} & 0 & 0 & 0 & 0 \\ \bar{C}_{12} & \bar{C}_{22} & 0 & 0 & 0 & 0 \\ 0 & 0 & 0 & 0 & 0 & 0 \\ 0 & 0 & 0 & \bar{C}_{44} & 0 & 0 \\ 0 & 0 & 0 & 0 & \bar{C}_{55} & 0 \\ 0 & 0 & 0 & 0 & 0 & \bar{C}_{66} \end{bmatrix}
\tag{4.57}
$$

in which

$$
\bar{C}_{11} = C_{11} - \frac{C_{13}^2}{C_{33}} = \frac{E}{1-\mu^2}
$$
$$
\bar{C}_{12} = C_{12} - \frac{C_{13}C_{23}}{C_{33}} = \frac{E\nu}{1-\mu^2}
\tag{4.58}
$$
$$
\bar{C}_{22} = C_{22} - \frac{C_{23}^2}{C_{33}} = \frac{E}{1-\mu^2}
$$
$$
\bar{C}_{44} = \kappa^2 C_{44}, \quad \bar{C}_{55} = \kappa^2 C_{55}, \quad \bar{C}_{66} = C_{66}
$$

where κ^2 is the shear correction factor.

4.6.3 QUADRATURE RECTANGULAR THICK PLATE ELEMENT

Assume that the thickness of the plate element h is constant. The strain energy for a thick plate element is given by

$$U = \frac{1}{2}\int_0^a\int_0^b \{\bar{\varepsilon}^A\}^T[\bar{D}]\{\bar{\varepsilon}^A\}\,dydx + \frac{1}{2}\int_0^a\int_0^b \{\bar{\varepsilon}^S\}^T[D]\{\bar{\varepsilon}^S\}\,dydx \tag{4.59}$$

where

$$\bar{D}_{ij} = \frac{h^3}{12}\bar{C}_{ij} \ (i,j=1,2,3,6), \quad \bar{D}_{ij} = h\bar{C}_{ij} \ (i,j=4,5)$$
$$D_{ij} = hC_{ij} \ (i,j=1,2,3,6), \quad D_{ij} = \frac{h^3}{12}C_{ij} \ (i,j=4,5) \tag{4.60}$$

and the elements in vectors $\{\bar{\varepsilon}^A\}$ and $\{\bar{\varepsilon}^S\}$ are defined by

$$\begin{cases}\bar{\varepsilon}_x^A = \dfrac{\partial\theta}{\partial x}, & \bar{\gamma}_{yz}^A = \phi + \dfrac{\partial w_0}{\partial y} \\[2mm] \bar{\varepsilon}_y^A = \dfrac{\partial\phi}{\partial y}, & \bar{\gamma}_{zx}^A = \theta + \dfrac{\partial w_0}{\partial x} \\[2mm] \bar{\varepsilon}_z^A = 0, & \bar{\gamma}_{xy}^A = \dfrac{\partial\theta}{\partial y} + \dfrac{\partial\phi}{\partial x}\end{cases} \quad \text{and} \quad \begin{cases}\bar{\varepsilon}_x^S = \dfrac{\partial u_0}{\partial x}, & \bar{\gamma}_{yz}^S = \dfrac{\partial\psi}{\partial y} \\[2mm] \bar{\varepsilon}_y^S = \dfrac{\partial v_0}{\partial y}, & \bar{\gamma}_{zx}^S = \dfrac{\partial\psi}{\partial x} \\[2mm] \bar{\varepsilon}_z^S = \psi_b, & \bar{\gamma}_{xy}^S = \dfrac{\partial u_0}{\partial y} + \dfrac{\partial v_0}{\partial x}\end{cases} \tag{4.61}$$

Note that z does not appear in Eq. (4.61), different from Eq. (4.52). The kinetic energy is given by

$$T = \frac{1}{2}\int_{-h/2}^{h/2}\int_0^a\int_0^b \rho\left[\left(\frac{\partial u(x,y,z,t)}{\partial t}\right)^2 + \left(\frac{\partial v(x,y,z,t)}{\partial t}\right)^2 + \left(\frac{\partial w(x,y,z,t)}{\partial t}\right)^2\right]dydxdz \tag{4.62}$$

An $N\times N$-node quadrature thick rectangular plate element is considered. For simplicity and illustration, GLL nodes are used for the element. Each node has six DOFs, that is, $u_0, v_0, w_0, \theta, \phi$, and ψ. A 6×6-node quadrature rectangular thick plate element is schematically shown in Fig. 4.4.

The three displacement components in Eq. (4.47) are assumed by

$$\{u,v,w\}^T = \sum_{i=1}^N\sum_{j=1}^N [L_{ij}(\xi,\eta)]\{q_k(\xi_i,\eta_j)\} = [L]\{q\} \tag{4.63}$$

where q_k represents the six generalized displacements, namely, $u_0(x,y,t)$, $v_0(x,y,t)$, $w_0(x,y,t)$, $\theta(x,y,t)$, $\phi(x,y,t)$, and $\psi(x,y,t)$, and $[L]$ is the Lagrange interpolation matrix defined by

$$[L_{ij}(\xi,\eta)] = \begin{bmatrix} N_{ij} & 0 & 0 & zN_{ij} & 0 & 0 \\ 0 & N_{ij} & 0 & 0 & zN_{ij} & 0 \\ 0 & 0 & N_{ij} & 0 & 0 & zN_{ij} \end{bmatrix} \tag{4.64}$$
$$(i,j=1,2,...,N)$$

in which $N_{ij}(\xi,\eta) = l_i(\xi)l_j(\eta)$ $(i,j=1,2,...,N)$ are shape functions, and $l_i(\xi)$ and $l_i(\eta)$ are Lagrange interpolation functions defined by Eq. (1.3). Note that $\xi\in[-1,1]$ and $\eta\in[-1,1]$.

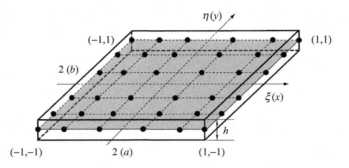

FIGURE 4.4

Sketch of a 6 × 6-Node Quadrature Rectangular Thick Plate Element

From Eq. (4.63), the six generalized displacements within the quadrature thick plate element can be expressed by

$$\{q_k\}^T = \sum_{i=1}^{N}\sum_{j=1}^{N}[N_{ij}(\xi,\eta)]\{q_k(\xi_i,\eta_j)\}^T \quad (k=1,2,...,6) \tag{4.65}$$

The strain fields can be expressed by

$$\{\varepsilon\} = \sum_{i=1}^{N}\sum_{j=1}^{N}[B_{ij}(\xi,\eta)]\{q_k(\xi_i,\eta_j)\} = [B(\xi,\eta)]\{q\} \tag{4.66}$$

Using Eq. (4.65), $\{\bar{\varepsilon}^A\}$ and $\{\bar{\varepsilon}^S\}$ can be expressed by

$$\begin{aligned}\{\bar{\varepsilon}^A\} &= [B^A(\xi,\eta)]\{q\} \\ \{\bar{\varepsilon}^S\} &= [B^S(\xi,\eta)]\{q\}\end{aligned} \tag{4.67}$$

where

$$[B_{ij}^A(\xi,\eta)] = \begin{bmatrix} 0 & 0 & 0 & \dfrac{\partial N_{ij}}{\partial x} & 0 & 0 \\ 0 & 0 & 0 & 0 & \dfrac{\partial N_{ij}}{\partial y} & 0 \\ 0 & 0 & 0 & 0 & 0 & 0 \\ 0 & 0 & \dfrac{\partial N_{ij}}{\partial y} & 0 & N_{ij} & 0 \\ 0 & 0 & \dfrac{\partial N_{ij}}{\partial x} & N_{ij} & 0 & 0 \\ 0 & 0 & 0 & \dfrac{\partial N_{ij}}{\partial y} & \dfrac{\partial N_{ij}}{\partial x} & 0 \end{bmatrix} \quad (i,j=1,2,...,N) \tag{4.68}$$

and

$$[B_{ij}^S(\xi,\eta)] = \begin{bmatrix} \dfrac{\partial N_{ij}}{\partial x} & 0 & 0 & 0 & 0 & 0 \\[2mm] 0 & \dfrac{\partial N_{ij}}{\partial y} & 0 & 0 & 0 & 0 \\[2mm] 0 & 0 & N_{ij} & 0 & 0 & 0 \\[2mm] 0 & 0 & 0 & 0 & 0 & \dfrac{\partial N_{ij}}{\partial y} \\[2mm] 0 & 0 & 0 & 0 & 0 & \dfrac{\partial N_{ij}}{\partial x} \\[2mm] \dfrac{\partial N_{ij}}{\partial y} & \dfrac{\partial N_{ij}}{\partial x} & 0 & 0 & 0 & 0 \end{bmatrix} \quad (i,j=1,2,...,N) \tag{4.69}$$

Substituting Eqs. (4.68) and (4.69) into Eq. (4.59) results

$$\begin{aligned} U &= \frac{1}{2}\{q\}^T \int_{-1}^{1}\int_{-1}^{1} \frac{ab}{4}[B^A(\xi,\eta)]^T \left[\bar{D}\right][B^A(\xi,\eta)] d\eta\, d\xi \{q\} \\ &\quad + \frac{1}{2}\{q\}^T \int_{-1}^{1}\int_{-1}^{1} \frac{ab}{4}[B^S(\xi,\eta)]^T [D][B^S(\xi,\eta)] d\eta\, d\xi \{q\} \\ &= \frac{1}{2}\{q\}^T [k]\{q\} \end{aligned} \tag{4.70}$$

Substituting Eq. (4.65) into Eq. (4.62) results

$$\begin{aligned} T &= \frac{1}{2} \int_{-h/2}^{h/2}\int_{0}^{a}\int_{0}^{b} \rho\left[\left(\frac{\partial u(x,y,z,t)}{\partial t}\right)^2 + \left(\frac{\partial v(x,y,z,t)}{\partial t}\right)^2 + \left(\frac{\partial w(x,y,z,t)}{\partial t}\right)^2\right] dzdydx \\ &= \frac{1}{2}\{\dot{q}\}^T \int_{-1}^{1}\int_{-1}^{1} \frac{ab}{4}[N]^T [\mu][N] d\eta d\xi \ \{\dot{q}\} \end{aligned} \tag{4.71}$$

where $[N]$ and $[\mu]$ are defined by

$$[N] = \begin{bmatrix} N_{ij} \end{bmatrix} \quad (i,j=1,2,...,N) \tag{4.72}$$

in which

$$[N_{ij}] = \begin{bmatrix} N_{ij} & 0 & 0 & 0 & 0 & 0 \\ 0 & N_{ij} & 0 & 0 & 0 & 0 \\ 0 & 0 & N_{ij} & 0 & 0 & 0 \\ 0 & 0 & 0 & N_{ij} & 0 & 0 \\ 0 & 0 & 0 & 0 & N_{ij} & 0 \\ 0 & 0 & 0 & 0 & 0 & N_{ij} \end{bmatrix} \tag{4.73}$$

$$[\mu] = \begin{bmatrix} \rho h & 0 & 0 & 0 & 0 & 0 \\ 0 & \rho h & 0 & 0 & 0 & 0 \\ 0 & 0 & \rho h & 0 & 0 & 0 \\ 0 & 0 & 0 & \rho I & 0 & 0 \\ 0 & 0 & 0 & 0 & \rho I & 0 \\ 0 & 0 & 0 & 0 & 0 & \rho I \end{bmatrix} \tag{4.74}$$

where $I = h^3/12$ is the moment of inertia.

The stiffness matrix and mass matrix are obtained by using the GLL quadrature, namely,

$$[k] = \sum_{m=1}^{N}\sum_{n=1}^{N}\left\{\frac{ab}{4}H_m H_n \left[B^A(\xi_m,\eta_n)\right]^T \left[\bar{D}\right]\left[B^A(\xi_m,\eta_n)\right]\right\}$$
$$+ \sum_{m=1}^{N}\sum_{n=1}^{N}\left\{\frac{ab}{4}H_m H_n \left[B^S(\xi_m,\eta_n)\right]^T [D]\left[B^S(\xi_m,\eta_n)\right]\right\} \tag{4.75}$$

$$[m] = \sum_{m=1}^{N}\sum_{n=1}^{N}\left\{\frac{ab}{4}H_m H_n \left[N(\xi_m,\eta_n)\right]^T [\mu]\left[N(\xi_m,\eta_n)\right]\right\} \tag{4.76}$$

The first-order derivatives, that is, $\partial N_{ij}/\partial\xi$ and $\partial N_{ij}/\partial\eta$, at the GLL points are calculated by using the DQ rule. For rectangular plate element, $\partial N_{ij}/\partial x = (2/a)\partial N_{ij}/\partial\xi$, $\partial N_{ij}/\partial y = (2/b)\partial N_{ij}/\partial\eta$. It is easy to show that the mass matrix is a diagonal matrix.

For quadrilateral quadrature thick plate elements, Eqs. (4.75) and (4.76) are modified by

$$[k] = \sum_{m=1}^{N}\sum_{n=1}^{N}\left\{H_m H_n \left[B^A(\xi_m,\eta_n)\right]^T \left[\bar{D}\right]\left[B^A(\xi_m,\eta_n)\right]|J(\xi_m,\eta_n)|\right\}$$
$$+ \sum_{m=1}^{N}\sum_{n=1}^{N}\left\{H_m H_n \left[B^S(\xi_m,\eta_n)\right]^T [D]\left[B^S(\xi_m,\eta_n)\right]|J(\xi_m,\eta_n)|\right\} \tag{4.77}$$

$$[m] = \sum_{m=1}^{N}\sum_{n=1}^{N}\left\{H_m H_n \left[N(\xi_m,\eta_n)\right]^T [\mu]\left[N(\xi_m,\eta_n)\right]|J(\xi_m,\eta_n)|\right\} \tag{4.78}$$

where $|J(\xi,\eta)|$ is the determinant of the Jacobian matrix, defined by

$$|J(\xi,\eta)| = \begin{vmatrix} \dfrac{\partial x}{\partial\xi} & \dfrac{\partial y}{\partial\xi} \\ \dfrac{\partial x}{\partial\eta} & \dfrac{\partial y}{\partial\eta} \end{vmatrix} \tag{4.79}$$

Quadrature plate elements based on other higher-order plate theories can be formulated in a similar way. Since the quadrature thick plate elements are the same as the time-domain spectral thick plate elements if GLL nodes and GLL quadrature are used, more details on the formulations may be found in [6,9,10]. For other types of nodes, Gauss quadrature and the method similar to the quadrature bar element should be used. The formulation is much simpler than the ones of the time-domain spectral thick plate element reported in literature.

For structural analysis by the QEM, the assemblage procedures are exactly the same as the FEM if more than one quadrature element is used in the analysis, since various quadrature elements are essentially the high-order finite elements. Physically, at each common nodal point, the generalized displacement components at the quadrature element level, that is, the generalized displacement components at the element boundary nodes, are the same as those after assemblage. If the generalized forces at nodal points are regarded as internal forces, then the sum of such internal forces at each common nodal point should be equal to the corresponding external applied generalized loads. Since generalized forces and generalized displacements are vectors, the vectorial sum is usually replaced by the scale sum of their components in a common coordinate system at the connecting nodal point. Similar to the FEM, the coordinate transformation may be necessary for the quadrature bar element and quadrature beam element during assemblage. For the quadrature plate element, similar to the isoparametric element in conventional FEM, the coordinate transformation is not necessary since the element stiffness matrix and mass matrix are formulated in the global coordinate system.

The derivatives at the integration points can also be computed by using the HDQ rule. This is equivalent to using Eq. (1.32) as the shape functions, instead of using the Lagrange interpolation functions defined by Eq. (1.3) as the shape functions. Even the derivatives at the integration points can be computed by using the HDQ rule in one direction and the DQ rule in the other direction. This is equivalent to using Eq. (1.32) as the shape functions in one direction and using Eq. (1.3) as the shape functions in the other direction. Therefore, the formulation of a quadrature element is more flexible and simpler, since explicit expressions of the derivatives of shape functions are not needed.

4.7 QUADRATURE THIN BEAM ELEMENT

The strain energy for the Euler–Bernoulli beam element with uniform cross-section is given by

$$U = \frac{1}{2} \int_0^L EI \left(\frac{\partial^2 v(x,t)}{\partial x^2} \right)^2 dx \tag{4.80}$$

The kinetic energy is given by

$$T = \frac{1}{2} \int_0^L \rho A \left(\frac{\partial v(x,t)}{\partial t} \right)^2 dx \tag{4.81}$$

and the work done by the axial compressive force $P(x)$ is given by

$$W_n = \frac{1}{2} \int_0^L P(x) \left(\frac{\partial v(x,t)}{\partial x} \right)^2 dx \tag{4.82}$$

where $v(x, t)$ is the transverse displacement, t is the time; L, A, and I are the elemental length, cross-sectional area, and the moment of inertia about the z-axis, respectively; E and ρ are the elasticity modulus and mass density, respectively; and $P(x)$ is the distributed axial force and positive when in compression. Equations (4.80)–(4.82) are used to derive the stiffness matrix, mass matrix, and geometric stiffness matrix by using the DQ rule. Thus the element belongs to the weak-form quadrature thin beam element or simply the quadrature beam element.

Consider an N-node Euler-Bernoulli beam element. For illustration, GLL points are used as the nodes of the quadrature beam element. Each inner node has one nodal DOF, $v_i (i = 2, 3, ...N - 1)$, and each end node has two nodal DOFs, $v_i, v_{xi} (i = 1, N)$, the deflection and its first-order derivative with respect to x. A five-node Euler–Bernoulli quadrature beam element is schematically shown in Fig. 4.5.

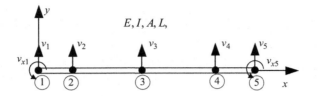

FIGURE 4.5

Sketch of a Five-Node Quadrature Thin Beam Element

The deflection for an N-node quadrature thin beam element is

$$v(x,t) = \sum_{i=1}^{N} l_i(x)v(x_i,t) = \sum_{i=1}^{N} l_i(x)v_i(t) = \sum_{i=1}^{N} l_i(\xi)v_i \quad (4.83)$$

where $v_i(t)$ is the nodal displacement, and the shape function $l_i(x)$ or $l_i(\xi)$ is the Lagrange interpolation function defined by Eq. (1.3) or Eq. (1.32), which are used for the ordinary DQM or the harmonic differential quadrature method (HDQM). Note that $\xi = 2x / L$ and $\xi \in [-1,1]$.

Alternatively, the shape function for an N-node quadrature beam element is the Hermite interpolation function defined by Eq. (2.7), namely,

$$v(x,t) \approx \varphi_1(x)v(x_1,t) + \cdots + \varphi_N(x)v(x_N,t) + \psi_1(x)v_x(x_1,t) + \psi_N(x)v_x(x_N,t)$$

$$= \sum_{j=1}^{N} \varphi_j(\xi)v(x_j,t) + \psi_1(\xi)v_x(x_1,t) + \psi_N(\xi)v_x(x_N,t) = \sum_{j=1}^{N+2} h_j(\xi)\bar{v}_j \quad (4.84)$$

where $\phi_j(x)$ and $\psi_j(x)$ are the $(N + 1)$th-order polynomials, $\bar{v}_j = v_j (j = 1,2,...,N)$, $\bar{v}_{N+1} = v_{x1}$, $\bar{v}_{N+2} = v_{xN}$.

Substituting the assumed deflection, Eq. (4.83), into Eqs. (4.80)–(4.82), respectively, results

$$U = \frac{1}{2}\int_0^L EI \left(\sum_{i=1}^{N+2} \frac{d^2 l_i(x)}{dx^2} \bar{v}_i(t) \right)^2 dx = \frac{1}{2}\{\bar{v}\}^T [k]\{\bar{v}\} \quad (4.85)$$

$$T = \frac{1}{2}\int_0^L \rho A \left(\sum_{i=1}^{N} l_i(x) \dot{v}_i(t) \right)^2 dx = \frac{1}{2}\{\dot{v}\}^T [m]\{\dot{v}\} \quad (4.86)$$

$$W_n = \frac{1}{2}\int_0^L P(x) \left(\sum_{i=1}^{N} \frac{\partial l_i(x)}{\partial x} v_i(t) \right)^2 dx = \frac{1}{2}\{\bar{v}\}^T [g]\{\bar{v}\} \quad (4.87)$$

where $[k]$, $[m]$, $[g]$ are the stiffness matrix, mass matrix, and geometric stiffness matrix, respectively, $\{\bar{v}\}$ and $\{\dot{v}\}$ are nodal displacement vector with derivative DOFs and nodal velocity vector without derivative DOFs, respectively, and the over dot denotes the first-order derivative with respect to time t.

Since the number of DOFs is usually large, thus $[k]$, $[m]$, and $[g]$ are obtained by numerical integration. For the element with GLL nodes, GLL quadrature is used in the numerical integration. For the uniform thin beam under constant axial compression, the elements k_{ij} in stiffness matrix are

$$k_{ij} = EI \int_0^L \frac{d^2 l_i(x)}{dx^2} \frac{d^2 l_j(x)}{dx^2} dx = EI \int_{-1}^{1} \frac{d^2 l_i(\xi)}{d\xi^2} \frac{d^2 l_j(\xi)}{d\xi^2} \frac{16}{L^4}|J(\xi)| d\xi$$

$$= \frac{16EI}{L^4} \sum_{k=1}^{N} H_k \frac{d^2 l_i(\xi_k)}{d\xi^2} \frac{d^2 l_j(\xi_k)}{d\xi^2} L/2 = \frac{8EI}{L^3} \sum_{k=1}^{N} H_k B_{ki} B_{kj} \quad (4.88)$$

$$(i,j = 1,2,...,N,N+1,N+2)$$

where $|J(x)| = |\partial x / \partial \xi| = L / 2$ is the determinant of Jacobian matrix, ξ_i and H_i are abscissas and weights, respectively, of the N-point GLL quadrature which can be found in Table I.1 in Appendix I for N up to 21, B_{ki} or B_{kj} are the weighting coefficients of the second-order derivative with respect to ξ in the modified DQM. The derivative DOF can be introduced by either MMWC-3, that is, Eq. (3.17), or MMWC-4, that is, Eq. (3.21). This formulation is different from the time-domain spectral element, high-order finite element, or quadrature Timoshenko beam element. The formulation is more flexible in quadrature thin beam element, since various methods in the DQM can be used to determine the weighting coefficients of the second-order derivative.

Care should be taken that in Eq. (4.88), terms in B_{ki} and B_{kj} related to the DOF of $\partial v / \partial \xi$ should be multiplied by a factor of $L/2$, since not the $\partial v / \partial \xi$, but the $\partial v / \partial x$ is the DOF. Otherwise, the assembled structural stiffness matrix by quadrature thin beam elements with different length would be incorrect. The reason is that the weighting coefficients of the second-order derivative are formulated in the region $[-1,1]$ and not in the region $[0,L]$, $\partial v / \partial x$ is the same at the common node jointed by two quadrature thin beam elements, but $\partial v / \partial \xi$ is not if the length of the two elements is different.

The elements m_{ij} in the mass matrix are

$$m_{ij} = \rho A \int_0^L l_i(x) l_j(x) dx = \rho A \int_{-1}^{1} l_i(\xi) l_j(\xi) |J(\xi)| d\xi$$
$$= \rho A \sum_{k=1}^{N} H_k l_i(\xi_k) l_j(\xi_k) L / 2 = \rho A L / 2 \sum_{k=1}^{N} H_k \delta_{ki} \delta_{kj} = \frac{\rho A L}{2} H_i \delta_{ij} \qquad (4.89)$$
$$(i, j = 1, 2, ..., N)$$

where

$$\delta_{ij} = \begin{cases} 1 & (i = j) \\ 0 & (i \neq j) \end{cases} \qquad (4.90)$$

The mass matrix is in diagonal form and only the deflection DOFs are included in Eq. (4.86) or Eq. (4.89), since terms related to the derivative DOF are zero by using either Eq. (4.83) or Eq. (4.84). This is quite different from the consistent mass matrix in the FEM. In the conventional FEM, the consistent mass matrix is a full matrix. This is the major advantage of using the GLL quadrature. Due to using the GLL quadrature to obtain the mass matrix, the mass matrix is not fully integrated, since the GLL quadrature is only accurate up to a polynomial of degree of $(2N - 3)$. However, the mass matrix is equivalent to the row or column summed fully integrated consistent mass matrix.

Assume the compressive force P is constant. Then the elements g_{ij} in the geometric stiffness matrix are given by

$$g_{ij} = P \int_0^L \frac{dl_i(x)}{dx} \frac{dl_j(x)}{dx} dx = P \int_{-1}^{1} \frac{dl_i(\xi)}{d\xi} \frac{dl_j(\xi)}{d\xi} \frac{4}{L^2} |J(\xi)| d\xi$$
$$= \frac{4P}{L^2} \sum_{k=1}^{N} H_k \frac{dl_i(\xi_k)}{d\xi} \frac{dl_j(\xi_k)}{d\xi} \frac{L}{2} = \frac{2P}{L} \sum_{k=1}^{N} H_k A_{ki} A_{kj} \qquad (4.91)$$
$$(i, j = 1, 2, ..., N)$$

It should be pointed out that the terms related to the derivative DOF are also zero in matrix $[g]$. In other words, only the deflection DOFs are included in Eqs. (4.87) and (4.91).

If Eq. (4.84) is used, simply change $l_j(\xi_k)$ and $l_i(\xi_k)$ in Eqs. (4.88)–(4.91) to $h_i(\xi_k)$ and $h_j(\xi_k)$, respectively. The weighting coefficients can be explicitly computed by using Eqs. (2.14)–(2.18). The difference is that the terms in matrix $[g]$ related to the derivative DOFs are no longer zero. The range of subscripts i and j in Eq. (4.91) should be changed from N to $N+2$, namely, $(i, j = 1, 2, ...N, N+1, N+2)$. In such cases, both matrix $[m]$ and matrix $[g]$ are not fully integrated due to the use of the GLL quadrature rule.

If other distributions of nodes, such as the Chebyshev points or the approximate Lebesgue-optimal grid points, that is, the expanded-Chebyshev grid points [5,6], are used, either GLL quadrature or Gauss quadrature can be used for numerical integration. Then the newly proposed method by the author and his research associates [7] should be used.

$\xi_k (k = 1, 2, ...N)$ denote the node points and $\varsigma_k (k = 1, 2, ...N+2)$ the Grid V which containing N Gauss integration points, the corresponding N weights are $\bar{H}_k (k = 2, 3, ..., N+1)$. Since the shape function of the N-node beam element, for example, $h_j(\xi)$ in Eq. (4.84), is an $(N+1)$th-order polynomial, it can be interpolated by Ref. [7]

$$h_j(\xi) = \sum_{k=1}^{N+2} \hat{l}_k(\xi) h_j(\varsigma_k) = \sum_{k=1}^{N+2} \hat{l}_k(\xi) h_{kj} \tag{4.92}$$

where $\hat{l}_k(\xi)$ are Lagrange interpolation functions based on grid points ς_k, defined by

$$\hat{l}_j(\xi) = \frac{(\xi - \varsigma_1)(\xi - \varsigma_2)...(\xi - \varsigma_{j-1})(\xi - \varsigma_{j+1})...(\xi - \varsigma_{N+2})}{(\varsigma_j - \varsigma_1)(\varsigma_j - \varsigma_2)...(\varsigma_j - \varsigma_{j-1})(\varsigma_j - \varsigma_{j+1})...(\varsigma_j - \varsigma_{N+2})} = \prod_{\substack{k=1 \\ k \neq j}}^{N+2} \frac{\xi - \varsigma_k}{\varsigma_j - \varsigma_k} \tag{4.93}$$

The h_{jk} in Eq. (4.92) are the values of shape function at grid points ς_k, namely,

$$h_j(\xi)_{\xi=\varsigma_k} = h_j(\varsigma_k) = h_{kj} \tag{4.94}$$

With the proposed method [7], the weighting coefficients of the first- and second-order derivatives of the shape functions at the integration points can be computed by

$$\begin{cases} \bar{A}_{ij} = h'_j(\varsigma_i) = \sum_{k=1}^{N+2} \hat{l}'_k(\varsigma_i) h_j(\varsigma_k) = \sum_{k=1}^{N+2} \hat{l}'_k(\varsigma_i) h_{kj} = \sum_{k=1}^{N+2} \hat{A}_{ik} h_{kj} \\ \bar{B}_{ij} = h''_j(\varsigma_i) = \sum_{k=1}^{N+2} \hat{l}''_k(\varsigma_i) h_j(\varsigma_k) = \sum_{k=1}^{N+2} \hat{l}''_k(\varsigma_i) h_{kj} = \sum_{k=1}^{N+2} \hat{B}_{ik} h_{kj} \end{cases} \tag{4.95}$$

where the over bar means that the weighting coefficients are computed on the basis of points ς_k and not on the element nodes, and $\hat{B}_{ik}$ can also be conveniently computed by

$$\hat{B}_{ik} = \sum_{l=1}^{N+2} \hat{A}_{il} \hat{A}_{lk} \tag{4.96}$$

in which $\hat{A}_{ij}$ can be explicitly computed by

$$\hat{A}_{ij} = \hat{l}'_j(\varsigma_i) = \begin{cases} \prod_{\substack{k=1 \\ k \neq i, j}}^{N+2} (\varsigma_i - \varsigma_k) \Big/ \prod_{\substack{k=1 \\ k \neq j}}^{N+2} (\varsigma_j - \varsigma_k) & (i \neq j) \\ \sum_{\substack{k=1 \\ k \neq i}}^{N+2} \frac{1}{(\varsigma_i - \varsigma_k)} & (i = j) \end{cases} \tag{4.97}$$

It is seen that the way to compute the weighting coefficients by the proposed method is far simpler than the explicit formulas presented in Chapter 2. Once $\bar{B}_{ij}$ are known, the elements in stiffness matrix can be computed by

$$k_{ij} = \frac{8EI}{L^3} \sum_{k=2}^{N+1} \bar{H}_k h_i''(\varsigma_k) h_j''(\varsigma_k) = \frac{8EI}{L^3} \sum_{k=2}^{N+1} \bar{H}_k \bar{B}_{ki} \bar{B}_{kj} \quad (i, j = 1, 2, ..., N+2) \tag{4.98}$$

The elements g_{ij} in the geometric stiffness matrix are given by

$$g_{ij} = \frac{2P}{L} \sum_{k=2}^{N+1} \bar{H}_k \bar{A}_{ki} \bar{A}_{kj} \quad (i, j = 1, 2, ..., N, N+1, N+2) \tag{4.99}$$

The technique of row summation should be used to obtain a diagonal mass matrix. If Eq. (4.83) is used to get the mass matrix, the diagonal term in the mass matrix can be computed by

$$m_{jj} = \frac{\rho AL}{2} \sum_{k=2}^{N+1} \bar{H}_k l_j(\varsigma_k) = \frac{\rho AL}{2} \sum_{k=2}^{N+1} \bar{H}_k l_{kj} \quad (j = 1, 2, ..., N) \tag{4.100}$$

Note that only terms related to the deflection DOFs are included in Eq. (4.100).

It should be mentioned that different from Eq. (4.10), Eq. (4.92) is not exact but accurate enough. If an exact method is required, simply replace $l_j(x)$ in Eqs. (2.8) and (2.11) by Eq. (4.10). The weighting coefficient of its first-order derivative at integration points can be calculated by Eq. (4.12) and the weighting coefficients of higher-order derivatives at integration points can be calculated in a similar way. Thus, Eqs. (2.14)–(2.16) can be used to explicitly compute the weighting coefficients at all integration points, simply replace x_i by ς_i and $l_j^{[k]}(x_i)$ by $l_j^{[k]}(\varsigma_i)$, where $\varsigma_i (i = 1, 2, ..., N)$ are the N integration points.

4.8 QUADRATURE THIN RECTANGULAR PLATE ELEMENT

For simplicity and illustration, a thin anisotropic rectangular plate element with uniform thickness h is considered.

A 25-node quadrature rectangular plate element is schematically shown in Fig. 4.3. GLL points are used as the nodes of the element, different from the early version of the quadrature plate element in Ref. [2]. The Cartesian coordinate system is set at the middle plane of the plate. Since the differential equation of an isotropic skew plate shows mathematically analogy to the one of an anisotropic rectangular plate, the quadrature thin plate element can also be used to analyze the isotropic skew thin plate problem.

The strain energy for a rectangular thin plate element with uniform thickness is given by

$$U = \frac{1}{2} \int_0^a \int_0^b \{\kappa\}^T [D] \{\kappa\} \, dy \, dx \tag{4.101}$$

where $[D]$ is a 3×3 stiffness matrix of the material, and $\{\kappa\}$ is the curvature vector defined by

$$\{\kappa\}^T = \left| \begin{array}{ccc} \dfrac{\partial^2 w}{\partial x^2} & \dfrac{\partial^2 w}{\partial y^2} & 2\dfrac{\partial^2 w}{\partial x \partial y} \end{array} \right| \tag{4.102}$$

where $w(x, y, t)$ is the deflection.

The kinetic energy of the plate element is given by

$$T = \frac{1}{2} \int_0^a \int_0^b \rho h \left(\frac{\partial w(x,y,t)}{\partial t} \right)^2 dy\, dx \tag{4.103}$$

where t is the time, and ρ the mass density of the material.

The work done by the in-plane forces, that is, $P_x(x, y) = \sigma_x(x, y)h$, $P_y(x, y) = \sigma_y(x, y)h$, and $P_{xy}(x, y) = \tau_y(x, y)h$, is given by

$$
\begin{aligned}
W_n &= \frac{1}{2} \int_0^a \int_0^b \left[P_x \left(\frac{\partial w}{\partial x} \right)^2 + P_y \left(\frac{\partial w}{\partial y} \right)^2 + 2P_{xy} \left(\frac{\partial w}{\partial x} \right) \left(\frac{\partial w}{\partial y} \right) \right] dy\, dx \\
&= \frac{1}{2} \int_0^a \int_0^b \left[\frac{\partial w}{\partial x}, \frac{\partial w}{\partial y} \right] \begin{bmatrix} P_x & P_{xy} \\ P_{xy} & P_y \end{bmatrix} \begin{Bmatrix} \dfrac{\partial w}{\partial x} \\ \dfrac{\partial w}{\partial y} \end{Bmatrix} dy\, dx \\
&= \frac{1}{2} \int_0^a \int_0^b \{e\}^T [P] \{e\}\ dy\, dx
\end{aligned}
\tag{4.104}
$$

where $\sigma_x(x, y)$, $\sigma_y(x, y)$, $\tau_{xy}(x, y)$ are in-plane stress components. For buckling analysis, the positive direction of the in-plane stress components is just opposite to the ones defined in the theory of elasticity.

Equations (4.101), (4.103), and (4.104) are used to derive the stiffness matrix, mass matrix, and the geometric stiffness matrix of the quadrature thin plate element, respectively. GLL quadrature is used for numerical integration. The DQ rule is used to compute the derivatives at all integration points. It is seen that the formulation is different from the early version of quadrature plate element presented in Ref. [2].

4.8.1 QUADRATURE RECTANGULAR PLATE ELEMENT WITH LAGRANGE INTERPOLATION

For the plate element with Lagrange interpolation, each inner node has one DOF (w), each boundary node has two DOFs (w, w_x or w, w_y), and each corner node has three DOFs (w, w_x, w_y). The unknowns are exactly the same as the ones in the strong form of the new version of the DQEM presented in Chapter 2. Similar to the strong-form DQEM, $N_\xi = N_\eta = N$, where N_ξ, N_η are the number of nodes in x, y (or ξ, η) directions.

The deflection function for the $N \times N$-node quadrature rectangular plate element is

$$
\begin{aligned}
w(x, y, t) &= \sum_{i=1}^{N} \sum_{j=1}^{N} l_i(x) l_j(y) w(x_i, y_j, t) \\
&= \sum_{i=1}^{N} \sum_{j=1}^{N} l_i(\xi) l_j(\eta) w_{ij}(t)
\end{aligned}
\tag{4.105}
$$

where $w_{ij}(t)$ is the nodal deflection, and the shape function $l_i(\xi)$ or $l_j(\eta)$ is the Lagrange interpolation function, defined by Eq. (1.3).

It is seen that the first-order derivative w_x or w_y at plate element boundaries are not included in Eq. (4.105) and will be introduced as the DOFs of the plate element during formulation of the weighting coefficients of the second-order derivative at the boundary points.

Substituting Eq. (4.105) into Eqs. (4.101), (4.103), and (4.104), respectively, results

$$U = \frac{1}{2}\{\overline{w}\}^T [k]\{\overline{w}\} \tag{4.106}$$

$$T = \frac{1}{2}\{\dot{w}\}^T [m]\{\dot{w}\} \tag{4.107}$$

$$W_n = \frac{1}{2}\{w\}^T [g]\{w\} \tag{4.108}$$

where $[k]$ is the stiffness matrix, $[m]$ is the mass matrix, $[g]$ is the geometric stiffness matrix, $\{\overline{w}\}$ and $\{\dot{w}\}$ are nodal displacement vector and nodal velocity vector, respectively, and the over dot denotes the first-order derivative with respect to time t. Note that $\{\overline{w}\}$ contains $w_{ij} (i, j = 1,2,...,N)$, $(w_x)_{i1}, (w_x)_{iN} (i = 1,2,...N), (w_y)_{1j}, (w_y)_{Nj} (j = 1,2,...N)$, where w_x, w_y are the first-order derivative with respect to x and y, respectively. It should be emphasized that w_x, w_y are not the first-order derivative with respect to ξ and η.

Since the number of DOFs is usually large, thus $[k]$, $[m]$, and $[g]$ are obtained by numerical integration. GLL quadrature is used since GLL points are used as the element nodes. The stiffness matrix $[k]$ is given by

$$[k] = \frac{ab}{4} \sum_{i=1}^{N} \sum_{j=1}^{N} H_i H_j [B(\xi_i, \eta_j)]^T [D][B(\xi_i, \eta_j)] \tag{4.109}$$

where (ξ_i, η_j) and H_i, H_j are abscissas and weights, respectively, of the $N \times N$-point GLL quadrature rule, $[B(\xi_i, \eta_j)]$ is the strain matrix at point (ξ_i, η_j), containing the weighting coefficients of the first- and the second-order derivatives with respect to x and y in the modified DQM, computed by using Eq. (1.30) and MMWC-3 or MMWC-4, that is, Eq. (3.17) or Eq. (3.21). In details, $[B(\xi_i, \eta_j)]$ is defined by

$$[B(\xi_i, \eta_j)]\{\overline{w}\} = \begin{bmatrix} \dfrac{4}{a^2} \displaystyle\sum_{k=1}^{N+2} B_{ik}^{\xi} \overline{w}_{kj} \\[2mm] \dfrac{4}{b^2} \displaystyle\sum_{k=1}^{N+2} B_{jk}^{\eta} \overline{w}_{ik} \\[2mm] \dfrac{8}{ab} \displaystyle\sum_{l=1}^{N+2}\sum_{k=1}^{N+2} A_{il}^{\xi} A_{jk}^{\eta} \overline{w}_{lk} \end{bmatrix} \quad (i, j = 1,2,...,N) \tag{4.110}$$

where superscripts ξ and η imply that the weighting coefficients of the corresponding derivative is taken with respect to ξ and η, the over bar in $\overline{w}_{ik}$ means that deflection as well as the first-order derivatives with respect to x and y along boundaries are included. A_{ik}, B_{jk} are the weighting coefficients of the first- and second-order derivatives with respect to ξ or η in the modified DQM, respectively. Care should be taken that terms related to the DOF of $\partial w / \partial \xi$ in B_{ik}^{ξ} should be multiplied by a factor of $a/2$, and terms related to the DOF of $\partial w / \partial \eta$ in B_{jk}^{η} should be multiplied by a factor of $b/2$, since the first-order partial derivatives $\partial w / \partial x$ and $\partial w / \partial y$ are the nodal DOFs. Otherwise, the structural stiffness matrix obtained by the assemblage of the stiffness matrix of elements with different length and width will be incorrect.

The mass matrix is in diagonal form and its diagonal terms m_{ll} are

$$m_{ll} = \frac{\rho hab}{4} H_i H_j \quad (i, j = 1,2,...,N) \tag{4.111}$$

where

$$I = N \times (i-1) + j \qquad (i, j = 1, 2, ..., N) \tag{4.112}$$

Note that only the deflection DOFs are included in Eq. (4.107) or Eq. (4.111), since terms related to the first-order derivative DOFs are zero. This is quite different from the consistent mass matrix in FEM. In the FEM, the consistent mass matrix is a full matrix. Due to the use of the GLL quadrature to obtain the mass matrix, the mass matrix is not fully integrated since the GLL quadrature is only accurate up to a polynomial of degree of $(2N - 3)$. However, the mass matrix is equivalent to the row or column summed fully integrated consistent mass matrix.

The geometric stiffness matrix $[g]$ is given by

$$[g] = \frac{ab}{4} \sum_{i=1}^{N} \sum_{j=1}^{N} H_i H_j [E(\xi_i, \eta_j)]^T [P(\xi_i, \eta_j)][E(\xi_i, \eta_j)] \tag{4.113}$$

where $[E(\xi_i, \eta_j)]$ is defined by

$$[E(\xi_i, \eta_j)]\{\overline{w}\} = \begin{bmatrix} \dfrac{2}{a} \sum_{k=1}^{N} A_{ik}^{\xi} w_{kj} \\[2mm] \dfrac{2}{b} \sum_{k=1}^{N} A_{jk}^{\eta} w_{ik} \end{bmatrix} \qquad (i = 1, 2, ..., N; j = 1, 2, ..., N) \tag{4.114}$$

in which A_{ik} is the weighting coefficients of the first-order derivative with respect to ξ or η in the modified DQM. Note that the derivative DOFs are not involved in Eq. (4.108) or Eq. (4.113). This is quite different from the FEM. Due to using the GLL quadrature, the stiffness matrix, mass matrix, and the geometrical stiffness matrix are all approximately integrated. In other words, reduced integration is used in one or both directions during the formulations of the quadrature plate element matrix equations with Lagrange interpolations.

4.8.2 QUADRATURE RECTANGULAR PLATE ELEMENT WITH HERMITE INTERPOLATION

For the plate element with Hermite interpolation, each inner node has one DOF (w), each boundary node has two DOFs (w, w_x or w, w_y), and each corner node has four DOFs (w, w_x, w_y, w_{xy}) different from the element with Lagrange interpolation. The unknowns are exactly the same as the ones in the strong-form DQEM.

The deflection function for an $N \times N$-node quadrature rectangular thin plate element in bending is given by

$$\begin{aligned} w(x, y, t) &= \sum_{i=1}^{N+2} \sum_{j=1}^{N+2} h_i(x) h_j(y) \overline{w}(x_i, y_j, t) \\ &= \sum_{i=1}^{N+2} \sum_{j=1}^{N+2} h_i(\xi) h_j(\eta) \overline{w}_{ij}(t) \end{aligned} \tag{4.115}$$

where $\overline{w}_{ij}(t)$ is the nodal deflection, the first-order derivative at boundary points and the mixed second-order derivative at four corner points, the shape function $h_i(\xi)$ or $h_j(\eta)$ is the Hermite interpolation function, defined by Eq. (2.7).

Substituting Eq. (4.115) into Eqs. (4.101), (4.103), and (4.104), respectively, results

$$U = \frac{1}{2}\{\overline{w}\}^T [k]\{\overline{w}\} \tag{4.116}$$

$$T = \frac{1}{2}\{\dot{w}\}^T [m]\{\dot{w}\} \tag{4.117}$$

$$W_n = \frac{1}{2}\{\overline{w}\}^T [g]\{\overline{w}\} \tag{4.118}$$

where $[k]$ is the stiffness matrix, $[m]$ is the mass matrix, $[g]$ is the geometric stiffness matrix, $\{\overline{w}\}$ and $\{\dot{w}\}$ are nodal displacement vector and nodal velocity vector, respectively, and the over dot denotes the first-order derivative with respect to time t. Note that $\{\overline{w}\}$ contains $w_{ij}(i, j = 1, 2,..., N)$, $(w_x)_{i1}$, $(w_x)_{iN}$ $(i = 1, 2,..., N)$, $(w_y)_{1j}$, $(w_y)_{Nj}$ $(j = 1, 2,..., N)$, and $(w_{xy})_{ij}$ $(i = 1, N; j = 1, N)$, where w_x, w_y are the first-order derivative with respect to x or y and w_{xy} is the mixed second-order derivative with respect to x and y.

Since the number of DOFs is usually large, $[k]$, $[m]$, and $[g]$ are obtained by numerical integration. GLL quadrature is used in the QEM with GLL nodes. The stiffness matrix $[k]$ is given by

$$[k] = \frac{ab}{4} \sum_{i=1}^{N} \sum_{j=1}^{N} H_i H_j [B(\xi_i, \eta_j)]^T [D][B(\xi_i, \eta_j)] \tag{4.119}$$

where (ξ_i, η_j) and H_i, H_j are abscissas and weights, respectively, of the $N \times N$-point GLL quadrature rule, and $[B(\xi_i, \eta_j)]$ is the strain matrix at point (ξ_i, η_j), containing the weighting coefficients of the first- and the second-order derivatives with respect to ξ and η in the DQEM, computed by using Eqs. (2.14)–(2.18). In details, $[B(\xi_i, \eta_j)]$ are defined by

$$[B(\xi_i, \eta_j)]\{\overline{w}\} = \begin{bmatrix} \dfrac{4}{a^2} \displaystyle\sum_{k=1}^{N+2} B_{ik}^{\xi} \overline{w}_{kj} \\[2ex] \dfrac{4}{b^2} \displaystyle\sum_{k=1}^{N+2} B_{jk}^{\eta} \overline{w}_{ik} \\[2ex] \dfrac{8}{ab} \displaystyle\sum_{l=1}^{N+2} \sum_{k=1}^{N+2} A_{il}^{\xi} A_{jk}^{\eta} \overline{w}_{lk} \end{bmatrix} \quad (i, j = 1, 2, ..., N) \tag{4.120}$$

where superscripts ξ and η imply that the weighting coefficients of the corresponding derivatives are taken with respect to ξ or η, the over bar in $\overline{w}_{ik}$ means that deflections, the first-order derivatives with respect to x or y along boundaries as well as the mixed second-order derivatives with respect to x and y at four corner points are included. A_{ik}, A_{jk} are the weighting coefficients of the first- and second-order derivatives with respect to ξ or η (depending on the superscript) in the DQEM.

Care should be taken that the terms related to the DOF of $\partial w/\partial \xi$ in $A_{il}^{\xi}, B_{ik}^{\xi}$ should be multiplied by a factor of $a/2$, and the terms related to the DOF of $\partial w/\partial \eta$ in $A_{il}^{\eta}, B_{ik}^{\eta}$ should be multiplied by a factor of $b/2$, since the DOFs are the first-order derivatives $\partial w/\partial x$ and $\partial w/\partial y$ and the mixed second-order derivative $\partial^2 w/\partial x \partial y$, but not the first-order derivatives $\partial w/\partial \xi$ and $\partial w/\partial \eta$ and the mixed second-order derivative $\partial^2 w/\partial \xi \partial \eta$. Otherwise, the structural stiffness matrix formulated by the assemblage of elemental stiffness matrixes for elements with different length and width would be incorrect.

Due to the property of Hermite interpolation functions at the nodal points (the GLL points), the mass matrix is in a diagonal form. Besides it does not contain terms related to the DOFs of the first-order derivatives and the mixed second-order derivatives.

The diagonal terms m_{II} in mass matrix are

$$m_{II} = \frac{\rho h a b}{4} H_i H_j \qquad (i, j = 1, 2, ..., N) \tag{4.121}$$

where

$$I = N \times (i - 1) + j \qquad (i, j = 1, 2, ..., N) \tag{4.122}$$

The geometric stiffness matrix $[g]$ is given by

$$[g] = \frac{ab}{4} \sum_{i=1}^{N} \sum_{j=1}^{N} H_i H_j [E(\xi_i, \eta_j)]^T [P(\xi_i, \eta_j)][E(\xi_i, \eta_j)] \tag{4.123}$$

where $[E(\xi_i, \eta_j)]$ are defined by

$$[E(\xi_i, \eta_j)]\{\overline{w}\} = \begin{bmatrix} \dfrac{2}{a} \sum_{k=1}^{N+2} A_{ik}^{\xi} \overline{w}_{kj} \\ \dfrac{2}{b} \sum_{k=1}^{N+2} A_{jk}^{\eta} \overline{w}_{ik} \end{bmatrix} \qquad (i = 1, 2, ..., N; j = 1, 2, ..., N) \tag{4.124}$$

in which A_{ik} is the weighting coefficients of the first-order derivative with respect to ξ or η (depending on the superscript) in the DQEM. For the same reason, the terms related to the DOF of $\partial w/\partial \xi$ in A_{ik}^{ξ} should be multiplied by a factor of $a/2$, and the terms related to the DOF of $\partial w/\partial \eta$ in A_{jk}^{η} should be multiplied by a factor of $b/2$.

Note that the derivative DOFs are involved in Eq. (4.124), different from Eq. (4.113). Due to the use of the GLL quadrature, the stiffness matrix, mass matrix, and the geometrical stiffness matrix are all approximately integrated. In other words, reduced integration is used in one or both directions during the formulations of the quadrature plate element matrix equations with Hermite interpolations.

4.8.3 QUADRATURE RECTANGULAR PLATE ELEMENT WITH MIXED INTERPOLATIONS

Due to the introduction of the mixed second-order derivative w_{xy} as the DOF at corner points, the element is neither convenient for applying the boundary conditions nor easy to be extended to a quadrilateral plate element.

In order not to include the mixed second-order derivative w_{xy} as the DOF at corner points, the weighting coefficients of the first-order derivative are determined by using the Lagrange interpolation but not by the Hermite interpolation [4]. In other words, the weighting coefficients of the first-order derivative in Eq. (4.110) are determined by using Lagrange interpolation and the weighting coefficients of the second-order derivatives in Eq. (4.110) are determined by using Hermite interpolation. It does not matter whether Lagrange interpolation functions or Hermite interpolation functions are used as the shape functions to formulate the mass matrix, since the results are the same. In this way, the displacement vector does not contain the mixed second-order derivative w_{xy}. All inner nodes have one DOF (w), each boundary node has two DOFs (w, w_x or w, w_y), and each corner node has three DOFs (w, w_x, w_y). The unknowns are exactly the same as the quadrature rectangular plate element with Lagrange interpolations. However, the geometric stiffness matrix $[g]$ may be different from Eq. (4.114), since weighting coefficients of the first-order derivative in Eq. (4.124) can be determined either by the Hermite interpolations or by the Lagrange interpolations. In other words, to remove the DOFs w_{xy}, the weighting coefficients of the first-order derivative in Eq. (4.120) are determined by using the Lagrange interpolations.

Alternatively, the weighting coefficients of the first- and second-order derivatives are determined by using the Lagrange interpolation in the ξ direction and by using the Hermite interpolation in the η direction, namely, the element is formulated with mixed Lagrange and Hermite interpolations [11]. Again each inner node has one DOF (w), each boundary node has two DOFs (w, w_x or w, w_y), and each corner node has three DOFs (w, w_x, w_y). The unknowns are exactly the same as the quadrature rectangular plate element with Lagrange interpolations.

Assume that Lagrange interpolation is used in the x or ξ direction and Hermite interpolation is used in the y or η direction. The deflection function for the $N \times N$-node mixed quadrature rectangular plate element is given by

$$
\begin{aligned}
w(x,y,t) &= \sum_{i=1}^{N}\sum_{j=1}^{N+2} l_i(x)h_j(y)\overline{w}(x_i,y_j,t) \\
&= \sum_{i=1}^{N}\sum_{j=1}^{N+2} l_i(\xi)h_j(\eta)\overline{w}_{ij}(t)
\end{aligned}
\tag{4.125}
$$

where $\overline{w}_{ij}(t)$ is the nodal deflection and the first-order derivative w_y taken with respect to y, different from Eq. (4.105) or Eq. (4.115), and the shape function $l_i(\xi)$ or $h_j(\eta)$ are the Lagrange interpolation functions and Hermite interpolation functions, defined by Eq. (1.3) and Eq. (2.7), respectively.

Substituting Eq. (4.125) into Eqs. (4.101), (4.103), and (4.104) results

$$
U = \frac{1}{2}\{\overline{w}\}^T [k]\{\overline{w}\}
\tag{4.126}
$$

$$
T = \frac{1}{2}\{\dot{w}\}^T [m]\{\dot{w}\}
\tag{4.127}
$$

$$
W_n = \frac{1}{2}\{\overline{w}\}^T [g]\{\overline{w}\}
\tag{4.128}
$$

where $[k]$ is the stiffness matrix, $[m]$ is the mass matrix, $[g]$ is the geometric stiffness matrix, $\{\overline{w}\}$ and $\{\dot{w}\}$ are nodal displacement vector and nodal velocity vector, respectively, and the over dot denotes the first-order derivative with respect to time t. Note that $\{\overline{w}\}$ contains w_{ij} ($i, j = 1,2,...N$), $(w_x)_{i1}$, $(w_x)_{iN}$ ($i = 1,2,...N$), $(w_j)_{1j}$, $(w_y)_{Nj}$ ($j = 1,2,...N$), where w_x, w_y are the first-order derivatives with respect to x or y, respectively. It should be emphasized again that w_x, w_y are not the first-order derivatives with respect to ξ and η, respectively, this is similar to the FEM. Besides, the first-order derivatives w_x at plate boundary ($x = 0$ and $x = a$ shown in Fig. 4.3) are introduced during formulation of the weighting coefficients of the second-order derivative at the boundary points, although they do not appear in Eq. (4.125).

Since the number of DOFs is usually large, $[k]$, $[m]$, and $[g]$ are obtained by numerical integration. GLL quadrature is used for the element with GLL nodes. The stiffness matrix $[k]$ is given by

$$
[k] = \frac{ab}{4}\sum_{i=1}^{N}\sum_{j=1}^{N} H_i H_j [B(\xi_i,\eta_j)]^T [D][B(\xi_i,\eta_j)]
\tag{4.129}
$$

where (ξ_i, η_j) and H_i, H_j are abscissas and weights, respectively, of the $N \times N$-point GLL quadrature rule, $[B(\xi_i,\eta_j)]$ is the strain matrix at point (ξ_i,η_j), containing the weighting coefficients of the first- and the second-order derivatives with respect to ξ in the modified DQM and the weighting coefficients of the first- and the second-order derivatives with respect to η in the DQEM. In other words, the weighting coefficients of the first-order and second-order derivatives with respect to ξ are computed by using Eq. (1.30) together with MMWC-3, that is, Eq. (3.17), the weighting coefficients of the first-order and second-order derivatives with respect to η are computed explicitly by using Eqs. (2.14)–(2.18). In details, $[B(\xi_i, \eta_j)]$ are defined by

$$[B(\xi_i, \eta_j)]\{\bar{w}\} = \begin{bmatrix} \dfrac{4}{a^2} \displaystyle\sum_{k=1}^{N+2} B_{ik}^{\xi} \bar{w}_{kj} \\ \dfrac{4}{b^2} \displaystyle\sum_{k=1}^{N+2} B_{jk}^{\eta} \bar{w}_{ik} \\ \dfrac{8}{ab} \displaystyle\sum_{l=1}^{N+2} \sum_{k=1}^{N+2} A_{il}^{\xi} A_{jk}^{\eta} \bar{w}_{lk} \end{bmatrix} \qquad (i, j = 1, 2, ..., N) \tag{4.130}$$

where superscripts ξ or η imply that the weighting coefficients of the corresponding derivatives are taken with respect to ξ or η; the over bar in $\bar{w}_{ik}$ means that deflection as well as the first-order derivatives with respect to x and y along boundaries are included. The mixed second-order derivatives w_{xy} at four corner nodes are not involved. $A_{il}^{\xi}, B_{ik}^{\xi}$ are the weighting coefficients of the first- and second-order derivatives with respect to ξ in the modified DQM and $A_{jk}^{\eta}, B_{jk}^{\eta}$ are the weighting coefficients of the first- and second-order derivatives with respect to η in the DQEM.

Care should be taken that terms related to $\partial w/\partial \xi$ in B_{ik}^{ξ} should be multiplied by a factor of $a/2$ and terms related to $\partial w/\partial \eta$ in $A_{jk}^{\eta}, B_{jk}^{\eta}$ should be multiplied by a factor of $b/2$, since $\partial w/\partial x$ and $\partial w/\partial y$ are the DOFs. Otherwise, the structural stiffness matrix obtained by the assemblage of elemental stiffness matrices of quadrature plate elements with different length and width would be incorrect.

Due to the property of Lagrange and Hermite interpolation functions at the nodal points (the GLL points), the mass matrix is in a diagonal form. Besides, the mass matrix does not contain terms related to the DOFs of the first-order derivatives.

The diagonal terms m_{II} in mass matrix are

$$m_{II} = \frac{\rho h a b}{4} H_i H_j \qquad (i, j = 1, 2, ..., N) \tag{4.131}$$

where

$$I = N \times (i-1) + j \qquad (i, j = 1, 2, ..., N) \tag{4.132}$$

The geometric stiffness matrix $[g]$ is given by

$$[g] = \frac{ab}{4} \sum_{i=1}^{N} \sum_{j=1}^{N} H_i H_j [E(\xi_i, \eta_j)]^T [P(\xi_i, \eta_j)][E(\xi_i, \eta_j)] \tag{4.133}$$

where $[E(\xi_i, \eta_j)]$ are defined by

$$[E(\xi_i, \eta_j)]\{\bar{w}\} = \begin{bmatrix} \dfrac{2}{a} \displaystyle\sum_{k=1}^{N+2} A_{ik}^{\xi} \bar{w}_{kj} \\ \dfrac{2}{b} \displaystyle\sum_{k=1}^{N+2} A_{jk}^{\eta} \bar{w}_{ik} \end{bmatrix} \qquad (i = 1, 2, ..., N; j = 1, 2, ..., N) \tag{4.134}$$

where A_{ik}^{ξ} and A_{jk}^{η} are the weighting coefficients of the first-order derivatives with respect to ξ in the modified DQM and to η in the DQEM, respectively. For the same reason, terms related to the DOF of $\partial w/\partial \eta$ in A_{jk}^{η} should be multiplied by a factor of $b/2$.

If Hermite interpolation functions are used as the shape functions in the x or ξ direction and Lagrange interpolation functions are used as the shape functions in the y or η direction, the stiffness matrix and mass matrix of the mixed quadrature rectangular plate element can be formulated in a similar way.

It should be pointed out that the formulation of various weak-form quadrature plate elements is more flexible than the FEM or the time-domain SPE, since the derivatives at the integration points can be expressed in terms of the DQ rule, the rule of DQEM, the HDQ rule, and their combinations. For quadrature rectangular plate element with HDQ rule, the formulation is essentially the same as the quadrature rectangular plate element with DQ rule; thus, details are omitted.

4.9 **EXTENSION TO QUADRILATERAL PLATE ELEMENT WITH CURVED EDGES**

Similar to the quadrature plane element, the quadrature plate elements in bending can be also extended to quadrilateral elements even with curved edges. Due to complexity in transformation of the DOF $\partial^2 w/\partial\xi\partial\eta$ to $\partial^2 w/\partial x\partial y$ and difficulty in applying the essential boundary conditions, the quadrature plate element with DOFs $\partial^2 w/\partial\xi\partial\eta$ is not considered in this section.

Since the curvature vector defined by Eq. (4.102) involves second-order partial derivatives, equations for transforming the second-order partial derivatives from $\xi-\eta$ coordinate system to Cartesian $x-y$ coordinate system are also necessary.

Using the chain rule of partial differentiation yields [1,4,12]

$$\left\{ \begin{array}{c} \dfrac{\partial w}{\partial x} \\ \dfrac{\partial w}{\partial y} \end{array} \right\} = \left[\begin{array}{cc} \dfrac{\partial \xi}{\partial x} & \dfrac{\partial \eta}{\partial x} \\ \dfrac{\partial \xi}{\partial y} & \dfrac{\partial \eta}{\partial y} \end{array} \right] \left\{ \begin{array}{c} \dfrac{\partial w}{\partial \xi} \\ \dfrac{\partial w}{\partial \eta} \end{array} \right\} = \left[\begin{array}{cc} \xi_x & \eta_x \\ \xi_y & \eta_y \end{array} \right] \left\{ \begin{array}{c} \dfrac{\partial w}{\partial \xi} \\ \dfrac{\partial w}{\partial \eta} \end{array} \right\} \tag{4.135}$$

Using Eq. (4.42), it is easy to see that

$$\left[\begin{array}{cc} \xi_x & \eta_x \\ \xi_y & \eta_y \end{array} \right] = \dfrac{1}{|J|} \left[\begin{array}{cc} y_\eta & -y_\xi \\ -x_\eta & x_\xi \end{array} \right] \tag{4.136}$$

Using Eqs. (4.135) and (4.136), one has

$$\left\{ \begin{array}{l} \dfrac{\partial w}{\partial x} = \dfrac{y_\eta}{|J|}\dfrac{\partial w}{\partial \xi} - \dfrac{y_\xi}{|J|}\dfrac{\partial w}{\partial \eta} \\ \dfrac{\partial w}{\partial y} = \dfrac{x_\xi}{|J|}\dfrac{\partial w}{\partial \eta} - \dfrac{x_\eta}{|J|}\dfrac{\partial w}{\partial \xi} \end{array} \right. \tag{4.137}$$

Thus,

$$
\begin{aligned}
\frac{\partial^2 w}{\partial x^2} &= \frac{\partial}{\partial x}\left(\frac{y_\eta}{|J|}\frac{\partial w}{\partial \xi} - \frac{y_\xi}{|J|}\frac{\partial w}{\partial \eta}\right) = \left(\frac{y_\eta}{|J|}\frac{\partial}{\partial \xi} - \frac{y_\xi}{|J|}\frac{\partial}{\partial \eta}\right)\left(\frac{y_\eta}{|J|}\frac{\partial w}{\partial \xi} - \frac{y_\xi}{|J|}\frac{\partial w}{\partial \eta}\right) \\
&= \frac{y_\eta}{|J|}\left(\frac{y_\eta}{|J|}\frac{\partial^2 w}{\partial \xi^2} + \frac{y_{\xi\eta}}{|J|}\frac{\partial w}{\partial \xi} - \frac{y_\eta}{|J|^2}\frac{\partial |J|}{\partial \xi}\frac{\partial w}{\partial \xi} - \frac{y_\xi}{|J|}\frac{\partial^2 w}{\partial \xi \partial \eta} - \frac{y_{\xi\xi}}{|J|}\frac{\partial w}{\partial \eta} + \frac{y_\xi}{|J|^2}\frac{\partial |J|}{\partial \xi}\frac{\partial w}{\partial \eta}\right) \\
&\quad - \frac{y_\xi}{|J|}\left(\frac{y_\eta}{|J|}\frac{\partial^2 w}{\partial \xi \partial \eta} + \frac{y_{\eta\eta}}{|J|}\frac{\partial w}{\partial \xi} - \frac{y_\eta}{|J|^2}\frac{\partial |J|}{\partial \eta}\frac{\partial w}{\partial \xi} - \frac{y_\xi}{|J|}\frac{\partial^2 w}{\partial \eta^2} - \frac{y_{\xi\eta}}{|J|}\frac{\partial w}{\partial \eta} + \frac{y_\xi}{|J|^2}\frac{\partial |J|}{\partial \eta}\frac{\partial w}{\partial \eta}\right)
\end{aligned}
\tag{4.138}
$$

where

$$
y_{\xi\xi} = \frac{\partial^2 y}{\partial \xi^2}, \; y_{\xi\eta} = \frac{\partial^2 y}{\partial \xi \partial \eta}, \; y_{\eta\eta} = \frac{\partial^2 y}{\partial \eta^2}
\tag{4.139}
$$

By the same token, one has

$$
\begin{aligned}
\frac{\partial^2 w}{\partial y^2} &= \frac{\partial}{\partial y}\left(-\frac{x_\eta}{|J|}\frac{\partial w}{\partial \xi} + \frac{x_\xi}{|J|}\frac{\partial w}{\partial \eta}\right) = \left(\frac{x_\eta}{|J|}\frac{\partial}{\partial \xi} - \frac{x_\xi}{|J|}\frac{\partial}{\partial \eta}\right)\left(\frac{x_\eta}{|J|}\frac{\partial w}{\partial \xi} - \frac{x_\xi}{|J|}\frac{\partial w}{\partial \eta}\right) \\
&= \frac{x_\eta}{|J|}\left(\frac{x_\eta}{|J|}\frac{\partial^2 w}{\partial \xi^2} + \frac{x_{\xi\eta}}{|J|}\frac{\partial w}{\partial \xi} - \frac{x_\eta}{|J|^2}\frac{\partial |J|}{\partial \xi}\frac{\partial w}{\partial \xi} - \frac{x_\xi}{|J|}\frac{\partial^2 w}{\partial \xi \partial \eta} - \frac{x_{\xi\xi}}{|J|}\frac{\partial w}{\partial \eta} + \frac{x_\xi}{|J|^2}\frac{\partial |J|}{\partial \xi}\frac{\partial w}{\partial \eta}\right) \\
&\quad - \frac{x_\xi}{|J|}\left(\frac{x_\eta}{|J|}\frac{\partial^2 w}{\partial \xi \partial \eta} + \frac{x_{\eta\eta}}{|J|}\frac{\partial w}{\partial \xi} - \frac{x_\eta}{|J|^2}\frac{\partial |J|}{\partial \eta}\frac{\partial w}{\partial \xi} - \frac{x_\xi}{|J|}\frac{\partial^2 w}{\partial \eta^2} - \frac{x_{\xi\eta}}{|J|}\frac{\partial w}{\partial \eta} + \frac{x_\xi}{|J|^2}\frac{\partial |J|}{\partial \eta}\frac{\partial w}{\partial \eta}\right)
\end{aligned}
\tag{4.140}
$$

and

$$
\begin{aligned}
\frac{\partial^2 w}{\partial x \partial y} &= \frac{\partial}{\partial x}\left(-\frac{x_\eta}{|J|}\frac{\partial w}{\partial \xi} + \frac{x_\xi}{|J|}\frac{\partial w}{\partial \eta}\right) = \left(\frac{y_\eta}{|J|}\frac{\partial}{\partial \xi} - \frac{y_\xi}{|J|}\frac{\partial}{\partial \eta}\right)\left(-\frac{x_\eta}{|J|}\frac{\partial w}{\partial \xi} + \frac{x_\xi}{|J|}\frac{\partial w}{\partial \eta}\right) \\
&= \frac{y_\eta}{|J|}\left(-\frac{x_\eta}{|J|}\frac{\partial^2 w}{\partial \xi^2} - \frac{x_{\xi\eta}}{|J|}\frac{\partial w}{\partial \xi} + \frac{x_\eta}{|J|^2}\frac{\partial |J|}{\partial \xi}\frac{\partial w}{\partial \xi} + \frac{x_\xi}{|J|}\frac{\partial^2 w}{\partial \xi \partial \eta} + \frac{x_{\xi\xi}}{|J|}\frac{\partial w}{\partial \eta} - \frac{x_\xi}{|J|^2}\frac{\partial |J|}{\partial \xi}\frac{\partial w}{\partial \eta}\right) \\
&\quad - \frac{y_\xi}{|J|}\left(-\frac{x_\eta}{|J|}\frac{\partial^2 w}{\partial \xi \partial \eta} - \frac{x_{\eta\eta}}{|J|}\frac{\partial w}{\partial \xi} + \frac{x_\eta}{|J|^2}\frac{\partial |J|}{\partial \eta}\frac{\partial w}{\partial \xi} + \frac{x_\xi}{|J|}\frac{\partial^2 w}{\partial \eta^2} + \frac{x_{\xi\eta}}{|J|}\frac{\partial w}{\partial \eta} - \frac{x_\xi}{|J|^2}\frac{\partial |J|}{\partial \eta}\frac{\partial w}{\partial \eta}\right)
\end{aligned}
\tag{4.141}
$$

where

$$
x_{\xi\xi} = \frac{\partial^2 x}{\partial \xi^2}, \; x_{\xi\eta} = \frac{\partial^2 x}{\partial \xi \partial \eta}, \; x_{\eta\eta} = \frac{\partial^2 x}{\partial \eta^2}
\tag{4.142}
$$

Thus, the curvature vector can be rewritten by

$$
\{\kappa\} = \left[E(\xi,\eta)\right]\left[\begin{array}{ccccc}\dfrac{\partial w}{\partial \xi} & \dfrac{\partial w}{\partial \eta} & \dfrac{\partial^2 w}{\partial \xi^2} & \dfrac{\partial^2 w}{\partial \eta^2} & \dfrac{\partial^2 w}{\partial \xi \partial \eta}\end{array}\right]^T
\tag{4.143}
$$

With Eq. (4.40), the expression of each term in $[C]$, defined in Eq. (4.144), can be obtained. In terms of DQ, the strain at point (ξ_i, η_j) is now given by

$$[B(\xi_i,\eta_j)]\{\bar{w}_L\} = \left[C\left(\xi_i,\eta_j\right)\right] \begin{bmatrix} \sum\limits_{k=1}^{N+2} A_{ik}^{\xi}\bar{w}_{kj} \\ \sum\limits_{k=1}^{N+2} A_{jk}^{\eta}\bar{w}_{ik} \\ \sum\limits_{k=1}^{N+2} B_{ik}^{\xi}\bar{w}_{kj} \\ \sum\limits_{k=1}^{N+2} B_{jk}^{\eta}\bar{w}_{ik} \\ \sum\limits_{l=1}^{N+2}\sum\limits_{k=1}^{N+2} A_{il}^{\xi}A_{jk}^{\eta}\bar{w}_{lk} \end{bmatrix} \qquad (i,j=1,2,...,N) \qquad (4.144)$$

where $\{\bar{w}_L\}$ is nodal displacement vector in local coordinate ξ–η system and contains w_{ij} $(i, j = 1, 2,..., N)$, $(w_{\xi})_{i1}$, $(w_{\xi})_{iN}$ $(i = 1, 2,..., N_{\eta})$, $(w_{\eta})_{1j}$, $(w_{\eta})_{jN}$ $(j = 1, 2,..., N_{\xi})$, respectively.

The strain energy for a quadrilateral thin plate element with uniform thickness is now given by

$$\begin{aligned} U &= \frac{1}{2}\iint\limits_{Area}\{\kappa(x,y)\}^T[D]\{\kappa(x,y)\}dy\,dx \\ &= \frac{1}{2}\int_{-1}^{1}\int_{-1}^{1}\{\kappa(\xi,\eta)\}^T[D]\{\kappa(\xi,\eta)\}|J(\xi,\eta)|d\eta\,d\xi \\ &= \frac{1}{2}\{\bar{w}_L\}^T[k_L]\{\bar{w}_L\} \end{aligned} \qquad (4.145)$$

where $[k_L]$ is the stiffness matrix in local coordinate ξ–η system.

The kinetic energy of the thin quadrilateral plate element is given by

$$\begin{aligned} T &= \frac{1}{2}\iint\limits_{Area}\rho h\left(\frac{\partial w(x,y,t)}{\partial t}\right)^2 dy\,dx \\ &= \frac{1}{2}\int_{-1}^{1}\int_{-1}^{1}\rho h\left(\frac{\partial w(\xi,\eta,t)}{\partial t}\right)^2|J(\xi,\eta)|d\eta\,d\xi \\ &= \frac{1}{2}\{\dot{w}\}^T[m]\{\dot{w}\} \end{aligned} \qquad (4.146)$$

where $[m]$ is the mass matrix, and $\{\dot{w}\}$ is the nodal velocity vector, the over dot denotes the first-order derivative with respect to time t.

The work done by the in-plane forces is given by

$$\begin{aligned} W_n &= \frac{1}{2}\iint\limits_{Area}\left[\frac{\partial w}{\partial x},\frac{\partial w}{\partial y}\right]\begin{bmatrix} P_x & P_{xy} \\ P_{xy} & P_y \end{bmatrix}\begin{Bmatrix} \dfrac{\partial w}{\partial x} \\ \dfrac{\partial w}{\partial y} \end{Bmatrix} dy\,dx \\ &= \frac{1}{2}\int_{-1}^{1}\int_{-1}^{1}\{e(\xi,\eta)\}^T[P(\xi,\eta)]\{e(\xi,\eta)\}|J(\xi,\eta)|\,d\eta\,d\xi = \frac{1}{2}\{\bar{w}_L\}^T[g_L]\{\bar{w}_L\} \end{aligned} \qquad (4.147)$$

where $[g_L]$ is the geometric stiffness matrix in local coordinate ξ–η system, $\{e\}$ is apparent.

The stiffness matrix $[k]$ can be obtained by GLL quadrature rule and is given by

$$[k_L] = \sum_{i=1}^{N} \sum_{j=1}^{N} H_i H_j [B(\xi_i, \eta_j)]^T [D][B(\xi_i, \eta_j)] |J(\xi_i, \eta_j)| \tag{4.148}$$

The mass matrix $[m]$ is also in diagonal form and its diagonal terms m_{II} are given by

$$m_{II} = \rho h H_i H_j |J(\xi_i, \eta_j)| \qquad (i, j = 1, 2, ..., N) \tag{4.149}$$

where

$$I = N \times (i - 1) + j \qquad (i, j = 1, 2, ..., N) \tag{4.150}$$

And the geometric stiffness matrix $[g_L]$ is given by

$$[g_L] = \sum_{i=1}^{N} \sum_{j=1}^{N} H_i H_j \{e(\xi_i, \eta_j)\}^T [P(\xi_i, \eta_j)]\{e(\xi_i, \eta_j)\} |J(\xi_i, \eta_j)| \tag{4.151}$$

Unlike the quadrature thick plate elements, coordinate transformation is necessary for assemblage, since nodal displacement vector $\{\overline{w}_L\}$ contains the first-order derivatives with respect to ξ and η and the local coordinate $(\xi-\eta)$ system is generally different for different quadrature plate element. The first-order derivatives w_ξ or w_η at points on the common boundary is to be transformed to the normal slope w_n at the corresponding point.

Let $[T]$ be the coordinate transformation matrix, and the nodal displacement vector in global coordinate system be $\{\overline{w}\}$. For a quadrilateral plate element, $\{\overline{w}\}$ contains w_{ij} $(i, j = 1, 2, ..., N)$, $(w_n)_{i1}$, $(w_n)_{iN}$ $(i = 1, 2, ..., N_\eta)$, and $(w_n)_{1j}$, $(w_n)_{Nj}$ $(j = 1, 2, ..., N_\xi)$, where w_n is the normal slope on the element edges. Thus, one has

$$\{\overline{w}_L\} = [T]\{\overline{w}\} \tag{4.152}$$

For the DOFs w_{ij} $(i, j = 1, 2, ..., N)$, the coordinate transform is not necessary. Thus simply put 1 in the corresponding diagonal terms in $[T]$. With Eq. (4.137) one has

$$w_n = n_x w_x + n_y w_y$$
$$= \frac{w_\xi (y_\eta n_x - x_\eta n_y)}{|J|} + \frac{w_\eta (x_\xi n_y - y_\xi n_x)}{|J|} \tag{4.153}$$

where n_x and n_y are the direction cosines of the normal at a nodal point on the element edge. From Eq. (4.153) one has

$$w_\xi = \frac{|J| w_n - w_\eta \left[x_\xi n_y - y_\xi n_x \right]}{y_\eta n_x - x_\eta n_y} \tag{4.154}$$

Equation (4.154) is used to transform w_ξ at $\xi = \pm 1$ to w_n. In Eq. (4.154), w_n is retained as the DOF and w_η is expressed in terms of the DQ rule along the edge, namely, as the weighted sum of deflections at all boundary points along edge $\xi = -1$ or $\xi = 1$.

By the same token, one can transform w_η at $\eta = \pm 1$ to w_n. From Eq. (4.153) one has

$$w_\eta = \frac{|J| w_n - w_\xi \left[y_\eta n_x - x_\eta n_y \right]}{x_\xi n_y - y_\xi n_x} \tag{4.155}$$

where w_n is retained as the DOF, and w_ξ is expressed in terms of the DQ rule along the edge, that is, as the weighted sum of deflections at all boundary points along edge $\eta = -1$ or $\eta = 1$. Equations (4.154) and (4.155) are also reported in Ref. [4], however, most quadrature plate elements presented herein are different from the one presented in Ref. [4].

As was mentioned previously, the coordinate transformation is not necessary for the deflection, or equivalently put one at the diagonal term and zeros at all other terms for the corresponding row in $[T]$. Therefore, the stiffness matrix in global coordinate system is given by

$$[k] = [T]^T [k_L][T] \tag{4.156}$$

where the superscript T denotes the matrix transpose.

Since the velocity vector $\{\dot{w}\}$ in Eq. (4.146) does not contain the first-order derivative DOFs with respect to time t, coordinate transformation is not necessary for the mass matrix $[m]$; this is similar to the quadrature thick plate element.

Transformation for the geometric matrix may be necessary if the method of mixed interpolations is used. In such cases, the geometric stiffness matrix in global coordinate system is given by

$$[g] = [T]^T [g_L][T] \tag{4.157}$$

As was mentioned earlier, due to difficulty in transformation of $w_{\xi\eta}$ to w_{xy} at corners, the quadrature plate element with Hermite interpolation is only limited to rectangular plate or skew plate. It is known that the skew plate with isotropic materials is analogue to a rectangular plate with anisotropic materials.

Different from the strong-form DQEM, the stiffness matrix and geometric stiffness matrix of the quadrature elements are symmetric, the mass matrix is a diagonal matrix but the magnitude of diagonal terms is not equal to each other.

If the nodes are not GLL points, then Gauss quadrature is used for efficiently numerical integration and the proposed method in Ref. [7] should be used to employ the DQ rule to compute the weighting coefficients at Gauss points.

4.10 **DISCUSSION**

Usually the number of the nodal points in each direction is greater than five thus larger size of element can be used as compared to the conventional FEM. The QEM has overcome the limitations existing in the ordinary DQM and the DQEM, thus is expected to be an alternative method to the FEM for efficiently analyzing problems in the area of structural mechanics.

4.10.1 **ASSEMBLAGE PROCEDURES**

For structural analysis by the QEM, the assemblage procedures are exactly the same as the FEM if more than one element is used in the analysis. Physically, at each common nodal point, the generalized displacement components (the deflection and derivatives) at the quadrature element level are the same as those after assemblage. The sum of internal forces (the shearing force and bending moment) at each common nodal point should be equal to the corresponding external applied generalized loads. Since generalized forces and generalized displacements are vectors, the vectorial sum is usually replaced by the scale sum of their components in the common coordinate system at the connecting nodal point.

Similar to the FEM, the coordinate transformation may be necessary. Remember that the coordinate transformation may be only necessary for the first-order derivatives at the boundary points.

Care should be taken on the direction of normal slope when quadrilateral quadrature thin plate elements are involved during assemblage. If n_x and n_y are the direction cosines of the out normal at a point on the element edge, then the normal direction of the edge is opposite for the two neighboring elements, and only one of them is retained as the DOF in the structural displacement vector. In programming, this is a drawback for the quadrature thin plate element with degrees of first-order derivatives. If the first-order derivatives are not used as the DOFs for a quadrature thin plate element [3], the performance of the element will be degraded, since nonconforming elements are formulated.

4.10.2 WORK EQUIVALENT LOAD VECTOR

For static analysis, the distributed load should be lumped to nodal points by using work equivalent method. This is similar to the consistent load vector in conventional FEM.

For beam element under distributed load $q(x)$, the deflection for an N-node quadrature beam element is

$$v(x) = \sum_{i=1}^{N} l_i(x) v(x_i) = \sum_{i=1}^{N} l_i(x) v_i = \sum_{i=1}^{N} l_i(\xi) v_i \tag{4.158}$$

Using the work equivalent method, one has

$$W = \frac{1}{2} \lfloor F \rfloor \{v\} = \frac{1}{2} \int_0^L q(x) v(x) \, dx \tag{4.159}$$

where $\{F\}$ is the work equivalent load vector and $\{v\}$ is the nodal displacement vector.

Using Eqs. (4.158) and (4.159), one has

$$F_i = \int_0^L q(x) l_i(x) \, dx = \frac{L}{2} \int_{-1}^{1} q\left(\frac{\xi L + L}{2}\right) l_i(\xi) \, d\xi = \frac{L}{2} H_i q\left(\frac{\xi_i L + L}{2}\right) \tag{4.160}$$

where ξ_i and H_i are abscissas and weights, respectively, of the N-point GLL quadrature which can be found in Table I.1 in Appendix I for N up to 21, and F_i is the nodal force component corresponding to nodal displacement component v_i.

If Eq. (4.84) is used as the displacement in the beam element, F_i remains the same and computed by Eq. (4.160), since the equivalent nodal bending moments at the two end points are zero due to the use of the GLL quadrature.

By the same token, the work equivalent load vector can be obtained for the quadrature plate element. For the quadrature rectangular plate element, the nodal force component corresponding to nodal displacement component w_{ij}, that is, F_{ij}, can be computed by

$$\begin{aligned}
F_{ij} &= \int_0^a \int_0^b q(x,y) l_i(x) l_j(y) \, dy \, dx \\
&= \frac{ab}{4} \int_{-1}^{1} \int_{-1}^{1} q\left(\frac{\xi a + a}{2}, \frac{\eta b + b}{2}\right) l_i(\xi) l_j(\eta) \, d\eta \, d\xi \\
&= \frac{ab}{4} \sum_{i=1}^{N} \sum_{j=1}^{N} H_i H_j q\left(\frac{\xi_i a + a}{2}, \frac{\eta_j b + b}{2}\right)
\end{aligned} \tag{4.161}$$

For quadrilateral quadrature plate element, the nodal force component corresponding to nodal displacement component w_{ij}, that is, F_{ij}, can be computed by

$$
\begin{aligned}
F_{ij} &= \iint_{Area} q(x,y)l_i(x)l_j(y)\,dy\,dx \\
&= \int_{-1}^{1}\int_{-1}^{1} \bar{q}(\xi,\eta)l_i(\xi)l_j(\eta)\left|J(\xi,\eta)\right|d\eta\,d\xi \\
&= \sum_{i=1}^{N}\sum_{j=1}^{N} H_i H_j \bar{q}(\xi_i,\eta_j)\left|J(\xi_i,\eta_j)\right|
\end{aligned}
\tag{4.162}
$$

where $|J(\xi_i,\eta_j)|$ is the Jacobian determinant at GLL point (ξ_i,η_j). The over bar on q is apparent.

It should be mentioned that due to the use of the GLL quadrature, the equivalent nodal force components corresponding to the first-order derivatives at boundary points are zero, whether Lagrange interpolation functions or Hermite interpolation functions are used as the shape functions.

4.10.3 QUADRATURE PLATE ELEMENTS WITH NODES OTHER THAN GLL POINTS

For simplicity in presentation, GLL nodes and GLL quadrature are used to derive the element matrix equations for most elements. If other distributions of nodes are to be used, the modifications are simple with the newly proposed method by the author and his research associates [7]. For illustration, two examples are given herein. The other elements can be treated in a similar way.

1. Quadrature plane stress (strain) rectangular plate element

$\varsigma_k = \iota_k$ ($k = 1, 2, ..., N$) denote the integration points in ξ and η directions and $\bar{H}_k (k = 1, 2, ..., N)$. For the quadrature plane stress (strain) rectangular plate element presented in Section 4.5, the stiffness matrix $[k]$, Eq. (4.36), is modified as

$$
[k] = \frac{ab}{4}\sum_{i=1}^{N}\sum_{j=1}^{N} \bar{H}_i \bar{H}_j [B(\varsigma_i,\iota_j)]^T [C][B(\varsigma_i,\iota_j)]
\tag{4.163}
$$

The strain matrix at the integration points, Eq. (4.37), is modified to

$$
[B(\varsigma_i,\iota_j)]\{\bar{u}\} =
\begin{bmatrix}
\dfrac{2}{a}\displaystyle\sum_{l=1}^{N}\sum_{k=1}^{N} \bar{A}_{il}^{\xi} l_{jk} u_{lk} \\[3mm]
\dfrac{2}{b}\displaystyle\sum_{l=1}^{N}\sum_{k=1}^{N} l_{il} \bar{A}_{jk}^{\eta} v_{lk} \\[3mm]
\dfrac{2}{b}\displaystyle\sum_{l=1}^{N}\sum_{k=1}^{N} l_{il} \bar{A}_{jk}^{\eta} u_{lk} + \dfrac{2}{a}\displaystyle\sum_{l=1}^{N}\sum_{k=1}^{N} \bar{A}_{il}^{\xi} l_{jk} v_{lk}
\end{bmatrix}
\tag{4.164}
$$
$$(i,j = 1, 2, ..., N)$$

where $l_{il} = l_i(\varsigma_i)$, $l_{jk} = l_k(\iota_j)$, $\bar{A}_{ij}$ is computed by Eq. (4.12), and superscripts ξ and η imply that the weighting coefficient of the corresponding derivative is taken with respect to ξ or η, respectively.

To ensure a diagonal mass matrix, the technique of row (or column) summation is used. The diagonal terms in the mass matrix, Eq. (4.38), is modified as

$$
m_{II} = m_{JJ} = \frac{\rho hab}{4}\sum_{k=1}^{N}\sum_{m=1}^{N} \bar{H}_k l_{ki} \bar{H}_m l_{mj}
\tag{4.165}
$$
$$(i,j = 1, 2, ..., N; I = N \times (i-1) + j ;\ J = I + N \times N)$$

where $l_{ki} = l_i(\varsigma_k)$ and $l_{mj} = l_j(\iota_m)$.

It is not difficult to show that these equations are the same as the corresponding equations if GLL points are used as the element nodes and GLL quadrature is used in the numerical integration.

2. Quadrature rectangular plate element with mixed interpolations

For the quadrature rectangular plate element with mixed interpolations presented in Section 4.8.3, the stiffness matrix $[k]$, Eq. (4.129), is modified as

$$[k] = \frac{ab}{4} \sum_{i=1}^{N} \sum_{j=1}^{N} \bar{H}_i \bar{H}_j [B(\varsigma_i, \iota_j)]^T [D][B(\varsigma_i, \iota_j)] \tag{4.166}$$

The strain matrix at the integration points, Eq. (4.130), is modified to

$$[B(\varsigma_i, \iota_j)]\{\bar{w}\} = \begin{bmatrix} \dfrac{4}{a^2} \displaystyle\sum_{l=1}^{N+2} \sum_{k=1}^{N} \bar{B}_{il}^{\xi} l_{jk} \bar{w}_{lk} \\[2ex] \dfrac{4}{b^2} \displaystyle\sum_{l=1}^{N} \sum_{k=1}^{N+2} l_{il} \bar{B}_{jk}^{\eta} \bar{w}_{lk} \\[2ex] \dfrac{8}{ab} \displaystyle\sum_{l=1}^{N+2} \sum_{k=1}^{N+2} \bar{A}_{il}^{\xi} \bar{A}_{jk}^{\eta} \bar{w}_{lk} \end{bmatrix} \qquad (i, j = 1, 2, ..., N) \tag{4.167}$$

where the superscripts ξ and η imply that the weighting coefficient of the corresponding derivative is taken with respect to ξ or η, $\bar{A}_{jk}^{\eta}, \bar{B}_{jk}^{\eta}$ are computed by Eq. (4.95), $\bar{A}_{il}^{\xi}$ is computed by Eq. (4.12), and $\bar{B}_{il}^{\xi}$ is computed by Eq. (4.168) [7], namely,

$$\bar{B}_{1k} = \sum_{l=2}^{N-1} \bar{A}_{1l} A_{lk}; \quad \bar{B}_{1(N+1)} = \bar{A}_{11}; \quad \bar{B}_{1(N+2)} = \bar{A}_{1N}$$

$$\bar{B}_{ik} = \sum_{l=1}^{N} \bar{A}_{il} A_{lk}; \quad \bar{B}_{i(N+1)} = \bar{B}_{i(N+2)} = 0 \quad (i = 2, 3, ..., N-1) \tag{4.168}$$

$$\bar{B}_{Nk} = \sum_{l=2}^{N-1} \bar{A}_{Nl} A_{lk}; \quad \bar{B}_{N(N+1)} = A_{N1}; \quad \bar{B}_{N(N+2)} = A_{NN}$$

$$(k = 1, 2, ..., N)$$

where $\bar{A}_{lk}$ is computed by Eq. (4.12) and A_{lk} is computed by Eq. (1.30). MMWC-3 is used to introduce the additional DOF.

To ensure a diagonal mass matrix, the technique of row (or column) summation is used. The diagonal terms m_{II} in mass matrix, Eq. (4.131), is modified as

$$m_{II} = \frac{\rho hab}{4} \sum_{k=1}^{N} \sum_{m=1}^{N} \bar{H}_k l_{ki} \bar{H}_m l_{mj} \qquad (i, j = 1, 2, ..., N; \; I = N \times (i-1) + j) \tag{4.169}$$

where $l_{ki} = l_i(\varsigma_k)$ and $l_{mj} = l_j(\iota_m)$.

The geometric stiffness matrix $[g]$, Eq. (4.133), is modified as

$$[g] = \frac{ab}{4} \sum_{i=1}^{N} \sum_{j=1}^{N} \bar{H}_i \bar{H}_j [E(\varsigma_i, \iota_j)]^T [P(\varsigma_i, \iota_j)][E(\varsigma_i, \iota_j)] \tag{4.170}$$

where $[E(\varsigma, \iota_j)]$ is modified as

$$[E(\varsigma_i, \iota_j)]\{\overline{w}\} = \begin{bmatrix} \dfrac{2}{a} \displaystyle\sum_{l=1}^{N+2} \sum_{k=1}^{N} \overline{A}_{il}^{\xi} l_{jk} \overline{w}_{lk} \\[4mm] \dfrac{2}{b} \displaystyle\sum_{l=1}^{N} \sum_{k=1}^{N+2} l_{il} \overline{A}_{jk}^{\eta} \overline{w}_{lk} \end{bmatrix} \qquad (i,j = 1,2,...,N) \tag{4.171}$$

In Eq. (4.171), $l_{il} = l_l(\varsigma_i)$, $l_{jk} = l_k(\iota_j)$, $\overline{A}_{il}^{\xi}$ is computed by Eq. (4.12) and $\overline{A}_{jk}^{\eta}$ is computed by Eq. (4.95).

4.10.4 NUMERICAL EXAMPLES

To show the accuracy and efficiency of the quadrature thin plate elements in bending presented in this chapter, two problems are analyzed by using the QEM, that is, the free vibration of SSFF isotropic rhombic plates with skew angles of 30° and 75°, shown in Fig. 4.6. SSFF denotes that edges 1 and 2 are simply supported and edges 3 and 4 are free. Poisson's ratio μ is 0.3. Only one 21×21-node quadrature plate element is used in the analysis for simplicity.

Table 4.1 lists the first six-mode frequency parameters ($\omega a^2 \sqrt{\rho h / D}$) for the SSFF rhombic plate with relatively small skew angle ($\theta = 30°$). Symbols ω, a, ρ, h, D denote the circular frequency, the length of the plate edge, mass density, thickness, and the bending rigidity, respectively.

In Table 4.1, symbols QEM-LL, QEM-HH, and QEM-LH denote that the quadrature thin plate element is formulated with Lagrange interpolation in both directions, Hermite interpolation in both directions, and Lagrange interpolation in the ξ direction and Hermite interpolation in the η direction. QEM-hL denotes that the quadrature thin plate element is formulated with harmonic interpolation in the ξ direction and Lagrange interpolation in the η direction. QEM-hh denotes that the quadrature thin plate element is formulated with harmonic interpolations in both directions. QEM-HHa denotes that the weighting coefficients of the second-order derivatives are computed by using Hermite interpolation, but

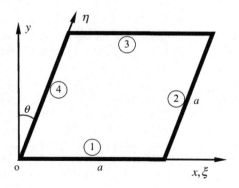

FIGURE 4.6

Sketch of a Rhombic Plate

Table 4.1 Comparison of Frequency Parameters $\omega a^2 \sqrt{\rho h / D}$ for SSFF Isotropic Rhombic Plates by Various Weak-Form Element Methods ($N = 21, \theta = 30°, \mu = 0.3$) [11]

Method	Mode Number				
	1	2	3	4	5
QEM-LL	3.2168	16.092	29.413	36.074	57.629
QEM-HH	3.2195	16.092	29.416	36.078	57.629
QEM-LH	3.2180	16.092	29.414	36.075	57.629
QEM-hL	3.2169	16.092	29.413	36.074	57.629
QEM-hh	3.2169	16.092	29.413	36.074	57.629
QEM-HHa	3.2169	16.092	29.413	36.074	57.629
MLS-Ritz [13]	3.2136	16.092	29.409	36.069	57.629
NASTRAN	3.2067	16.080	29.384	36.020	57.561
Upper bound [14]	3.2186	16.092	29.416	36.084	57.652

the weighting coefficients of the first-order derivatives are computed by using Lagrange interpolation [4]. In this way the DOFs w_{xy} at corners are not used. MLS-Ritz means the method of moving least square Ritz [13]. NASTRAN denotes that the results are obtained by the commercial software NASTRAN with meshes of 100×100. The accurate upper-bound solution is obtained by Rayleigh–Ritz method with corner functions to take into consideration corner stress singularities [14]. Most of the results listed in Table 4.1 are directly cited from reference [11].

It can be seen that except for the QEM-HH, all solutions are slightly lower than the accurate upper-bound solutions given by McGee et al. [14], the accuracy with only one 21×21-node quadrature thin plate element is excellent. Among the various methods, the QEM-LH yields the most accurate fundamental frequency.

Table 4.2 lists the first five-mode frequency parameters ($\omega a^2 \sqrt{\rho h / D}$) for the SSFF isotropic rhombic plate with very large skew angle ($\theta = 75°$). Most of the results listed in Table 4.2 are directly cited from Ref. [11]. Due to the severe stress singularity at the obtuse angles of the rhombic plates, this is a very difficult problem and many approximate and numerical methods, including the modified DQM, have encountered serious convergence problems. It can be seen that the fundamental frequencies obtained by various methods differ a lot from each other; however, all other four frequencies are close to each other. The fifth mode frequency cited from Ref. [11] is obviously incorrect and might be caused by a misprint. Again the QEM-LH yields the most accurate fundamental solutions as compared to the upper-bound solution. When the number of element QEM-LH increases to 31×31, denoted by QEM-LH (31) in Table 4.2, the accuracy of fundamental frequency is further improved, but all other four frequencies change very little.

It is interesting to see that the conventional FEM also yields poor fundamental frequency. Increasing the mesh size to 150×150 or using triangular element does not improve the accuracy of the fundamental frequency. The results are similar to the ones listed in Table 4.2 and thus are omitted.

Table 4.2 Comparison of Frequency Parameters $\omega a^2 \sqrt{\rho h / D}$ **for SSFF Isotropic Rectangular Plates by Various Weak-Form Element Methods** ($N = 21, \theta = 75°, \mu = 0.3$) [11]

Method	Mode Number				
	1	2	3	4	5
QEM-LL	7.0635	39.925	68.993	104.78	143.97
QEM-HH	10.618	39.926	69.436	104.78	144.16
QEM-LH	8.0420	39.925	69.096	104.78	144.01
QEM-LH [31]	8.1106	39.925	69.104	104.78	144.02
QEM-hL	7.1043	39.925	68.997	104.78	143.97
QEM-hh	7.1353	39.925	69.000	104.78	143.97
QEM-HHa	7.0835	39.925	68.995	104.78	143.97
MLS-Ritz [13]	6.4018	39.925	68.931	104.78	143.94
NASTRAN	6.6022	39.701	68.494	104.06	142.90
Upper bound [14]	8.2184	39.935	69.306	105.34	**157.64**

The value given in bold is to emphasize the datum is not accurate enough.

4.11 SUMMARY

Another type of the DQ-based element methods, the weak-form differential QEM or simply called the QEM, is presented in this chapter. It is seen that the formulations of the QEM are different from the DQEM presented in Chapter 2. The QEM is essentially the same as the high-order FEM. The difference from the FEM is that the distribution of nodal points in the QEM is not uniform. The difference from the SEM is that the formulation of stiffness matrix is simpler and more flexible, since various explicit formulas to calculate the weighting coefficients at integration points can be used and the explicit expression of the derivative of the shape functions is not needed. This also makes the implementation of an N-node quadrature element simpler than the SEM. To use the available explicit formulas of computing the weighting coefficients in the DQM, a simple way is proposed to formulate quadrature elements with nodes other than GLL nodes. Various quadrature elements, such as quadrature bar element, quadrature beam elements, and quadrature plate elements, are presented. For illustration, numerical examples are given. It is shown that quadrature thin plate elements can yield similar accurate solutions as the modified DQM.

Since the formulations are essentially the same as the FEM and SEM in principle, the difficulty existing in the DQM, such as the difficulty for applying the multiple boundary conditions and dealing with the loading and geometric discontinuities and complex geometry, does not exist in the QEM. Besides, spurious eigenvalues do not exist since the stiffness matrix and mass matrix are symmetric. Thus, QEM is more suitable for dynamic analysis than the DQEM, such as simulating the wave propagation in plate-like structures, and would find more applications in practice.

Attention should be paid to the novel idea on the constitutive equation in formulations of the quadrature thick plate element. Thickness locking phenomenon does not exist in the element although simple plate theory is employed. It is shown [15] that the simple plate theory can provide reasonably accurate results at relatively high frequency up to 1000 kHz (or more precisely up to a frequency to thickness ratio of 1.0 MHz mm). Thus the quadrature thick plate element can be efficiently used in simulations of wave propagation in plate-like structure in the frequency to thickness ratio of 1.0 MHz mm.

REFERENCES

[1] C.W. Bert, M. Malik, The differential quadrature method for irregular domains and application to plate vibration, Int. J. Mech. Sci. 38 (6) (1996) 589–606.

[2] W.L. Chen, A.G. Striz, C.W. Bert, High-accuracy plane stress and plate elements in the quadrature element method, Int. J. Solids Struct. 37 (2000) 627–647.

[3] Y. Xing, B. Liu, High-accuracy differential quadrature finite element method and its application to free vibrations of thin plate with curvilinear domain, Int. J. Numer. Meth. Eng. 80 (2009) 1718–1742.

[4] H.Z. Zhong, Z.G. Yue, Analysis of thin plates by the weak form quadrature element method, Sci. China Phys. Mech. 55 (5) (2012) 861–871.

[5] A. Żak, A novel formulation of a spectral plate element for wave propagation in isotropic structures, Finite Elem. Anal. Des. 45 (2009) 650–658.

[6] X. Wang, F. Wang, C. Xu, et al. New spectral plate element for simulating Lamb wave propagations in plate structures, J. NUAA 44 (5) (2012) 645–651 (in Chinese).

[7] C. Jin, X. Wang, L. Ge, Novel weak form quadrature element method with expanded Chebyshev nodes, Appl. Math. Lett. 34 (2014) 51–59.

[8] E. Carrera, S. Brischetto, Analysis of thickness locking in classical, refined and mixed multilayered plate theories, Compos. Struct. 82 (2008) 549–562.

[9] F. Wang, X. Wang, Z. Feng, Simulation of wave propagation in plate structures by using new spectral element with piezoelectric coupling, J. Vibroeng. 15 (1) (2013) 268–276.

[10] L. Ge, X. Wang, C. Jin, Numerical modeling of PZT-induced Lamb wave-based crack detection in plate-like structures, Wave Motion 51 (2014) 867–885.

[11] X. Wang, Z. Wu, Differential quadrature analysis of free vibration of rhombic plates with free edges, Appl. Math. Comput. 225 (2013) 171–183.

[12] C. Shu, Differential Quadrature and Its Application in Engineering, Springer-Verlag, London, 2000.

[13] L. Zhou, W.X. Zheng, Vibration of skew plates by the MLS-Ritz method, Int. J. Mech. Sci. 50 (2008) 1133–1141.

[14] G. McGee, J.W. Kim, A.W. Leissa, The influence of corner stress singularities on the vibration characteristics of rhombic plates with combinations of simply supported and free edges, Int. J. Mech. Sci. 41 (1999) 17–41.

[15] L. Ge, X. Wang, F. Wang, Accurate modeling of PZT-induced Lamb wave propagation in structures by using a novel spectral finite element method, Smart Mater. Struct. 23 (2014) 095018.

IN-PLANE STRESS ANALYSIS

5.1 INTRODUCTION

This chapter demonstrates the in-plane stress analysis of rectangular plate by using the differential quadrature method (DQM). Two formulations are used: one is based on the displacements and the other is based on the Airy stress functions. The equivalent boundary conditions are derived if the method of Airy stress function is used, which is more convenient for plates with all edges having only stress boundary conditions. Since only second-order partial differential equations are involved if unknowns are displacement components, there is no difficulty in applying the displacement boundary conditions. For the force boundary conditions; however, care should be taken at the corner points, since only two out of the three boundary conditions can be applied. If Airy stress function is used, a fourth-order compatibility equation, similar to the one of plate in bending, is solved. Therefore, multiple boundary conditions are encountered. The two formulations are worked out in detail and examples are given.

5.2 FORMULATION-I

Consider the in-plane elasticity problem. An isotropic thin rectangular plate with length a and width b is subjected to a uniaxial nonuniform distributed edge load, shown in Fig. 5.1. The plate thickness is h, and the edge load is $\sigma_{x0} = -\sigma_0 \cos(\pi y/b)$. The modulus of elasticity and Poisson's ratio are denoted by E and μ, respectively.

In terms of displacement components u and v, the governing differential equations are given by

$$
\begin{cases}
\dfrac{E}{1-\mu^2}\left(\dfrac{\partial^2 u}{\partial x^2} + \dfrac{(1-\mu)}{2}\dfrac{\partial^2 u}{\partial y^2} + \dfrac{(1+\mu)}{2}\dfrac{\partial^2 v}{\partial x \partial y}\right) + f_x = 0 \\[4mm]
\dfrac{E}{1-\mu^2}\left(\dfrac{\partial^2 v}{\partial y^2} + \dfrac{(1-\mu)}{2}\dfrac{\partial^2 v}{\partial x^2} + \dfrac{(1+\mu)}{2}\dfrac{\partial^2 u}{\partial x \partial y}\right) + f_y = 0
\end{cases}
\tag{5.1}
$$

where f_x, f_y are body force components in the x and y directions.

The displacement boundary conditions are

$$
u = u_0, \; v = v_0
\tag{5.2}
$$

where u_0, v_0 are specified.

The force boundary conditions along $x = \pm\, a/2$ are

Differential Quadrature and Differential Quadrature Based Element Methods. 978-0-12-803081-3

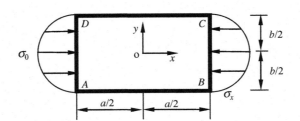

FIGURE 5.1

Sketch of a Rectangular Plate Under the Uniaxial Edge Compression

$$\frac{E}{1-\mu^2}\left(\frac{\partial u}{\partial x}+\mu\frac{\partial v}{\partial y}\right)=\sigma_{x0}, \quad \frac{E}{2(1+\mu)}\left(\frac{\partial v}{\partial x}+\frac{\partial u}{\partial y}\right)=\tau_{xy0} \tag{5.3}$$

and the force boundary conditions along $y=\pm b/2$ are

$$\frac{E}{1-\mu^2}\left(\mu\frac{\partial u}{\partial x}+\frac{\partial v}{\partial y}\right)=\sigma_{y0}, \quad \frac{E}{2(1+\mu)}\left(\frac{\partial v}{\partial x}+\frac{\partial u}{\partial y}\right)=\tau_{xy0} \tag{5.4}$$

where $\sigma_{x0}, \sigma_{y0}, \tau_{xy0}$ are given.

For the problem shown in Fig. 5.1, only force boundary conditions are involved, i.e.,

$$\begin{cases} \sigma_{x0}=-\sigma_0\cos\left(\dfrac{\pi y}{b}\right) \\ \sigma_{y0}=\tau_{xy0}=0 \end{cases} \tag{5.5}$$

Let N be the total number of grid points in the x and y directions. A 7×7 grid spacing is schematically shown in Fig. 5.2. Each point has two nodal degrees of freedom, u_{ij}, v_{ij} $(i, j=1,2,...,N)$.

In terms of differential quadrature, Eq. (5.1) at all inner grid points is given by

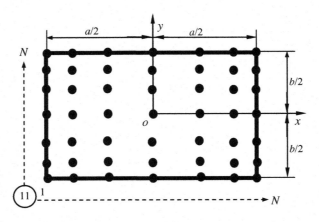

FIGURE 5.2

Sketch of a Rectangular Plate with 7×7 Grid Spacing

$$\begin{cases} \dfrac{E}{1-\mu^2}\left(\displaystyle\sum_{k=1}^{N} B_{ik}^x u_{kl} + \dfrac{(1-\mu)}{2}\sum_{k=1}^{N} B_{lk}^y u_{ik} + \dfrac{(1+\mu)}{2}\sum_{j=1}^{N}\sum_{k=1}^{N} A_{ij}^x A_{lk}^y v_{jk}\right) = 0 \\[4mm] \dfrac{E}{1-\mu^2}\left(\displaystyle\sum_{k=1}^{N} B_{lk}^y v_{ik} + \dfrac{(1-\mu)}{2}\sum_{k=1}^{N} B_{ik}^x v_{kl} + \dfrac{(1+\mu)}{2}\sum_{j=1}^{N}\sum_{k=1}^{N} A_{ij}^x A_{lk}^y u_{jk}\right) = 0 \end{cases} \tag{5.6}$$

$$(i,l = 2,3,...,N-1)$$

where $B_{ij}^x, B_{ij}^y, A_{ij}^x, A_{ij}^y$ are the weighting coefficients of the second-order derivative with respect to the variable of x or y and the weighting coefficients of the first-order derivative with respect to the variable of x or y, $f_x = f_y = 0$ for the problem considered.

In terms of differential quadrature, the force boundary conditions along $x = \pm\, a/2$ are

$$\begin{cases} \dfrac{E}{1-\mu^2}\left(\displaystyle\sum_{k=1}^{N} A_{ik}^x u_{kl} + \mu\sum_{k=1}^{N} A_{lk}^y v_{ik}\right) = -\sigma_0 \cos\left(\dfrac{\pi y_i}{b}\right) (i=1,2,...,N;\ l=1,N) \\[4mm] \dfrac{E}{2(1+\mu)}\left(\displaystyle\sum_{k=1}^{N} A_{ik}^x v_{kl} + \sum_{k=1}^{N} A_{lk}^y u_{ik}\right) = 0 \quad (i=2,3,...,N-1;\ l=1,N) \end{cases} \tag{5.7}$$

and the force boundary conditions along $y = \pm\, b/2$ are

$$\begin{cases} \dfrac{E}{1-\mu^2}\left(\mu\displaystyle\sum_{k=1}^{N} A_{ik}^x u_{kl} + \sum_{k=1}^{N} A_{lk}^y v_{ik}\right) = 0 \qquad (i=1,N;\ l=1,2,...,N) \\[4mm] \dfrac{E}{2(1+\mu)}\left(\displaystyle\sum_{k=1}^{N} A_{ik}^x v_{kl} + \sum_{k=1}^{N} A_{lk}^y u_{ik}\right) = 0 \quad (i=1,N;\ l=2,3,...,N-1) \end{cases} \tag{5.8}$$

Note that the first equation in Eq. (5.7) is applied at the degree of freedom (DOF) u_{il} and the other one is applied at v_{il}. The first equation in Eq. (5.8) is applied at the DOF v_{il} and the other one is applied at u_{il}. It is seen that shear force boundary condition is not used at corners. This is one of the keys to success in using the DQM.

Equations (5.6) and (5.8) contain $2(N \times N)$ algebraic equations; thus, the number of equations is equal to the number of unknowns. In matrix form, the following equation is obtained,

$$[K]\{U\} = \{F\} \tag{5.9}$$

or in partitioned form as

$$\begin{bmatrix} [k_{uu}] & [k_{uv}] \\ [k_{vu}] & [k_{vv}] \end{bmatrix} \begin{Bmatrix} \{u\} \\ \{v\} \end{Bmatrix} = \begin{Bmatrix} \{f\} \\ \{0\} \end{Bmatrix} \tag{5.10}$$

For the problem considered, the inverse of $[K]$ does not exist, since all boundary conditions are force boundary conditions. Thus, appropriate constraints on the displacements should be found and applied.

It is seen that the problem shown in Fig. 5.1 is doubly symmetric in both geometry and load; thus, constraints $u = 0$ at $x = 0$ and $v = 0$ at $y = 0$ can be applied if the entire plate is modeled and analyzed. In doing so, N should be an odd number. After applying the zero displacement conditions, Eq. (5.9) is modified as

$$[\bar{K}]\{U\} = \{F\} \tag{5.11}$$

Then the displacements at all grid points can be obtained by

$$\{U\} = \left[\bar{K}\right]^{-1}\{F\} \tag{5.12}$$

Once the displacements at all grid points are known, the stress components at any grid point can be calculated by

$$
\begin{cases}
(\sigma_x)_{il} = \dfrac{E}{1-\mu^2}\left(\displaystyle\sum_{k=1}^{N} A_{ik}^x u_{kl} + \mu \sum_{k=1}^{N} A_{lk}^y v_{ik}\right) \\[4mm]
(\sigma_y)_{il} = \dfrac{E}{1-\mu^2}\left(\mu \displaystyle\sum_{k=1}^{N} A_{ik}^x u_{kl} + \sum_{k=1}^{N} A_{lk}^y v_{ik}\right) & (i,l=1,2,\ldots,N) \\[4mm]
(\tau_{xy})_{il} = \dfrac{E}{2(1+\mu)}\left(\displaystyle\sum_{k=1}^{N} A_{ik}^x v_{kl} + \sum_{k=1}^{N} A_{lk}^y u_{ik}\right)
\end{cases} \tag{5.13}
$$

Applying the zero displacement condition is very simple in programming. Just put a big number, say, 10^{30}, at the diagonal term of matrix $[K]$, which corresponds to the zero displacement. This method is the same as the one used in the conventional finite element software.

5.3 FORMULATION-II

Since all boundary conditions are stress boundary conditions for the problem shown in Fig. 5.1, the method based on stress function can be used. The well-known Airy stress function (φ) without body forces are given by

$$\sigma_x = \frac{\partial^2 \varphi}{\partial y^2}; \quad \sigma_y = \frac{\partial^2 \varphi}{\partial x^2}; \quad \tau_{xy} = -\frac{\partial^2 \varphi}{\partial x \partial y} \tag{5.14}$$

Airy stress function defined by Eq. (5.14) satisfies the governing equations automatically. However, it should satisfy following compatibility equation, namely,

$$\frac{\partial^4 \varphi}{\partial x^4} + 2\frac{\partial^4 \varphi}{\partial x^2 \partial y^2} + \frac{\partial^4 \varphi}{\partial y^4} = 0 \tag{5.15}$$

Once φ is obtained, the in-plane stress components $\sigma_x, \sigma_y, \tau_{xy}$ can be computed by Eq. (5.14). In order to obtain φ numerically by the DQM, appropriate boundary conditions should be applied. It should be emphasized that one cannot directly apply the stress boundary conditions in terms of differential quadrature. Otherwise no unique solution would be resulted, since superimposing a linear function to Airy stress function will not affect the stress values. Therefore, equivalent boundary conditions in terms of φ and its first-order derivatives with respect to x and y should be sought first.

Let X and Y be the known in-plane load components acting on the boundary in the x and y directions, then φ and its first-order partial derivatives along boundaries can be computed by Ref. [1]

$$\varphi_B = \int_A^B (y_B - y)X\,ds - \int_A^B (x_B - x)Y\,ds$$

$$(\varphi_x)_B = \left(\frac{\partial \varphi}{\partial x}\right)_B = -\int_A^B Y ds \qquad (5.16)$$

$$(\varphi_y)_B = \left(\frac{\partial \varphi}{\partial y}\right)_B = \int_A^B X ds$$

where A and B are two arbitrary distinguished points on the plate boundary. Since superimposing a linear function to φ will not affect the stress components, the following assumption can be made, namely,

$$\varphi_A = 0; \quad \left(\frac{\partial \varphi}{\partial x}\right)_A = 0; \quad \left(\frac{\partial \varphi}{\partial y}\right)_A = 0 \qquad (5.17)$$

The left lower corner point is set as point A. For the loading shown in Fig. 5.1, $Y = 0$ and only the X along the plate boundary $(x = \pm a/2)$ is nonzero, i.e.,

$$X\left(x = -\frac{a}{2}\right) = \sigma_0 \cos\left(\frac{\pi y}{b}\right), \ X\left(x = \frac{a}{2}\right) = -\sigma_0 \cos\left(\frac{\pi y}{b}\right) \qquad (5.18)$$

Substituting the X and Y into Eq. (5.16) and performing the integration result the equivalent boundary conditions, which are as follows,

At point $x_i\left(-a/2 \leq x_i \leq a/2\right)$ on edge $y = -b/2$,

$$\varphi_i = \left(\frac{\partial \varphi}{\partial y}\right)_i = 0 \qquad (5.19)$$

At point $y_i\left(-b/2 \leq y_i \leq b/2\right)$ on edge $x = a/2$,

$$\varphi_i = \frac{\sigma_0 b}{2\pi^2}\left[2b\cos\left(\frac{\pi y_i}{b}\right) - 2\pi y_i - \pi b\right], \left(\frac{\partial \varphi}{\partial x}\right)_i = 0 \qquad (5.20)$$

At point $x_i\left(-a/2 \leq x_i \leq a/2\right)$ on edge $y = b/2$,

$$\varphi_i = -\frac{\sigma_0 b^2}{\pi}, \left(\frac{\partial \varphi}{\partial y}\right)_i = -\frac{2\sigma_0 b}{\pi} \qquad (5.21)$$

At point $y_i\left(-b/2 \leq y_i \leq b/2\right)$ on edge $x = -a/2$,

$$\varphi_i = \frac{\sigma_0 b}{2\pi^2}\left[2b\cos\left(\frac{\pi y_i}{b}\right) - 2\pi y_i - \pi b\right], \left(\frac{\partial \varphi}{\partial x}\right)_i = 0 \qquad (5.22)$$

It is noticed that Eq. (5.19) is not the same as Eq. (5.21), although both geometry and load are symmetry about the x axis. The reason is that in the derivation, point A is set at $(-a/2, -b/2)$. If another appropriate boundary point is used as the starting point for the integration, Eq. (5.19) and Eq. (5.21) may be exactly the same. However, the final results will not be altered; since this is equivalent to superimpose a linear function to Airy stress function; thus, the stress values will not be affected. It should also be mentioned that the third boundary condition derived by Eq. (5.16) is omitted in Eqs. (5.19) to (5.22), since only two boundary conditions along each boundary are necessary for a unique solution of a fourth-order partial differential equation.

The physical meaning of the first expression in Eq. (5.16) is the moment about point B by the edge applied forces between points A and B. The positive direction of the moment is counterclockwise for the right-hand coordinate system shown in Fig. 5.1 or in the positive z direction. The second expression in Eq. (5.16) is the negative resultant force in y direction for the edge applied forces between points A and B. And the third expression in Eq. (5.16) is the resultant force in x direction for the edge applied forces between points A and B.

Since the compatibility equation is a fourth-order partial differential equation, the modified DQM is used for solutions. In other words, either MMWC-3 or MMWC-4 is used to introduce the first-order derivative at the boundary points. In terms of differential quadrature, Eq. (5.15) at all inner grid points is given by

$$\sum_{k=1}^{N+2} D_{ik}^x \overline{\varphi}_{kl} + 2\sum_{j=1}^{N}\sum_{k=1}^{N} B_{ij}^x B_{lk}^y \varphi_{jk} + \sum_{k=1}^{N+2} D_{lk}^y \overline{\varphi}_{ik} = 0 \tag{5.23}$$
$$(i=2,3,...,N-1; l=2,3,...,N-1)$$

where superscripts x and y mean that the weighting coefficient of the corresponding derivative is taken with respect to x or y; the over bar in $\overline{\varphi}_{ik}$ means that φ as well as its first-order derivative with respect to x or y along boundaries is included, i.e., $\overline{\varphi}_{ik}$ contains $\varphi_{ij}(i,j=1,2,...,N)$, $(\varphi_x)_{i1},(\varphi_x)_{iN}(i=1,2,...,N)$ and $(\varphi_y)_{1j},(\varphi_y)_{Nj}(j=1,2,...,N)$.

Equation (5.23) can be written in the following partitioned matrix form,

$$\begin{bmatrix} K_{ib} & K_{ii} \end{bmatrix} \begin{Bmatrix} \varphi_b \\ \varphi_i \end{Bmatrix} = \{0\} \tag{5.24}$$

where subscripts b and i represent the quantities related to the boundary points and inner points, respectively.

From Eqs. (5.19) to (5.22), $\{\varphi_b\}=\{\varphi_{b0}\}$, where $\{\varphi_{b0}\}$ is given. Thus, $\{\varphi_i\}$ can be obtained by

$$\{\varphi_i\}=-[K_{ii}]^{-1}[K_{ib}]\{\varphi_{b0}\} \tag{5.25}$$

Alternatively, after applying the boundary conditions, Eq. (5.24) can be rewritten by

$$\begin{bmatrix} I & 0 \\ K_{ib} & K_{ii} \end{bmatrix} \begin{Bmatrix} \varphi_b \\ \varphi_i \end{Bmatrix} = \begin{Bmatrix} \varphi_{b0} \\ 0 \end{Bmatrix} \tag{5.26}$$

where $[I]$ is a unit submatrix.

Equation (5.26) can be further simplified as

$$[K]\{\overline{\varphi}\}=\{F\} \tag{5.27}$$

Therefore, $\{\overline{\varphi}\}$ can be found by

$$\{\overline{\varphi}\}=[K]^{-1}\{F\} \tag{5.28}$$

Once $\{\bar{\varphi}\}$ is obtained, the stress components at any inner grid point can be computed by

$$(\sigma_x)_{il} = \sum_{k=1}^{N} B_{lk}^y \varphi_{ik}, (\sigma_y)_{il} = \sum_{k=1}^{N} B_{ik}^x \varphi_{kl}, (\tau_{xy})_{il} = -\sum_{j=1}^{N}\sum_{k=1}^{N} A_{ij}^x A_{lk}^y \varphi_{jk}$$
$$(i = 2, 3, ...N-1; \quad l = 2, 3, ...N-1)$$

(5.29)

Note that for MMWC-3 or MMWC-4, the weighting coefficients of the first- and second-order derivatives at all inner points are the same as the ones in ordinary DQM; thus, the summation range in Eq. (5.29) and in the middle term of Eq. (5.23) is 1 to N and not 1 to $N+2$.

If different stress boundary conditions are applied, only Eqs. (5.19) to (5.22) are changed. In other words, only $\{\varphi_{b0}\}$ is changed. Therefore, it is very convenient to use the DQM for analyzing the in-plane stress of a rectangular plate with stress boundary conditions only. Besides, the DQ equations at all inner points are similar if buckling problems are to be analyzed by the modified DQM.

5.4 **RESULTS AND DISCUSSION**

Consider an isotropic square plate ($b=a$) under cosine distributed compression, shown in Fig. 5.1. Two FORTRAN programs, namely, programs 5 and 6, are developed and listed in Appendix VIII for readers' reference. The converted MATLAB files are also provided.

Figures 5.3–5.7 show the diagrams of the displacement components, $Eu/a\sigma_0$, $Ev/a\sigma_0$, and distribution of three stress components. Nondimensional stresses, i.e., σ_x/σ_0, σ_y/σ_0, and τ_{xy}/σ_0, are plotted. In the DQ analysis, Grid III is used and $N=15$.

The results shown in Figs. 5.3–5.7 are obtained by using the DQM with formulation I. Poisson's ratio is 0.3. Formulation I is more general than formulation II since it can be used for any type of boundary

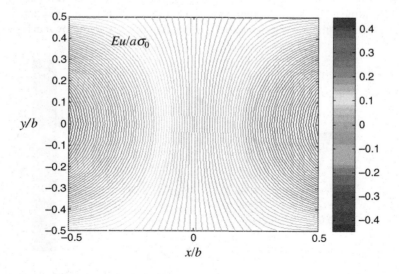

FIGURE 5.3

Distribution of Displacement $Eu/a\sigma_0$

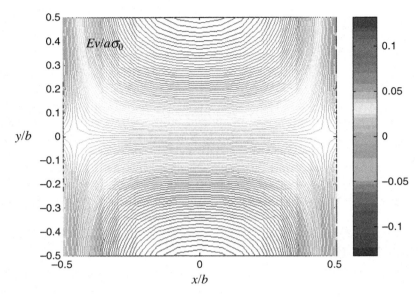

FIGURE 5.4

Distribution of Displacement $Ev/a\sigma_0$

conditions. Also, both the stress components and the displacement components can be obtained by the DQM. Formulation II is used for rectangular plates with stress boundary conditions only.

From Figs. 5.3 and 5.4 it is clearly seen that the deformation is symmetric about $x = 0$ and $y = 0$ but not uniform. From Fig. 5.5 [1], it is also seen that the distribution of stress varies and is not the same as the one applied on the edge. Besides, the other two stress components are not zero, clearly seen in Figs. 5.6–5.7 [1]. However, their magnitudes are much smaller than the one shown in Fig. 5.5, especially

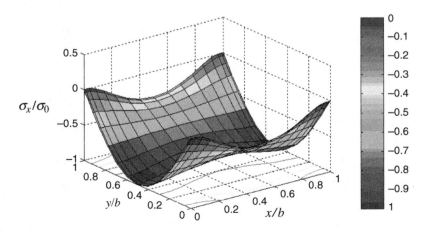

FIGURE 5.5

Distribution of In-plane Stress σ_x/σ_0

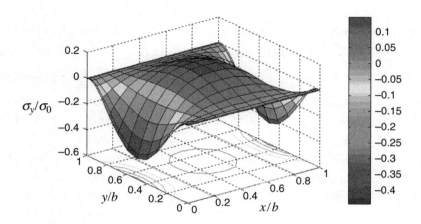

FIGURE 5.6

Distribution of In-plane Stress σ_y/σ_0

for the shear stress. If the variation of σ_x is not considered and the other two stress components are neglected in the buckling analysis, then serious error may be introduced in the calculated buckling load.

Table 5.1 lists the stress components σ_x/σ_0 obtained by the DQM with two different formulations. Results at two locations, i.e., $x = 0$ and $x = 0.391a$, are listed. It is seen that the accuracy of the results obtained by the two methods is similar. If larger N is used, the difference between the two formulations will decrease further. Unlike the finite element method, the accuracy of DQ solutions for both displacement and stress is similar. Accurate results can be obtained by using relatively small number of grid points.

When the aspect ratio a/b changes from 1 to 3, more uniform stress distribution of σ_x/σ_0 is observed. Figs. 5.8–5.10 [1] show the diagrams of three stress components for the case of $a/b = 3$. Again, nondimensional stresses, i.e., σ_x/σ_0, σ_y/σ_0, and τ_{xy}/σ_0, are plotted. The stress distributions shown

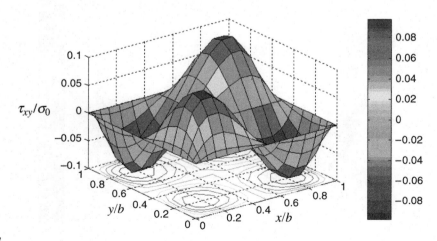

FIGURE 5.7

Distribution of In-plane Stress τ_{xy}/σ_0

Table 5.1 Stress Components σ_x/σ_0 Obtained by Two Different Formulations (15 × 15)

	x = 0.390915741a		x = 0	
y/b	Airy	Displacement	Airy	Displacement
−0.500000000	−4.94E-02	−4.97E-02	−0.370816984	−0.370757009
−0.487463956	−9.05E-02	−9.05E-02	−0.377846182	−0.377842531
−0.450484434	−0.201390851	−0.201447288	−0.406579755	−0.406613833
−0.390915741	−0.363632352	−0.363618468	−0.471503626	−0.471490816
−0.311744901	−0.561366654	−0.561363325	−0.575562305	−0.575559421
−0.216941870	−0.761879263	−0.761877614	−0.697836441	−0.697837577
−0.111260467	−0.913999163	−0.914000717	−0.798263616	−0.798264895
0.000000000	−0.970981115	−0.970979440	−0.837255837	−0.837255129
0.111260467	−0.913999163	−0.914000717	−0.798263616	−0.798264895
0.216941870	−0.761879263	−0.761877614	−0.697836441	−0.697837577
0.311744901	−0.561366654	−0.561363325	−0.575562305	−0.575559421
0.390915741	−0.363632352	−0.363618468	−0.471503626	−0.471490816
0.450484434	−0.201390851	−0.201447288	−0.406579755	−0.406613833
0.487463956	−9.05E-02	−9.05E-02	−0.377846182	−0.377842531
0.500000000	−4.94E-02	−4.97E-02	−0.370816984	−0.370757009

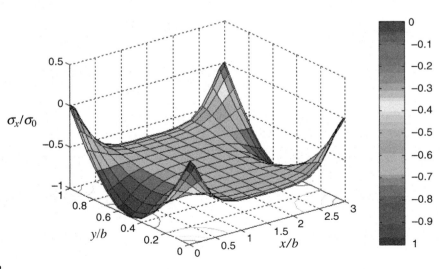

FIGURE 5.8

Distribution of In-plane Stress σ_x/σ_0 ($a/b = 3$)

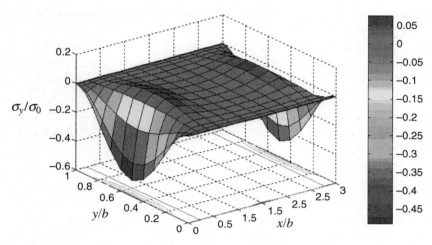

FIGURE 5.9

Distribution of In-plane Stress σ_y/σ_0 ($a/b = 3$)

in Figs. 5.8–5.10 are obtained by the DQM with formulation II. If displacement components are needed, however, formulation I is more convenient and recommended.

5.5 EQUIVALENT BOUNDARY CONDITIONS

In order to use the method of Airy stress function, the equivalent boundary conditions should be obtained to solve the compatibility equation by using the DQM. For illustrations, the equivalent boundary conditions for other uniaxial compressive loadings ($Y = 0$) are given below. Different combinations of applied loadings can be worked out following the same procedures easily.

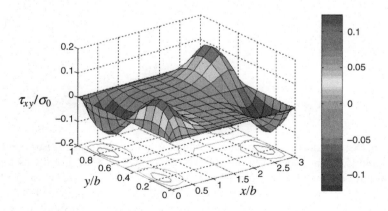

FIGURE 5.10

Distribution of In-plane Stress τ_{xy}/σ_0 ($a/b = 3$)

CASE 1

Rectangular plate subjected to uniaxial uniformly distributed compression (σ_0) as shown in Fig. 5.11. The nonzero forces are applied at a part of plate edges ($x = \pm a/2$) and given by

$$X\left(x = -\frac{a}{2}\right) = \sigma_0, \ X\left(x = \frac{a}{2}\right) = -\sigma_0 \quad \left(-\frac{b_1}{2} \leq y \leq \frac{b_1}{2}\right) \tag{5.30}$$

Using Eq. (5.17) and substituting the X expressed by Eq. (5.30) and $Y = 0$ into Eq. (5.16) result the equivalent boundary conditions are as follows:

At point $x_i \left(-a/2 \leq x_i \leq a/2\right)$ on edge $y = -b/2$

$$\varphi_i = \left(\frac{\partial \varphi}{\partial y}\right)_i = 0 \tag{5.31}$$

At point $y_i \left(-b/2 \leq y_i \leq b/2\right)$ on edge $x = a/2$

$$\begin{cases} \varphi_i = 0, \ \left(\frac{\partial \varphi}{\partial x}\right)_i = 0 \quad \left(-\frac{b}{2} \leq y_i \leq -\frac{b_1}{2}\right) \\ \varphi_i = -\frac{\sigma_0}{2}\left(y_i + \frac{b_1}{2}\right)^2, \ \left(\frac{\partial \varphi}{\partial x}\right)_i = 0 \quad \left(-\frac{b_1}{2} \leq y_i \leq \frac{b_1}{2}\right) \\ \varphi_i = -\sigma_0 b_1 y_i, \ \left(\frac{\partial \varphi}{\partial x}\right)_i = 0 \quad \left(\frac{b_1}{2} \leq y_i \leq \frac{b}{2}\right) \end{cases} \tag{5.32}$$

At point $x_i \left(-a/2 \leq x_i \leq a/2\right)$ on edge $y = b/2$,

$$\varphi_i = -\frac{\sigma_0 b_1 b}{2}, \ \left(\frac{\partial \varphi}{\partial y}\right)_i = -\sigma_0 b_1 \tag{5.33}$$

At point $y_i \left(-b/2 \leq y_i \leq b/2\right)$ on edge $x = -a/2$,

$$\begin{cases} \varphi_i = 0, \ \left(\frac{\partial \varphi}{\partial x}\right)_i = 0 \quad \left(-\frac{b}{2} \leq y_i \leq -\frac{b_1}{2}\right) \\ \varphi_i = -\frac{\sigma_0}{2}\left(y_i + \frac{b_1}{2}\right)^2, \ \left(\frac{\partial \varphi}{\partial x}\right)_i = 0 \quad \left(-\frac{b_1}{2} \leq y_i \leq \frac{b_1}{2}\right) \\ \varphi_i = -\sigma_0 b_1 y_i, \ \left(\frac{\partial \varphi}{\partial x}\right)_i = 0 \quad \left(\frac{b_1}{2} \leq y_i \leq \frac{b}{2}\right) \end{cases} \tag{5.34}$$

Although Eq. (5.16) yields three equivalent boundary conditions, only the two conditions listed in Eqs. (5.31) and (5.34) are needed in the stress analysis by the DQM. For the rectangular plate under uniaxial uniformly distributed compressive load, $b_1 = b$ in Eqs. (5.31) to (5.34). With the provided FORTRAN program or the converted MATLAB file in Appendix VIII, it is easy to show that the stress distributions obtained by the DQM are the same as the distributions of edge applied loads if Eqs. (5.31) to (5.34) with $b_1 = b$ are used. Thus, the written program has been verified.

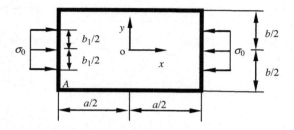

FIGURE 5.11

Sketch of a Rectangular Under Uniaxial Compressive Load

CASE 2

Rectangular plate subjected to uniaxial linearly distributed compression; the nonzero forces applied at plate edges $(x = \pm a/2)$ are given by

$$X\left(x = -\frac{a}{2}\right) = \frac{2\sigma_0}{3}\left(1 - \frac{y}{b}\right), \; X\left(x = \frac{a}{2}\right) = -\frac{2\sigma_0}{3}\left(1 - \frac{y}{b}\right) \tag{5.35}$$

Using Eq. (5.17) and substituting the X expressed by Eq. (5.35) and $Y = 0$ into Eq. (5.16) result the equivalent boundary conditions which are as follows:

At point $x_i \, (-a/2 \le x_i \le a/2)$ on edge $y = -b/2$

$$\varphi_i = \left(\frac{\partial\varphi}{\partial y}\right)_i = 0 \tag{5.36}$$

At point $y_i \, (-b/2 \le y_i \le b/2)$ on edge $x = a/2$

$$\varphi_i = \frac{\sigma_0}{36b}\left(4y_i^3 - 4b^3 - 15b^2 y_i - 12by_i^2\right), \left(\frac{\partial\varphi}{\partial x}\right)_i = 0 \tag{5.37}$$

At point $x_i \, (-a/2 \le x_i \le a/2)$ on edge $y = b/2$

$$\varphi_i = -\frac{7b^2}{18}\sigma_0, \left(\frac{\partial\varphi}{\partial y}\right)_i = -\frac{2b}{3}\sigma_0 \tag{5.38}$$

At point $y_i \, (-b/2 \le y_i \le b/2)$ on edge $x = -a/2$,

$$\varphi_i = \frac{\sigma_0}{36b}\left(4y_i^3 - 4b^3 - 15b^2 y_i - 12by_i^2\right), \left(\frac{\partial\varphi}{\partial x}\right)_i = 0 \tag{5.39}$$

Note that there are three conditions along each boundary by using Eq. (5.16); only two conditions are given and the third-one is not needed. With the provided FORTRAN program or its converted MATLAB file in Appendix VIII, it is easy to show that the stress distributions obtained by the DQM are exact and are the same as the distributions of edge applied loads. Thus, the written program has been verified further.

CASE 3

Rectangular plate subjected to uniaxial parabolically distributed compression; the nonzero forces applied at plate edges $(x=\pm a/2)$ are given by

$$X\left(x=\frac{a}{2}\right)=\frac{4\sigma_0}{b^2}\left(y^2-\frac{b^2}{4}\right), X\left(x=-\frac{a}{2}\right)=-\frac{4\sigma_0}{b^2}\left(y^2-\frac{b^2}{4}\right)$$

(5.40)

Using Eq. (5.17) and substituting the X expressed by Eq. (5.40) and $Y=0$ into Eq. (5.16) result the equivalent boundary conditions which are as follows [2]:

At point $x_i\,(-a/2\le x_i\le a/2)$ on edge $y=-b/2$,

$$\varphi_i=\left(\frac{\partial\varphi}{\partial y}\right)_i=0$$

(5.41)

At point $y_i\,(-b/2\le y_i\le b/2)$ on edge $x=a/2$,

$$\varphi_i=\frac{\sigma_0}{48b^2}\left(16y_i^4-3b^4-16b^3y_i-24b^2y_i^2\right),\left(\frac{\partial\varphi}{\partial x}\right)_i=0$$

(5.42)

At point $x_i\,(-a/2\le x_i\le a/2)$ on edge $y=b/2$,

$$\varphi_i=-\frac{\sigma_0 b^2}{3},\left(\frac{\partial\varphi}{\partial y}\right)_i=-\frac{2\sigma_0 b}{3}$$

(5.43)

At point $y_i\,(-b/2\le y_i\le b/2)$ on edge $x=-a/2$,

$$\varphi_i=\frac{\sigma_0}{48b^2}\left(16y_i^4-3b^4-16b^3y_i-24b^2y_i^2\right),\left(\frac{\partial\varphi}{\partial x}\right)_i=0$$

(5.44)

Again only two conditions along each boundary are given, since the third one is not used in the stress analysis by using the modified DQM.

For illustration, Eqs. (5.19) to (5.22) are used in the program presented in Appendix VIII, i.e., in Program 5. Example is also given to show how to modify the program for **Case** 3, namely, if Eqs. (5.41) to (5.44) are to be used. For other loadings, the modifications can be done in a similar way.

With the provided FORTRAN program or the converted MATLAB file in Appendix VIII, it is seen that the stress distributions are no longer the same as the ones of edge applied loads.

Although the general analytical procedure for determining the exact stress distribution within a rectangular plate, loaded by in-plane arbitrary external load, is available, its implementation seems not easy [3]. The implementation of the modified DQM is simpler. With formulation II (Program 5) and the equivalent boundary conditions, accurate stress components can be conveniently obtained by the modified DQM for a variety of force boundary conditions. Besides, the DQ solution accuracy is similar to the analytical solutions, which has also been demonstrated for the case of the rectangular plate under uniaxial cosine distributed edge compressive load [3].

5.6 SUMMARY

In this chapter, the differential quadrature method is used to perform the in-plane stress analysis for rectangular plates under nonlinearly distributed edge loads. Two ways are used. One way is based on the displacements; thus, two second-order partial differential equations are solved simultaneously by the ordinary DQM. The other way is based on the Airy stress function; thus, one fourth-order partial differential equation is solved by using the modified DQM. Numerical results show that the two ways are equivalent and the DQM can yield accurate results with relatively small number of grid points. Thus, the attractive features of rapid convergence, high accuracy, and computational efficiency are demonstrated.

Although the method based on the displacement is simple and more general, numerical experience shows that the shear stress condition at a plate corner should not be used, especially when singularity exists at the corner point [4]. With the help of MMWC-3 or MMWC-4, the method based on the Airy stress function is also convenient for plates with all edges having only stress boundary conditions.

REFERENCES

[1] X. Wang, L. Gan, Y. Zhang, Differential quadrature analysis of the buckling of thin rectangular plates with cosine-distributed compressive loads on two opposite sides, Adv. Eng. Softw. 39 (2008) 497–504.

[2] X. Wang, X. Wang, X. Shi, Accurate buckling loads of thin rectangular plates under parabolic edge compressions by differential quadrature method, Int. J. Mech. Sci. 49 (2007) 447–453.

[3] O. Mijušković, B. Ćorić, B. Šćepanović, Exact stress functions implementation in stability analysis of plates with different boundary conditions under uni and biaxial compression, Thin Wall. Struct. 80 (2014) 192–206.

[4] Z.B. Zhou, Application of differential quadrature element method to fracture analysis. Master Thesis. Nanjing University of Aeronautics and Astronautics, China, 2008 (in Chinese).

6
STATIC ANALYSIS OF THIN PLATE

6.1 INTRODUCTION

This chapter demonstrates the static analysis of isotropic and anisotropic rectangular thin plate in bending by using the modified DQM. Three types of loading, including distributed load in the entire plate, distributed load along a line, and a point load, are considered. Since a fourth-order partial differential equation is solved, either MMWC-3 or MMWC-4 can be used to apply the multiple boundary conditions. For the distributed load along a line or the concentrated load, a new approach is proposed to deal with the Dirac-delta function. Formulations are worked out in detail and six examples are given for demonstrations.

6.2 RECTANGULAR THIN PLATE UNDER GENERAL LOADING
6.2.1 BASIC EQUATIONS

Consider first the problem of a rectangular thin plate under general loading. Three types of load, i.e., distributed load, line distributed load along $x = x_L$, and point load at (x_p, y_q), are considered. The plate has length a, width b, and uniform thickness h. Anisotropic materials are considered.

The governing equation is given by

$$
D_{11}\frac{\partial^4 w}{\partial x^4} + 4D_{16}\frac{\partial^4 w}{\partial x^3 \partial y} + 2(D_{12} + 2D_{66})\frac{\partial^4 w}{\partial x^2 \partial y^2} + 4D_{26}\frac{\partial^4 w}{\partial x \partial y^3} + D_{22}\frac{\partial^4 w}{\partial y^4}
$$
$$
= q(x, y) + q^*(y)\delta(x - x_L) + P\delta(x - x_p)\delta(y - y_q) \tag{6.1}
$$

where D_{ij} are the flexural rigidity of the plate, w is the deflection, $\delta(.)$ is the Dirac-delta function, and $q(x,y)$, $q^*(y)$, and P are the distributed, line distributed, and point load, respectively.

The boundary conditions are summarized as follows:

1. Simply supported edge (S):

$$
w = M_x = 0 \text{ at } x = 0, a \text{ or } w = M_y = 0 \text{ at } y = 0, b \tag{6.2}
$$

2. Clamped edge (C):

$$
w = \frac{\partial w}{\partial x} = 0 \text{ at } x = 0, a \text{ or } w = \frac{\partial w}{\partial y} = 0 \text{ at } y = 0, b \tag{6.3}
$$

3. Free edge (F):

$$
Q_x = M_x = 0 \text{ at } x = 0, a \text{ or } Q_y = M_y = 0 \text{ at } y = 0, b \tag{6.4}
$$

where

$$M_x = D_{11}\frac{\partial^2 w}{\partial x^2} + D_{12}\frac{\partial^2 w}{\partial y^2} + 2D_{16}\frac{\partial^2 w}{\partial x \partial y} \tag{6.5}$$

$$M_y = D_{12}\frac{\partial^2 w}{\partial x^2} + D_{22}\frac{\partial^2 w}{\partial y^2} + 2D_{26}\frac{\partial^2 w}{\partial x \partial y} \tag{6.6}$$

$$Q_x = D_{11}\frac{\partial^3 w}{\partial x^3} + 4D_{16}\frac{\partial^3 w}{\partial x^2 \partial y} + \left(D_{12} + 4D_{66}\right)\frac{\partial^3 w}{\partial x \partial y^2} + 2D_{26}\frac{\partial^3 w}{\partial y^3} \tag{6.7}$$

$$Q_y = D_{22}\frac{\partial^3 w}{\partial y^3} + 4D_{26}\frac{\partial^3 w}{\partial x \partial y^2} + \left(D_{12} + 4D_{66}\right)\frac{\partial^3 w}{\partial x^2 \partial y} + 2D_{16}\frac{\partial^3 w}{\partial x^3} \tag{6.8}$$

If a corner is free, three boundary conditions, i.e., $M_x = M_y = R = 0$, should be applied, where the concentrated force R is given by

$$R = 2D_{16}\frac{\partial^2 w}{\partial x^2} + 2D_{26}\frac{\partial^2 w}{\partial y^2} + 4D_{66}\frac{\partial^2 w}{\partial x \partial y} \tag{6.9}$$

6.2.2 DIFFERENTIAL QUADRATURE FORMULATION

Let N be the total number of grid points in the x and y directions. A 7×7 grid spacing is schematically shown in Fig. 6.1. Each inner point has one degree of freedom (DOF), i.e., w_{ij} $(i, j = 2, 3, ..., N-1)$; each corner node has three DOFs, i.e., w_{ij}, w_{xij}, w_{yij} $(i, j = 1, N)$; each boundary node at $x = 0$ and a has two DOFs, i.e., w_{ij}, w_{xij} $(i = 2, ..., N-1, j = 1, N)$; and each boundary node at $y = 0$ and b has two DOFs, i.e., w_{ij}, w_{yij} $(i = 1, N, j = 2, ..., N-1)$.

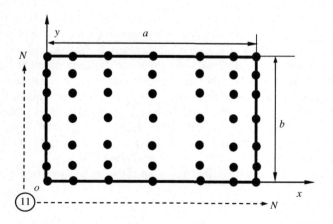

FIGURE 6.1

Sketch of a Rectangular Plate with 7 × 7 Grid Spacing

Either MMWC-3 or MMWC-4 can be used to introduce the additional DOF at the two end points in one dimension, i.e., $w_1' = (\partial w / \partial x)_{x=0}$ and $w_N' = (\partial w / \partial x)_{x=a}$. The first- to the fourth-order derivatives of the solution function w with respect to x at a grid point i can be symbolically written by

$$w_i^{(1)} = \sum_{j=1}^{N+2} A_{ij}^x \, \delta_j \qquad (i = 1, 2, ..., N)$$

$$w_i^{(2)} = \sum_{j=1}^{N+2} B_{ij}^x \, \delta_j \qquad (i = 1, 2, ..., N)$$

$$w_i^{(3)} = \sum_{j=1}^{N+2} C_{ij}^x \, \delta_j \qquad (i = 1, 2, ..., N)$$

$$w_i^{(4)} = \sum_{j=1}^{N+2} D_{ij}^x \, \delta_j \qquad (i = 1, 2, ..., N)$$

$$(6.10)$$

where $A_{ij}^x, B_{ij}^x, C_{ij}^x, D_{ij}^x$ are called the weighting coefficients of the first- to the fourth-order derivatives with respect to x, N is the total number of grid points in the x direction, and δ_j is defined by

$$\delta_j = w_j \left(j = 1, 2, ..., N \right), \ \delta_{N+1} = w_1', \ \delta_{N+2} = w_N' \qquad (6.11)$$

The weighting coefficients of the first- to the fourth-order derivatives with respect to y can be formulated in a similar way. In terms of the modified differential quadrature, Eq. (6.1) at all inner grid points can be expressed as

$$D_{11} \sum_{k=1}^{N+2} D_{ik}^x \bar{w}_{kl} + 4D_{16} \sum_{j=1}^{N+2} \sum_{k=1}^{N+2} C_{ij}^x A_{lk}^y \bar{w}_{jk} + 2\left(D_{12} + 2D_{66}\right) \sum_{j=1}^{N+2} \sum_{k=1}^{N+2} B_{ij}^x B_{lk}^y \bar{w}_{jk}$$

$$+ 4D_{26} \sum_{j=1}^{N+2} \sum_{k=1}^{N+2} A_{ij}^x C_{lk}^y \bar{w}_{jk} + D_{22} \sum_{k=1}^{N+2} D_{lk}^y \bar{w}_{ik} = p_{il} \qquad (6.12)$$

$$(i = 2, 3, ..., N-1; l = 2, 3, ..., N-1)$$

where superscripts x and y imply that the weighting coefficients of the corresponding derivatives are taken with respect to x or y; the over bar in $\bar{w}_{ik}$ means that deflection as well as the first-order derivatives with respect to x and y along boundaries are included. In detail, $\bar{w}_{ik}$ contains $w_{ij} (i, j = 1, 2, ..., N)$, $(w_x)_{i1}, (w_x)_{iN} (i = 1, 2, ..., N)$, and $(w_y)_{1j}, (w_y)_{Nj} (j = 1, 2, ..., N)$; p_{il} is the load at point (x_i, y_l), respectively.

To formulate a complete set of algebraic equations, the bending moment equation is placed at the position where the DOF of $\partial w / \partial x$ or $\partial w / \partial y$ is and the shear force equation is placed at the position where the boundary deflection w is. In detail, the bending moment equation M_x is placed at the position of $(w_x)_{i1}, (w_x)_{iN} (i = 1, 2, ..., N)$, the bending moment equation M_y is placed at the position of $(w_y)_{1j}, (w_y)_{Nj} (j = 1, 2, ..., N)$, the shear force equation Q_x is placed at the position of $w_{i1}, w_{iN} (i = 2, 3, ..., N-1)$, the shear force equation Q_y is placed at the position of $w_{1j}, w_{Nj} (j = 2, 3, ..., N-1)$, and the concentrated force R is placed at the position of four corner points $w_{ij} (i, j = 1, N)$.

More precisely, at the positions where $(\partial w / \partial x)_{x=0}$ and $(\partial w / \partial x)_{x=a}$ are, the following DQ equations are placed:

$$D_{11} \sum_{k=1}^{N+2} B_{ik}^x \bar{w}_{kl} + D_{12} \sum_{k=1}^{N+2} B_{ik}^y \bar{w}_{ik} + 2D_{16} \sum_{j=1}^{N+2} \sum_{k=1}^{N+2} A_{ij}^x A_{lk}^y \bar{w}_{jk} = \left(M_x\right)_{il} \qquad (6.13)$$

$$(i = 1, 2, ..., N; l = 1, N)$$

At the positions w_{il} $(i = 2, 3, ..., N - 1;\ l = 1,\ N)$, the following DQ equations are applied:

$$D_{11} \sum_{k=1}^{N+2} C_{ik}^x \bar{w}_{kl} + 4D_{16} \sum_{j=1}^{N+2} \sum_{k=1}^{N+2} B_{ij}^x A_{lk}^y \bar{w}_{jk} + \left(D_{12} + 4D_{66} \right) \sum_{j=1}^{N+2} \sum_{k=1}^{N+2} A_{ij}^x B_{lk}^y \bar{w}_{jk}$$

$$+ 2D_{26} \sum_{k=1}^{N+2} C_{lk}^y \bar{w}_{ik} = \left(Q_x \right)_{il} \qquad (i = 2, 3, ..., N - 1;\ l = 1,\ N) \tag{6.14}$$

Note that the four corner points, w_{il} $(i = 1, N;\ l = 1, N)$, are not included in Eq. (6.14).

Similarly, at the positions where $\left(\partial w / \partial y \right)_{y=0}$ and $\left(\partial w / \partial y \right)_{y=b}$ are, the following DQ equations are placed:

$$D_{12} \sum_{k=1}^{N+2} B_{ik}^x \bar{w}_{kl} + D_{22} \sum_{k=1}^{N+2} B_{lk}^y \bar{w}_{ik} + 2D_{26} \sum_{j=1}^{N+2} \sum_{k=1}^{N+2} A_{ij}^x A_{lk}^y \bar{w}_{jk} = \left(M_y \right)_{il} \tag{6.15}$$

$$(i = 1, N;\ l = 1,\ 2, ..., N)$$

At the positions w_{il} $(i = 1, N;\ l = 2, 3, ..., N - 1)$, the following DQ equations are applied:

$$D_{22} \sum_{k=1}^{N+2} C_{lk}^y \bar{w}_{ik} + 4D_{26} \sum_{j=1}^{N+2} \sum_{k=1}^{N+2} A_{ij}^x B_{lk}^y \bar{w}_{jk} + \left(D_{12} + 4D_{66} \right) \sum_{j=1}^{N+2} \sum_{k=1}^{N+2} B_{ij}^x A_{lk}^y \bar{w}_{jk}$$

$$+ 2D_{16} \sum_{k=1}^{N+2} C_{ik}^x \bar{w}_{kl} = \left(Q_y \right)_{il} \qquad (i = 1, N;\ l = 2, 3, ..., N - 1) \tag{6.16}$$

Again the four corner points, w_{il} $(i = 1, N;\ l = 1, N)$, are not included in Eq. (6.16).

Finally, at the four corner points, w_{il} $(i = 1, N;\ l = 1, N)$, the following DQ equation is applied:

$$2D_{16} \sum_{k=1}^{N+2} B_{ik}^x \bar{w}_{kl} + 2D_{26} \sum_{k=1}^{N+2} B_{lk}^y \bar{w}_{ik} + 4D_{66} \sum_{j=1}^{N+2} \sum_{k=1}^{N+2} A_{ij}^x A_{lk}^y \bar{w}_{jk} = R_{il} \tag{6.17}$$

$$(i = 1, N;\ l = 1, N)$$

Equations (6.12)–(6.17) can be written in the following partitioned matrix form:

$$\begin{bmatrix} K_{bb} & K_{bi} \\ K_{ib} & K_{ii} \end{bmatrix} \begin{Bmatrix} \Delta_b \\ W_i \end{Bmatrix} = \begin{Bmatrix} F_b \\ F_i \end{Bmatrix} \tag{6.18}$$

where subscripts b and i denote the quantities related to the boundary and inner points, respectively. Equation (6.18) contains $(N + 2) \times (N + 2) - 4$ algebraic equations; the number of equations is the same as the number of unknowns. The dimension of vector $\{W_i\}$ is $(N - 2)^2 \times 1$ and the dimension of vector $\{\Delta_b\}$ is $(8N - 4) \times 1$.

It is seen that any combinations of boundary conditions can be easily applied by using Eqs. (6.13)–(6.17), since either the deflection or the shear force, as well as either the slope or the bending moment, is specified. In other words, the element in $\{F_b\}$ is either zero if the corresponding DOF (either deflection or the first-order derivative) in $\{\Delta_b\}$ is unknown or unknown if the corresponding DOF in $\{\Delta_b\}$ is given.

After applying the boundary conditions and eliminating the zero DOFs in $\{\Delta_b\}$, Eq. (6.18) can be modified as

$$\left[K_{ii} - \bar{K}_{ib} \bar{K}_{bb}^{-1} \bar{K}_{bi} \right] \{W_i\} = \{F_i\} \tag{6.19}$$

where the over bar means that the rows and columns corresponding to the zero DOFs have been removed. Equation (6.19) can be further simplified as

$$[\bar{K}]\{W_i\} = \{F_i\} \tag{6.20}$$

Solving Eq. (6.20) yields the solutions, namely,

$$\{W_i\} = [\bar{K}]^{-1}\{F_i\} \tag{6.21}$$

If all edges of the rectangular plate are clamped, then $\{\Delta_b\} = 0$, thus Eq. (6.19) becomes

$$[K_{ii}]\{W_i\} = \{F_i\} \tag{6.22}$$

In other words, $[\bar{K}] = [K_{ii}]$. Once $\{W_i\}$ is known, the nonzero $\{\Delta_b\}$ can be computed by

$$\{\bar{\Delta}_b\} = -[\bar{K}_{bb}^{-1}][\bar{K}_{bi}]\{W_i\} \tag{6.23}$$

For static analysis of isotropic or orthotropic rectangular plate without a free edge, the boundary conditions can be directly built in during formulation of weighting coefficients by using the combination of MMWC-1 and MMWC-3. Therefore, only DQ equations at all inner grid points are needed, i.e.,

$$D_{11}\sum_{k=1}^{N}D_{ik}^{x}w_{kl} + 4D_{16}\sum_{j=1}^{N}\sum_{k=1}^{N}C_{ij}^{x}A_{lk}^{y}w_{jk} + 2(D_{12} + 2D_{66})\sum_{j=1}^{N}\sum_{k=1}^{N}B_{ij}^{x}B_{lk}^{y}w_{jk}$$
$$+ 4D_{26}\sum_{j=1}^{N}\sum_{k=1}^{N}A_{ij}^{x}C_{lk}^{y}w_{jk} + D_{22}\sum_{k=1}^{N}D_{lk}^{y}w_{ik} = p_{il} \tag{6.24}$$
$$(i = 2, 3, ..., N-1; l = 2, 3, ..., N-1)$$

Note that Eq. (6.24) is different from Eq. (6.12). The upper summation index changes from $N + 2$ to N and the first-order derivative DOFs are not involved. Also, the weighting coefficients of the third- and the fourth-order derivatives are more or less different, although the symbols are the same. Since MMWC-3 is used for building in the clamped boundary conditions, C_{ij}^{x} and D_{ij}^{x} in Eq. (6.24) and Eq. (6.12) are the same if edges $x = 0$ and a are clamped. Similarly, C_{ij}^{y} and D_{ij}^{y} in Eq. (6.24) and Eq. (6.12) are the same if edges $y = 0$ and b are clamped.

After applying the zero deflection condition at all boundary points, i.e., the degrees of deflection at all boundary points are removed, Eq. (6.24) can be rewritten as

$$[K_{ii}]\{W_i\} = \{F_i\} \tag{6.25}$$

Note that except for the case of rectangular plate with all edges fixed, Eq. (6.25) is different from Eq. (6.22), although the form is exactly the same. Due to building in the boundary conditions during formulation of weighting coefficients of higher-order derivatives, the programming is very simple. Subroutines are provided in Appendix II.2 for the one-dimensional cases of SS, CC, SC, and CS. They are called in Program 1 (see Appendix III). The isotropic or orthotropic rectangular plate problems with any combinations of simply supported and clamped boundary conditions under general loadings can be formulated by using Eq. (6.24).

Solving Eq. (6.25), the deflections at all inner points can be obtained as

$$\{W_i\} = [K_{ii}]^{-1}\{F_i\}$$

(6.26)

6.2.3 EQUIVALENT LOAD

If line distributed or point loads are involved, appropriately dealing with the Dirac-delta function appearing in Eq. (6.1) is one of the keys to success by using the DQM. A general method is proposed to deal with such cases.

Consider first the plate under line distributed load alone. Assume x_L is a grid point in the x direction. Since the Dirac-delta function is a distribution, Eq. (6.1) is integrated from 0 to a. Then the right-hand side of Eq. (6.1) becomes $q*(y)$. The left-hand side of Eq. (6.1) is integrated numerically. One has

$$\left(D_{11}\frac{\partial^4 w}{\partial x^4} + 4D_{16}\frac{\partial^4 w}{\partial x^3 \partial y} + 2(D_{12}+2D_{66})\frac{\partial^4 w}{\partial x^2 \partial y^2} + 4D_{26}\frac{\partial^4 w}{\partial x \partial y^3} + D_{22}\frac{\partial^4 w}{\partial y^4}\right)_i = 0$$
$$(i = 2,3,...,L-1,L+1,...,N-1)$$

(6.27)

$$\left(D_{11}\frac{\partial^4 w}{\partial x^4} + 4D_{16}\frac{\partial^4 w}{\partial x^3 \partial y} + 2(D_{12}+2D_{66})\frac{\partial^4 w}{\partial x^2 \partial y^2} + 4D_{26}\frac{\partial^4 w}{\partial x \partial y^3} + D_{22}\frac{\partial^4 w}{\partial y^4}\right)_i$$
$$= 2q*(y)/(aH_L) \qquad (i = L)$$

(6.28)

where H_L is the weight corresponding to the integration point ξ_L defined as $\xi_L = (2x_L - a)/a$.

There are many methods for numerical integrations. The simplest one is trapezoidal rule and the most accurate one is Gauss quadrature. GLL quadrature can also be used. If trapezoidal rule is used, $H_L = (\xi_{L+1} - \xi_{L-1})/2$. If Gauss quadrature is used, H_L can be calculated by the subroutine GRULE(N,X,W,Y) in Appendix II.1. For GLL quadrature, H_L can be found in Appendix I.

Consider next a point load applied at a grid point (x_p,y_q). Then double numerical integrations from 0 to a and 0 to b are performed. After doing so, one has

$$\left(D_{11}\frac{\partial^4 w}{\partial x^4} + 4D_{16}\frac{\partial^4 w}{\partial x^3 \partial y} + 2(D_{12}+2D_{66})\frac{\partial^4 w}{\partial x^2 \partial y^2} + 4D_{26}\frac{\partial^4 w}{\partial x \partial y^3} + D_{22}\frac{\partial^4 w}{\partial y^4}\right)_{il} = 0$$
$$(i = 2,3,...,p-1,p+1,...,N-1; \ l = 2,3,...,q-1,q+1,...,N-1)$$

(6.29)

$$\left(D_{11}\frac{\partial^4 w}{\partial x^4} + 4D_{16}\frac{\partial^4 w}{\partial x^3 \partial y} + 2(D_{12}+2D_{66})\frac{\partial^4 w}{\partial x^2 \partial y^2} + 4D_{26}\frac{\partial^4 w}{\partial x \partial y^3} + D_{22}\frac{\partial^4 w}{\partial y^4}\right)_{il}$$
$$= 4P/(abH_pH_q) \qquad (i = p, \ l = q)$$

(6.30)

where H_p and H_q are the weights corresponding to the integration point of (ξ_p, η_q). ξ_i, η_l are defined as $\xi_p = (2x_p - a)/a$ and $\eta_q = (2y_q - b)/b$.

If trapezoidal rule is used, $H_p = (\xi_{p+1} - \xi_{p-1})/2$ and $H_q = (\xi_{q+1} - \xi_{q-1})/2$. If Gauss quadrature is used, H_p and H_q can be calculated by the subroutine GRULE(N,X,W,Y) in Appendix II.1. For GLL quadrature, H_p and H_q can be found in Appendix I.

When x_L is not a grid point but is located between grid points of x_k and x_{k+1}, two ways can be used to get the equivalent line loads applied at grid points. In the first approach the line load is equivalent to line loads applied at x_k and x_{k+1}. If q_k^* and q_{k+1}^* denote the equivalent line loads applied at x_k and x_{k+1}, one has

$$q_k^*(y) = \frac{x_{k+1} - x_L}{x_{k+1} - x_k} q_L^*(y), \quad q_{k+1}^*(y) = \frac{x_L - x_k}{x_{k+1} - x_k} q_L^*(y) \tag{6.31}$$

Alternatively the line load can work equivalently to line loads applied at all nodes along x direction $(x_1, x_2, \ldots x_N)$, namely,

$$\frac{q^*(y)w(x_L, y)}{2} = \frac{1}{2}\sum_{j=1}^{N} q^*(y)l_j(x_L)w(y)_j = \frac{1}{2}\sum_{j=1}^{N} q_j^*(y)w(y)_j \tag{6.32}$$

Thus, the work equivalent line loads are given by

$$q_j^*(y) = q^*(y)l_j(x_L) \quad (j = 1, 2, \ldots, N) \tag{6.33}$$

where $l_j(x)$ are the Lagrange interpolation functions defined by Eq. (1.3).

Once the equivalent line loads are obtained, integration of Eq. (6.1) is performed. If Eq. (6.33) is used, one has

$$\left(D_{11}\frac{\partial^4 w}{\partial x^4} + 4D_{16}\frac{\partial^4 w}{\partial x^3 \partial y} + 2(D_{12} + 2D_{66})\frac{\partial^4 w}{\partial x^2 \partial y^2} + 4D_{26}\frac{\partial^4 w}{\partial x \partial y^3} + D_{22}\frac{\partial^4 w}{\partial y^4} \right)_i$$
$$= 2q_i^*(y)/(aH_i) \qquad\qquad (i = 2, 3, \ldots, N-1) \tag{6.34}$$

where H_i is the weight corresponding to the integration point ξ_i defined as $\xi_i = (2x_i - a)/a$.

If the concentrated loading point is not coincided with a grid point, similar procedures can be followed. For illustration, the second approach is used and the point load can work equivalently to point loads applied at all node points. The work equivalent point loads are given by

$$P_{il} = Pl_i(x_p)l_l(y_q) \quad (i, l = 1, 2, \ldots, N) \tag{6.35}$$

where $l_j(x)$ and $l_j(y)$ are the Lagrange interpolation functions defined by Eq. (1.3). It is easy to see that $P_{pq} = P$ and all others are zero if the point loading point (x_p, y_q) is coincided with a grid point.

By the same token, performing the double integration on Eq. (6.1) and using Eq. (6.35) yields

$$\left(D_{11}\frac{\partial^4 w}{\partial x^4} + 4D_{16}\frac{\partial^4 w}{\partial x^3 \partial y} + 2(D_{12} + 2D_{66})\frac{\partial^4 w}{\partial x^2 \partial y^2} + 4D_{26}\frac{\partial^4 w}{\partial x \partial y^3} + D_{22}\frac{\partial^4 w}{\partial y^4} \right)_{il}$$
$$= 4P_{il}/(abH_iH_l) \qquad\qquad (i, l = 2, 3, \ldots, N-1) \tag{6.36}$$

where H_i and H_l are the weights corresponding to integration point of (ξ_i, η_l). ξ_i, η_l are defined as $\xi_i = (2x_i - a)/a$ and $\eta_l = (2y_l - b)/b$.

Therefore, the definition of the load p_{il} in Eq. (6.12) or Eq. (6.24) is obvious if the line load and/or point load are also involved besides the load distributed in the entire plate.

Care should be taken that Grid V should be used in the DQM if Gauss quadrature is used in the numerical integration. Fortunately, this is not a limitation on the application of the DQM, since Grid V is the most reliable grid distribution to solve structure mechanics problems by using the DQM. Similarly, Grid VII should be used if GLL quadrature is used in the numerical integration. Any grids could be used if trapezoidal rule is used for the numerical integration. For moving point load problem; however, the second method is more convenient than the first method in programming.

6.3 APPLICATIONS

There exist two edge numbering sequences in literature. Both are used in this book for comparison purposes. The number sequence in this chapter is different from the one shown in Fig. 4.6. In this chapter, the edge numbering sequence starts from $x = 0, y = 0, x = a$, and ends at $y = b$. Symbol SFCC denotes that the plate is simply supported at $x = 0$, free at $y = 0$, and clamped at $x = a$ and $y = b$.

A FORTRAN program, named Program 1, is written for analyzing SSSS anisotropic rectangular plate without a free edge and is included in Appendix III. The name of the converted MATLAB file is Program1_Main. Currently Program 1 is for free vibration analysis. With only a few modifications, however, the FORTRAN program or MATLAB file can do static analysis of rectangular plate under general loadings and with six combinations of boundary conditions, i.e., SSSS, CCCC, SCCS, SSCS, SCCC, and CSCS.

6.3.1 RECTANGULAR PLATE UNDER UNIFORMLY DISTRIBUTED LOAD

EXAMPLE 6.1

SSSS isotropic rectangular plate under uniformly distributed load q is analyzed by the modified DQM. Two aspect ratios ($b/a = 1$ and 2) are considered. Since the Dirac-delta function is not involved, numerical integration is not required and $p_{il} = q$. The nondimensional center deflection and bending moment, defined by Eq. (6.37), are listed in Table 6.1. Poisson's ratio is taken as 0.3. Grid III is used and $N = 11$. From the data listed in Table 6.1, it can be seen that the rate of convergence is very high; the modified DQM with 11×11 grid spacing can yield very accurate deflection and bending moments. This is quite different from the finite element method; the accuracy of the stress obtained by the FEM is not as good as the displacement.

The nondimensional quantities appearing in Table 6.1 are defined by

$$\bar{w}_c = \frac{w_c D}{qa^4}, \quad \bar{M}_{xc} = \frac{M_{xc}}{qa^2} \tag{6.37}$$

where $D = Eh^3/(1 - \mu^2)$ is the bending rigidity of the plate, and h, E, and μ are plate thickness, modulus of elasticity, and Poisson's ratio, respectively.

Table 6.1 Nondimensional Center Deflection and Bending Moment Obtained by the Modified DQM

	SSSS				CCCC			
	$\bar{w}_c$		$\bar{M}_{xc}$		$\bar{w}_c$		$\bar{M}_{xc}$	
	$b/a = 1$	$b/a = 2$	$b/a = 1$	$b/a = 2$	$b/a = 1$	$b/a = 2$	$b/a = 1$	$b/a = 2$
DQM	0.004062	0.01013	0.04789	0.10168	0.001265	0.002533	0.02291	0.04116
Exact	0.004062	0.01013	0.04788	0.10168	0.001265	0.002533	-	-

EXAMPLE 6.2

CCCC isotropic rectangular plate under uniformly distributed load q ($p_{il} = q$) is analyzed by the modified DQM. Two aspect ratios ($b/a = 1$ and 2) are considered. The nondimensional center deflection and bending moment, defined by Eq. (6.37), are also listed in Table 6.1. Grid III is used and $N = 11$. From the data listed in Table 6.1, it can be seen that the rate of convergence is also very high; the modified DQM with 11×11 grid spacing can yield very accurate deflection and bending moment.

EXAMPLE 6.3

A cantilever isotropic rectangular plate under uniformly distributed load q ($p_{il} = q$) is analyzed by the modified DQM. The plate with aspect ratio $b/a = 2$ shown in Fig. 6.2 is considered, where a and b are the plate length and width, respectively. Poisson's ratio is taken as 0.3. Grid III is used. Due to symmetry of both load and geometry about the x-axis, the top half plate is analyzed by using the modified DQM with the grid spacing of 15×15. The nondimensional center deflections at points A and B, and bending moments at points C and D are listed in Table 6.2. Results obtained by the least squares boundary collocation method [1] and by the DQEM with grid spacing of 15×15 [2] are also included for comparisons. The rate of convergence is also very high because the results change only little when the number of grid points changes from 11 to 25. From the data listed in Table 6.2, it can be seen that except the bending moment at point D, all other results are close to the data obtained by the least squares boundary collocation method [1] and by the DQEM [2].

The nondimensional quantities appearing in Table 6.2 are defined by

$$\bar{w} = \frac{wD}{qa^4}, \quad \bar{M}_x = \frac{M_x}{qa^2} \tag{6.38}$$

where $D = Eh^3/(1 - \mu^2)$ is the bending rigidity of the plate, and h, E, and μ are plate thickness, modulus of elasticity, and Poisson's ratio, respectively.

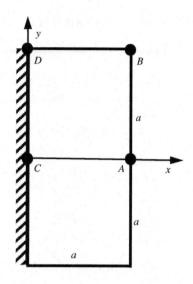

FIGURE 6.2

Sketch of a Cantilever Plate

Table 6.2 Nondimensional Center Deflection and Bending Moment Obtained by the Modified DQM

Methods	$A(\bar{w})$	$B(\bar{w})$	$C(\bar{M}x)$	$D(\bar{M}x)$
DQM	0.1278	0.1243	0.5136	0.0179
DQEM [2]	0.1284	0.1253	0.5174	0.6823
Collocation [1]	0.1282	0.1248	0.5329	0.3564

Theoretically speaking, the bending moment at point D should be zero. With N varying from 11 to 25, $\bar{M}_x$ at point D obtained by the modified DQM varies from positive to negative but is close to zero. On the other hand, the results obtained by the least squares boundary collocation method and the DQEM are not accurate. Perhaps due to difficulty in finding the fourth boundary condition at the plate corner point, the results obtained by the DQEM should be less accurate as compared to the data obtained by the modified DQM. Both MMWC-3 and MMWC-4 are tried for this problem; the results remain the same. Besides, the results are insensitive to the grid spacing for the problem considered. More details may be found in Ref. [3].

6.3.2 SKEW PLATE UNDER UNIFORMLY DISTRIBUTED LOAD

EXAMPLE 6.4

Isotropic rhombic ($b = a$) plate under uniformly distributed load q ($p_{il} = q$) is analyzed by the modified DQM, shown in Fig. 6.3. The skew plate is simply supported at all four edges. Five skew angles varying from 0° to 36° are considered.

It is known that the isotropic skew plate is equivalent to an anisotropic rectangular plate mathematically, namely [4],

$$
\begin{aligned}
D_{11} &= D_{22} = D = \frac{Eh^3}{12\left(1-\mu^2\right)} \\
D_{16} &= D_{26} = -D\sin\theta \\
D_{12} &= D\left(\mu\cos^2\theta + \sin^2\theta\right) \\
D_{66} &= D\left(1+\sin^2\theta - \mu\cos^2\theta\right)/2
\end{aligned}
\tag{6.39}
$$

where E, μ, and h are the modulus of elasticity, Poisson's ratio, and thickness of the skew plate, respectively. For isotropic rectangular plate, $\theta = 0°$.

The DQM with MMWC-3 is used. Table 6.3 lists the nondimensional center deflection $\bar{w}_c$ of the rhombic plate under uniformly distributed load. $\bar{w}_c$ is also defined by Eq. (6.37). It is seen that the modified DQM with the grid spacing of 11×11 can yield accurate center deflection.

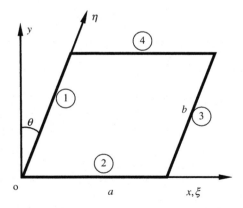

FIGURE 6.3

Sketch of a Skew Plate

Table 6.3 Nondimensional Center Deflection $\bar{w}_c$ and Bending Moment Obtained by the Modified DQM

θ	0	10	27	30	36
DQM	0.00406	0.00411	0.00445	0.00454	0.00478
Ashton [3]	0.00406	0.00411	0.00444	0.00452	0.00476

6.3.3 SQUARE PLATE UNDER LINE DISTRIBUTED LOAD

EXAMPLE 6.5

Isotropic square plate under a constant line load q is analyzed by the modified DQM. For illustration and simplicity, the line load is applied along $x_L = a/2$. Three combinations of boundary conditions, i.e., SFSF, SFSC, and CFCF, are considered. Poisson's ratio is taken as 0.0 and 0.3. When Poisson's ratio is 0.0, the problem is equivalent to the beam problem under point load; thus, analytical solutions are available.

Single numerical integration is performed by using Gauss quadrature since the Dirac-delta function is involved. The equivalent load p_{il} is given by

$$\begin{cases} p_{il} = 2q/[aH_{(N+1)/2}] & (i = (N+1)/2, l = 2, 3, ..., N-1) \\ p_{il} = 0 & \text{all others} \end{cases} \tag{6.40}$$

Grid V is used and $N = 11$. The nondimensional center deflection, $\bar{w}_c = 100(Dw_c/qla^3)$, is listed in Table 6.4.

From the data listed in Table 6.4, it can be seen that the rate of convergence is also very high; the modified DQM with 11×11 grid spacing can yield accurate central deflection. Poisson's ratio has appreciated effect on the central deflection.

Table 6.4 Center Deflection Coefficient $\bar{w}_c = 100(Dw_c/qla^3)$ Obtained by the Modified DQM

Boundary Condition	Exact ($\mu = 0.0$)	DQM ($\mu = 0.0$)	Error (%)	DQM ($\mu = 0.3$)	$(\bar{w}_c)_{0.3}/(\bar{w}_c)_{0.0}$
SFSF	2.083333	2.081776	−0.07	2.093060	1.0054
CFCF	0.520833	0.521175	0.07	0.512832	0.9840
SFCF	0.911458	0.911325	−0.01	0.901027	0.9887

6.3.4 SQUARE PLATE UNDER CONCENTRATED LOAD

EXAMPLE 6.6

Isotropic square plate under a concentrated load P is analyzed by the modified DQM. For illustration and simplicity, the point load is applied at the plate center. Three combinations of boundary conditions, i.e., SFSF, SFSC, and CFCF, are considered. Poisson's ratio is taken as 0.3. Since analytical solutions are not available, data obtained by FEM with fine meshes are included for comparisons.

Double numerical integrations are performed by using Gauss quadrature since the Dirac-delta function is involved. The equivalent load p_{il} is given by

$$\begin{cases} p_{il} = 4P/[aH_{(N+1)/2}]/[aH_{(N+1)/2}] & (i,l = (N+1)/2) \\ p_{il} = 0 & \text{all others} \end{cases} \tag{6.41}$$

Grid V is used. Since the convergence rate is not as high as the previous one, data obtained by DQM with $N = 21$ are listed in Table 6.5. The nondimensional center deflection is defined as $\bar{w}_c = 100(Dw_c/Pa^2)$.

From the data listed in Table 6.5, it can be seen that the rate of convergence is relatively high; the modified DQM with 21×21 grid spacing can yield accurate central deflection. Although the DQM with the method to deal with the Dirac-delta function may yield less accurate results than the DQEM for problems of beam and rectangular plate under a point load, the method presented in this chapter is more convenient than DQEM or FEM for solving problems of beam and rectangular plate under moving point or line distributed loads.

6.4 SUMMARY

In this chapter, the modified differential quadrature method is used to perform static analysis successfully for thin plates under various loadings, including the distributed load in the entire plate, the distributed load along a line, and a concentrated load. For the distributed load along a line and a concentrated load, a new approach is proposed to deal with the Dirac-delta function. Also, the formulation and

Table 6.5 Center Deflection Coefficient $\bar{w}_c = 100(Dw_c/Pa^2)$ Obtained by the Modified DQM

Boundary Condition	FEM	DQM	Difference (%)
SFSF	2.3217	2.317223	−0.19
CFCF	0. 76486	0.760162	−0.61
SFCF	1.1566	1.151957	−0.40

solution procedures can be used for rectangular plates with anisotropic materials. Several numerical examples are given.

Numerical results show that the modified DQM can yield accurate deflection as well as bending moment with relatively small number of grid points for all cases considered. Thus, the attractive features of rapid convergence, high accuracy, and computational efficiency of the DQM are demonstrated. Although only static analysis has been presented, the proposed approach to deal with the Dirac-delta function can also be used to solve the moving point load problem by using the modified DQM.

REFERENCES

[1] C.D. Xu, The Weighted Residual Method in Solid Mechanics, Shanghai, China, Tong Ji University Press, 1987 (in Chinese).

[2] X. Wang, Y. Wang, R. Chen, Static and free vibrational analysis of rectangular plates by the differential quadrature element method, Commun. Numer. Meth. Eng. 14 (1998) 1133–1141.

[3] Y. Wang, X. Wang, Y. Zhou, Static and free vibration analyses of rectangular plates by the new version of differential quadrature element method, Int. J. Numer. Meth. Eng. 59 (9) (2004) 1207–1226.

[4] J.E. Ashton, An analogy for certain anisotropic plates, J. Compos. Mater. 3 (1969) 355–358.

7

LINEAR BUCKLING ANALYSIS OF THIN PLATE

7.1 INTRODUCTION

This chapter demonstrates the linear buckling analysis of isotropic and anisotropic thin rectangular plates by using the modified (DQM) differential quadrature method. Since a fourth-order partial differential equation is solved, MMWC-3 or MMWC-4 should be used to apply the multiple boundary conditions. The stress components at all grid points should be obtained by the DQM or the modified DQM first if the stress distributions are unknown. Formulations are worked out in detail and examples are given for illustrations. The importance of using an appropriate way to apply the multiple boundary conditions is illustrated.

7.2 BUCKLING OF RECTANGULAR THIN PLATE

Consider the problem of linear buckling of the anisotropic rectangular thin plate. The plate has length a, width b, and thickness h. Anisotropic materials are considered. Uniform thickness is assumed.

The governing equation is given by

$$
\begin{aligned}
D_{11}\frac{\partial^4 w}{\partial x^4} + 4D_{16}\frac{\partial^4 w}{\partial x^3 \partial y} &+ 2(D_{12}+2D_{66})\frac{\partial^4 w}{\partial x^2 \partial y^2} + \\
4D_{26}\frac{\partial^4 w}{\partial x \partial y^3} + D_{22}\frac{\partial^4 w}{\partial y^4} &= \sigma_x h\frac{\partial^2 w}{\partial x^2} + 2\sigma_{xy} h\frac{\partial^2 w}{\partial x \partial y} + \sigma_y h\frac{\partial^2 w}{\partial y^2}
\end{aligned}
\tag{7.1}
$$

where D_{ij} are the flexural rigidity of the plate, w is the deflection, and $\sigma_x, \sigma_y, \tau_{xy}$ are the in-plane stress components, respectively. Note that the in-plane stress components in Eq. (7.1) may not be constant within the plate if the edge load is nonuniformly distributed. In such cases, the stress components at grid points can be obtained by using the modified DQM, as presented in Chapter 5. Because the same grid spacing is used for both in-plane stress analysis and buckling analysis by the modified DQM, it is very convenient when formulation II is used to obtain the in-plane stress components.

Introduce stress resultant P_x, P_y, P_{xy}, which are defined by

$$P_x = \sigma_x h, \ P_y = \sigma_y h, \ P_{xy} = \tau_{xy} h \tag{7.2}$$

The boundary conditions are summarized as follows:

1. Simply supported edge (S):

$$w = M_x = 0 \text{ at } x = 0, a \ \text{ or } \ w = M_y = 0 \text{ at } y = 0, b \tag{7.3}$$

2. Clamped edge (C):

$$w = \frac{\partial w}{\partial x} = 0 \ \text{ at } \ x = 0, a \ \text{ or } \ w = \frac{\partial w}{\partial y} = 0 \ \text{ at } \ y = 0, b \tag{7.4}$$

3. Free edge (F):

$$Q_x = M_x = 0 \ \text{ at } \ x = 0, \ a \ \text{ or } \ Q_y = M_y = 0 \ \text{ at } \ y = 0, b \tag{7.5}$$

where

$$M_x = D_{11} \frac{\partial^2 w}{\partial x^2} + D_{12} \frac{\partial^2 w}{\partial y^2} + 2 D_{16} \frac{\partial^2 w}{\partial x \partial y} \tag{7.6}$$

$$M_y = D_{12} \frac{\partial^2 w}{\partial x^2} + D_{22} \frac{\partial^2 w}{\partial y^2} + 2 D_{26} \frac{\partial^2 w}{\partial x \partial y} \tag{7.7}$$

$$Q_x = D_{11} \frac{\partial^3 w}{\partial x^3} + 4 D_{16} \frac{\partial^3 w}{\partial x^2 \partial y} + \left(D_{12} + 4 D_{66} \right) \frac{\partial^3 w}{\partial x \partial y^2} + 2 D_{26} \frac{\partial^3 w}{\partial y^3} \\ - P_x \frac{\partial w}{\partial x} - P_{xy} \frac{\partial w}{\partial y} \tag{7.8}$$

$$Q_y = D_{22} \frac{\partial^3 w}{\partial y^3} + 4 D_{26} \frac{\partial^3 w}{\partial x \partial y^2} + \left(D_{12} + 4 D_{66} \right) \frac{\partial^3 w}{\partial x^2 \partial y} + 2 D_{16} \frac{\partial^3 w}{\partial x^3} \\ - P_{xy} \frac{\partial w}{\partial x} - P_y \frac{\partial w}{\partial y} \tag{7.9}$$

If a corner is free, three boundary conditions, i.e., $M_x = M_y = R = 0$, should be applied, where the concentrated force R is given by

$$R = 2 D_{16} \frac{\partial^2 w}{\partial x^2} + 2 D_{26} \frac{\partial^2 w}{\partial y^2} + 4 D_{66} \frac{\partial^2 w}{\partial x \partial y} \tag{7.10}$$

Comparing Eqs. (7.8) and (7.9) with Eqs. (6.7) and (6.8) reveals that the expressions of shear force are slightly different. The effect of in-plane forces is considered in Eqs. (7.8) and (7.9) and will be reflected in the solutions when free edges are involved.

Let N_x and N_y be the total number of grid points in the x and y directions. Usually, $N_x = N_y = N$. A 9×9 grid spacing is schematically shown in Fig. 7.1. Each inner point has one degree of freedom (DOF), i.e., $w_{ij} (i = 2, 3, ..., N_y - 1, j = 2, 3, ..., N_x - 1)$, each corner node has three DOF, i.e., w_{ij}, w_{xij}, w_{yij} $(i = 1, N_y, j = 1, N_x)$, each boundary node on edges $x = 0$ and a has two DOF, i.e., $w_{ij}, w_{xij} (i = 2, ..., N_y - 1, j = 1, N_x)$, and each boundary node on edges $y = 0$ and b has two DOF, i.e., $w_{ij}, w_{yij} (i = 1, N_y, j = 2, ..., N_x - 1)$.

MMWC-3 or MMWC-4 can be used to introduce the additional DOF at the two end points in one dimension, i.e., $w_1' = (\partial w / \partial x)_{x=0}$ and $w_N' = (\partial w / \partial x)_{x=a}$. The first- to the fourth-order derivatives of the solution function w at a grid point i can be symbolically written by

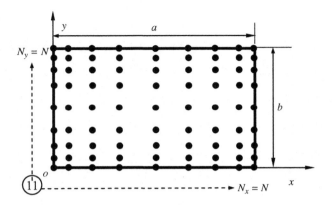

FIGURE 7.1

Sketch of a Rectangular Plate with 9 × 9 Grid Spacing

$$w_i^{(1)} = \sum_{j=1}^{N+2} A_{ij}^x \, \delta_j \qquad (i = 1, 2, ..., N)$$

$$w_i^{(2)} = \sum_{j=1}^{N+2} B_{ij}^x \, \delta_j \qquad (i = 1, 2, ..., N)$$

$$w_i^{(3)} = \sum_{j=1}^{N+2} C_{ij}^x \, \delta_j \qquad (i = 1, 2, ..., N)$$

$$w_i^{(4)} = \sum_{j=1}^{N+2} D_{ij}^x \, \delta_j \qquad (i = 1, 2, ..., N)$$

(7.11)

where $A_{ij}^x, B_{ij}^x, C_{ij}^x, D_{ij}^x$ are called the weighting coefficients of the first- to fourth-order derivatives with respect to x, N is the total number of grid points, and δ_j is defined by

$$\delta_j = w_j \, (j = 1, 2, ..., N), \ \ \delta_{N+1} = w_1', \ \ \delta_{N+2} = w_N'$$

(7.12)

The weighting coefficients of the first- to fourth-order derivatives with respect to y can be formulated in a similar way. Assume the total number of grid points in the x and y directions the same and denoted by N. For clarity, N_x and N_y are also used sometimes in this chapter, however.

In terms of the differential quadrature, Eq. (7.1) at all inner grid points can be expressed as

$$D_{11} \sum_{k=1}^{N+2} D_{ik}^x \bar{w}_{kl} + 4D_{16} \sum_{j=1}^{N+2} \sum_{k=1}^{N+2} C_{ij}^x A_{lk}^y \bar{w}_{jk} + 2\left(D_{12} + 2D_{66}\right) \sum_{j=1}^{N+2} \sum_{k=1}^{N+2} B_{ij}^x B_{lk}^y \bar{w}_{jk}$$

$$+ 4D_{26} \sum_{j=1}^{N+2} \sum_{k=1}^{N+2} A_{ij}^x C_{lk}^y \bar{w}_{jk} + D_{22} \sum_{k=1}^{N+2} D_{lk}^y \bar{w}_{ik}$$

$$= P_0 \left(\sum_{k=1}^{N} B_{lk}^x w_{kl} (P_x / P_0)_{il} + 2 \sum_{j=1}^{N} \sum_{k=1}^{N} A_{ij}^x A_{lk}^y w_{jk} (P_{xy} / P_0)_{il} + \sum_{k=1}^{N} B_{lk}^y w_{ik} (P_y / P_0)_{il} \right)$$

$$(i = 2, 3, ..., N_y - 1; l = 2, 3, ..., N_x - 1)$$

(7.13)

where P_0 is a constant to be found and the over bar on w means that the derivative DOFs are included.

To formulate a complete set of algebraic equations, the bending moment equation is placed at the position where the DOF of $\partial w / \partial x$ or $\partial w / \partial y$ is, and the shear force equation is placed at the position where the boundary deflection w is, the same as the ones presented in Chapter 6.

In detail, at the positions where $\left(\partial w / \partial x\right)_{x=0}$ and $\left(\partial w / \partial x\right)_{x=a}$ are, the following DQ equations, i.e., the bending moment M_x at points il, are placed:

$$D_{11}\sum_{k=1}^{N+2} B_{ik}^x \overline{w}_{kl} + D_{12}\sum_{k=1}^{N+2} B_{lk}^y \overline{w}_{ik} + 2D_{16}\sum_{j=1}^{N+2}\sum_{k=1}^{N+2} A_{ij}^x A_{lk}^y \overline{w}_{jk} = \left(M_x\right)_{il}$$

$$(i = 1, 2, ..., N_y; l = 1, N_x)$$

$$(7.14)$$

At the positions $w_{il}(i = 2, 3, ..., N_y - 1; \ l = 1, \ N_x)$, the following DQ equations, i.e., the shear force equations Q_x at points il, are applied:

$$D_{11}\sum_{k=1}^{N+2} C_{ik}^x \overline{w}_{kl} + 4D_{16}\sum_{j=1}^{N+2}\sum_{k=1}^{N+2} B_{ij}^x A_{lk}^y \overline{w}_{jk} + \left(D_{12} + 4D_{66}\right)\sum_{j=1}^{N+2}\sum_{k=1}^{N+2} A_{ij}^x B_{lk}^y \overline{w}_{jk}$$

$$+ 2D_{26}\sum_{k=1}^{N+2} C_{lk}^y \overline{w}_{ik} - \left(P_x\right)_{il}\sum_{k=1}^{N+2} A_{ik}^x \overline{w}_{kl} - \left(P_{xy}\right)_{il}\sum_{k=1}^{N+2} A_{lk}^y \overline{w}_{ik} = \left(Q_x\right)_{il}$$

$$(i = 2, 3, ..., N_y - 1; l = 1, N_x)$$

$$(7.15)$$

Note that the four corner points, w_{il} $(i = 1, N_y; \ l = 1, N_x)$, are not included in Eq. (7.15), and that Eq. (7.15) is slightly different from Eq. (6.14).

Similarly, at the positions where the DOF of $\left(\partial w / \partial y\right)_{y=0}$ and $\left(\partial w / \partial y\right)_{y=b}$ are, the following DQ equations, i.e., the bending moment M_y at points il, are placed:

$$D_{12}\sum_{k=1}^{N+2} B_{ik}^x \overline{w}_{kl} + D_{22}\sum_{k=1}^{N+2} B_{lk}^y \overline{w}_{ik} + 2D_{26}\sum_{j=1}^{N+2}\sum_{k=1}^{N+2} A_{ij}^x A_{lk}^y \overline{w}_{jk} = \left(M_y\right)_{il}$$

$$(i = 1, N_y; \ l = 1, 2, ..., N_x)$$

$$(7.16)$$

And at the positions w_{il} $(i = 1, N_y; \ l = 2, 3, ..., N_x - 1)$, the following DQ equations, i.e., the shear force equations Q_y at points il, are applied:

$$D_{22}\sum_{k=1}^{N+2} C_{lk}^y \overline{w}_{ik} + 4D_{26}\sum_{j=1}^{N+2}\sum_{k=1}^{N+2} A_{ij}^x B_{lk}^y \overline{w}_{jk} + \left(D_{12} + 4D_{66}\right)\sum_{j=1}^{N+2}\sum_{k=1}^{N+2} B_{ij}^x A_{lk}^y \overline{w}_{jk}$$

$$+ 2D_{16}\sum_{k=1}^{N+2} C_{ik}^x \overline{w}_{kl} - \left(P_{xy}\right)_{il}\sum_{k=1}^{N+2} A_{ik}^x \overline{w}_{kl} - \left(P_y\right)_{il}\sum_{k=1}^{N+2} A_{lk}^y \overline{w}_{ik} = \left(Q_y\right)_{il}$$

$$(i = 1, N_y; \ l = 2, 3, ..., N_x - 1)$$

$$(7.17)$$

Again the four corner points, w_{il} $(i = 1, N_y; \ l = 1, N_x)$, are not included in Eq. (7.17), and Eq. (7.17) is slightly different from Eq. (6.16).

Finally, at the four corner points, w_{il} $(i = 1, N_y; \ l = 1, N_x)$, the following DQ equations, i.e., the concentrated force R at points il, are applied:

$$2D_{16}\sum_{k=1}^{N+2} B_{ik}^x \overline{w}_{kl} + 2D_{26}\sum_{k=1}^{N+2} B_{lk}^y \overline{w}_{ik} + 4D_{66}\sum_{j=1}^{N+2}\sum_{k=1}^{N+2} A_{ij}^x A_{lk}^y \overline{w}_{jk} = R_{il}$$

$$(i = 1, N_y; \ l = 1, N_x)$$

$$(7.18)$$

For rectangular plates without free edges, the modification on the expression of the shear force is not important because these equations are not used in the analysis.

Combining Eqs. (7.13)–(7.18) together results in the following partitioned matrix equation:

$$\begin{bmatrix} K_{mm} & K_{mi} \\ K_{im} & K_{ii} \end{bmatrix} \begin{Bmatrix} \Delta_m \\ W_i \end{Bmatrix} = \begin{Bmatrix} F_m \\ P_0\left([P_{im}]\{\Delta_m\} + [P_{ii}]\{W_i\} \right) \end{Bmatrix} \tag{7.19}$$

where subscripts m and i denote the equations without and with nonzero in-plane stress component, and P_0 is apparent. In other words, the value of subscripts m and i varies depending on whether the plate has loaded free edges or not.

If the rectangular plate does not have loaded free edges, subscripts m and i are the same as the subscripts b and i in Eq. (6.18). When the rectangular plate has loaded free edges, however, the vector $\{W_i\}$ contains not only the displacement at all inner nodes but also the displacement at nodes on the loaded free boundary except to the corner points because the last two terms on the left-hand side of Eq. (7.15) or Eq. (7.17) are not zero. Therefore, vectors $\{\Delta_m\}$ and $\{F_m\}$ change accordingly. This is different from the ones in Eq. (6.18).

Again, any combinations of boundary conditions can be easily applied by using Eqs. (7.14)–(7.18), because either the deflection or the shear force, as well as either the slope or the bending moment, is specified. In other words, the element in $\{F_m\}$ is either zero if the corresponding DOF in $\{\Delta_m\}$ is unknown or unknown if the corresponding DOF in $\{\Delta_m\}$ is given. After imposing the regular boundary conditions, i.e., the zero generalized displacement conditions, the DQ equations for the linear buckling analysis of a rectangular plate become

$$\begin{bmatrix} \bar{K}_{mm} & \bar{K}_{mi} \\ \bar{K}_{im} & \bar{K}_{ii} \end{bmatrix} \begin{Bmatrix} \bar{\Delta}_m \\ \bar{W}_i \end{Bmatrix} = P_0 \begin{bmatrix} 0 & 0 \\ \bar{P}_{im} & \bar{P}_{ii} \end{bmatrix} \begin{Bmatrix} \bar{\Delta}_m \\ \bar{W}_i \end{Bmatrix} \tag{7.20}$$

where the over bar means that the quantities have been modified due to dropping of the zero generalized displacements.

After eliminating the nonzero $\{\bar{\Delta}_m\}$, Eq. (7.20) can be written as

$$\left[\bar{K}_{ii} - \bar{K}_{im}\bar{K}_{mm}^{-1}\bar{K}_{mi} \right]\{\bar{W}_i\} = P_0 \left[\bar{P}_{ii} - \bar{P}_{im}\bar{K}_{mm}^{-1}\bar{K}_{mi} \right]\{\bar{W}_i\} \tag{7.21}$$

Equation (7.21) can be further simplified to

$$\left[\bar{K} \right]\{\bar{W}_i\} = P_0 \left[\bar{P} \right]\{\bar{W}_i\} \tag{7.22}$$

Equation (7.22) is a generalized eigenvalue equation and can be solved by a generalized eigenvalue solver. The lowest eigenvalue corresponds to the buckling load. Alternatively, Eq. (7.22) can be transformed to a standard eigenvalue equation, namely:

$$\left[\bar{K} \right]^{-1}\left[\bar{P} \right]\{\bar{W}_i\} = 1/P_0\{\bar{W}_i\} \tag{7.23}$$

or in short,

$$[E]\{\bar{W}_i\} = \lambda_0\{\bar{W}_i\} \tag{7.24}$$

Equation (7.24) is a standard eigenvalue equation and can be solved by any standard eigenvalue solver. Alternatively, Eq. (7.24) can be efficiently solved by using the Power method because only the largest eigenvalue is required. Knowing the largest eigenvalue λ_0, the buckling load can be found accordingly.

If an isotropic or orthotropic rectangular plate without a free edge is considered, the boundary conditions can be directly built in during formulation of weighting coefficients by using the combination of MMWC-1 and MMWC-3 (or MMWC-4), similar to the static cases presented in Chapter 6. In such cases, only DQ equations at all inner grid points are needed, namely:

$$
D_{11} \sum_{k=1}^{N} D_{ik}^{x} w_{kl} + 4 D_{16} \sum_{j=1}^{N} \sum_{k=1}^{N} C_{ij}^{x} A_{lk}^{y} w_{jk} + 2 \left(D_{12} + 2 D_{66} \right) \sum_{j=1}^{N} \sum_{k=1}^{N} B_{ij}^{x} B_{lk}^{y} w_{jk}
$$

$$
+ 4 D_{26} \sum_{j=1}^{N} \sum_{k=1}^{N} A_{ij}^{x} C_{lk}^{y} w_{jk} + D_{22} \sum_{k=1}^{N} D_{lk}^{y} w_{ik}
$$

$$
= P_0 \left(\sum_{k=1}^{N} B_{ik}^{x} w_{kl} (P_x / P_0)_{il} + 2 \sum_{j=1}^{N} \sum_{k=1}^{N} A_{ij}^{x} A_{lk}^{y} w_{jk} (P_{xy} / P_0)_{il} + \sum_{k=1}^{N} B_{lk}^{y} w_{ik} (P_y / P_0)_{il} \right)
$$

$$
(i = 2, 3, ..., N_y - 1; l = 2, 3, ..., N_x - 1)
$$

(7.25)

Note that Eq. (7.25) is different from Eq. (7.13). The upper summation index is N and the first-order derivative DOF is not involved. Besides, the weighting coefficients of the third- and fourth-order derivatives are more or less different from the ones shown in Eq. (7.13), although the symbols are the same. Since MMWC-3 (or MMWC-4) is used for the clamped boundary conditions, C_{ij}^{x} and D_{ij}^{x} in Eq. (7.25) and Eq. (7.13) are the same if edges at $x = 0$ and a are clamped. Similarly, C_{ij}^{y} and D_{ij}^{y} in Eq. (7.25) and Eq. (7.13) are the same if edges at $y = 0$ and b are clamped.

In matrix form, Eq. (7.25) can be rewritten as

$$
[K_{ii}]\{W_i\} = P_0 [P_{ii}]\{W_i\}
$$

(7.26)

or in short,

$$
[K]\{W_i\} = \lambda [I]\{W_i\}
$$

(7.27)

where

$$
[K] = [K_{ii}]^{-1} [P_{ii}], \quad \lambda = \frac{1}{P_0}
$$

(7.28)

Solving either Eq. (7.26) directly by using a generalized eigenvalue solver or Eq. (7.27) by using a standard eigenvalue solver yields the buckling load. If Eq. (7.28) is solved, the inverse of the largest eigenvalue corresponds to the buckling load.

7.3 APPLICATIONS

A FORTRAN program is written for analyzing SSSS and CCCC anisotropic rectangular plates. The program is called Program 4 and included in Appendix VII. It has been converted to MATLAB file, too. With only a few modifications, the modified program can analyze the buckling of rectangular plates with different materials or applied in-plane loads.

7.3.1 RECTANGULAR PLATE UNDER UNIFORM EDGE COMPRESSIVE LOAD

EXAMPLE 7.1

An isotropic square plate with length a and uniform thickness h is under uniaxial compression. The edge applied load is uniformly distributed. Find the buckling load coefficient (simply called the buckling coefficient) for SSSS and CCCC plates. Poisson's ratio is 0.3. For comparisons, the buckling coefficient $\bar{P}_x$ is introduced and defined by

$$\bar{P}_x = \frac{P_x a^2}{D} \tag{7.29}$$

where D is the bending rigidity of the isotropic square plate and P_x is the buckling load.

The results obtained by the modified DQM with grid spacing of 9×9 are listed in Table 7.1. Grid III is used. It is seen that the buckling coefficients obtained by the modified DQM with MMWC-3 to apply the boundary conditions are more accurate than the ones obtained by the conventional DQM with the δ method to apply the boundary conditions [1], especially for the clamped boundary conditions.

7.3.2 RECTANGULAR PLATE UNDER LINEAR VARYING EDGE COMPRESSIVE LOAD

EXAMPLE 7.2

Isotropic rectangular plates with length a, width b, and uniform thickness h are under uniaxial compression. Find the buckling coefficient for SCSC plate. The symbol SCSC means that the edges at $x = 0$ and $x = a$ are simply supported and the other two edges are clamped. Poisson's ratio is 0.3. The edge applied load is a linear varying distributed load defined by

$$P_x = -P(1 - \alpha y / b) \tag{7.30}$$

where α is a constant and to be specified during analysis.
For comparisons, the buckling coefficient $\bar{P}_x$ is defined by

$$\bar{P}_x = \frac{P_x b^2}{D} \tag{7.31}$$

where D is the bending rigidity of the isotropic rectangular plate.

Convergence study [2] shows that the modified DQM with grid spacing of 13×13 can yield very accurate results. Therefore, the results obtained by the modified DQM with grid spacing of 13×13 are listed in Table 7.2 for the case $\alpha = 0$. The plate is under uniaxial compression with constant stress. Various aspect ratios are considered. It can be seen that very accurate results are obtained and the results are the same as the ones obtained by the power series method [3] and given by Timoshenko and Gere [4].

From Table 7.2, it is seen that the buckling coefficients for $a/b = 0.5$ and $a/b = 1$ are exactly the same. Buckling mode shapes shown in Figs. 7.2 and 7.3 [2] can explain this fact. It is seen clearly from Figs. 7.2 and 7.3 that $m = n = 1$ for $a/b = 0.5$ and $m = 2, n = 1$ for $a/b = 1$, where m and n are the number of half waves in the x and y directions. For $a/b = 1$, there are two half-waves in the x direction.

Table 7.1 Buckling Coefficient of Rectangular Plate Under Uniaxial Compression

Boundary Conditions	SSSS	CCCC
Modified DQM	39.4784	99.4726
DQM-δ [1]	40.0607	111.230
Exact	39.4784	99.3869

Table 7.2 Buckling Coefficient of SCSC Rectangular Plate Under Uniaxial Compression ($\alpha = 0$)

a/b	0.4	0.5	0.6	0.7	0.8	0.9	1.0
DQM (N = 13)	93.247	75.910	69.632	69.095	72.084	77.545	75.910
[4]	93.2	75.9	69.6	69.1	71.9	77.3	75.9
[3]	93.247	75.910	69.632	69.095	72.084	77.545	75.910

Tables 7.3 and 7.4 list the buckling coefficient for rectangular plates under liner varying load ($\alpha = 1$) and pure bending ($\alpha = 2$). The results obtained by the modified DQM with grid spacing of 15×15 are the same as the ones obtained by the power series method [3]. The energy method yields less accurate buckling coefficient. From Tables 7.3 and 7.4, it is seen that the modified DQM can yield very accurate buckling coefficient. Modifications of the provided FORTRAN program or MATLAB file (see Appendix VII) are simple to analyze the buckling for rectangular plates with other type of applied loads, materials, and combinations of boundary conditions.

It is seen again that the buckling coefficients for $a/b = 0.5$ and $a/b = 1$ are exactly the same for $\alpha = 1$ and $\alpha = 2$. The reason should be the same as the one for $\alpha = 0$.

EXAMPLE 7.3

An isotropic rectangular plate with length a, width b, and uniform thickness h is under uniaxial compression. Find the buckling coefficient for the SSSF plate. The symbol SSSF means that the edge at $y = b$ is free and the remaining three edges are simply supported. Poisson's ratio is 0.3. The edge compressive load is a linear varying distributed load, defined by

$$P_x = -P(1 - y/b) \tag{7.32}$$

For comparisons, the buckling coefficient $\bar{P}_x$ is defined by

$$\bar{P}_x = \frac{P_x b^2}{\pi^2 D} \tag{7.33}$$

In the DQ analysis, Grid III and grid spacing of 13×13 are used. The DQ results are listed in Table 7.5. Various aspect ratios are analyzed. Results are compared with the data listed in the *Handbook for Aircraft Design* [5] and the results obtained by the NASTRAN with fine meshes. It can be seen that the DQ results are close to the finite element results, but quite different from the ones listed in Ref. [5]. The possible reason is that the data listed in Ref. [5] is not accurate enough because the rate of the convergence of the Ritz method is low when free edge is involved.

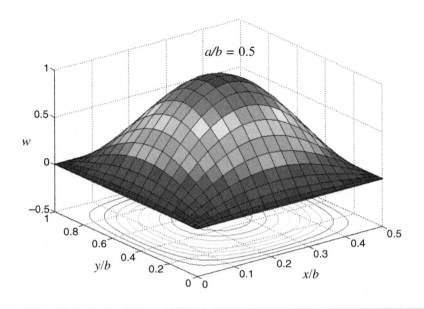

FIGURE 7.2

Buckling Mode of SCSC Rectangular Plate Under Uniaxial Compression ($\alpha = 0$)

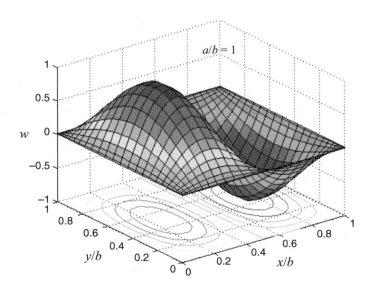

FIGURE 7.3

Buckling Mode of SCSC Square Plate Under Uniaxial Compression ($\alpha = 0$)

Table 7.3 Buckling Coefficient of SCSC Rectangular Plate Under Uniaxial Compression ($\alpha = 1$)

a/b	0.4	0.5	0.6	0.7	0.8	0.9	1.0
DQM ($N = 15$)	174.4	145.2	134.8	134.6	141.0	152.0	145.2
Energy[3]	175	145	135	134.7	141.0	152.1	145
Series [3]	174.4	145.2	134.8	134.6	141.0	152.0	145.2

Table 7.4 Buckling Coefficient of SCSC Rectangular Plate Under Uniaxial Compression ($\alpha = 2$)

a/b	0.4	0.5	0.6	0.7	0.8	0.9	1.0
DQM ($N = 15$)	400.4	391.5	411.8	422.5	400.4	391.3	391.5
Energy[3]	402	382.2	412.2	424	402	-	392
Series [3]	400.4	391.5	411.8	422.5	400.4	-	391.5

Table 7.5 Buckling Coefficient of SSSF Rectangular Plate Under Uniaxial Compression

a/b	Manual [5]	NASTRAN	DQM
0.8	6.72	6.18	6.20
1.0	5.02	4.758	4.77
1.5	3.36	3.15	3.16
2.0	2.80	2.53	2.55
3.0	2.40	2.07	2.08

EXAMPLE 7.4

An isotropic rectangular plate with length a, width b, and uniform thickness h is under uniaxial compression. Find the buckling coefficient for the SFSS plate. The symbol SFSS means that the edge at $x = a$ is free and the remaining three edges are simply supported. Poisson's ratio is 0.3. The load is a linear varying distributed load given by Eq. (7.32).

The results obtained by the modified DQM with grid spacing of 15×15 are listed in Table 7.6. Various aspect ratios are considered. The results are compared with the data listed in the *Handbook for Aircraft Design* [5] and data obtained by NASTRAN with fine meshes. It can be seen that the DQ results are close to the finite element data, but quite different from the ones listed in Ref. [5]. The possible reason is that the data listed in Ref. [5] are not accurate enough because the rate of the convergence of the Ritz method is low when a free edge is involved. It is believed that the DQ results listed in Tables 7.5 and 7.6 are the most accurate ones and can be used in practice.

Table 7.6 Buckling Coefficient of SFSS Rectangular Plate Under Uniaxial Compression

a/b	Manual [5]	NASTRAN	DQM
0.4	9.86	8.31	8.33
0.5	6.58	5.61	5.63
0.6	4.80	4.11	4.12
0.8	3.02	2.58	2.59
1.0	2.18	1.86	1.87
1.5	1.36	1.14	1.15
2.0	1.08	0.89	0.89
3.0	0.86	0.71	0.71

7.3.3 RECTANGULAR PLATE UNDER COSINE DISTRIBUTED EDGE COMPRESSIVE LOAD

EXAMPLE 7.5

Consider an SSSS rectangular plate under nonuniformly distributed edge compression, i.e., $\sigma_{x0} = -\sigma_0 \cos(\pi y / b)$, $\sigma_{y0} = \tau_{xy0} = 0$, shown in Fig. 5.1. For this loading case, $P_x \neq -\sigma_0 h \cos(\pi y / b)$, $P_y \neq 0$, and $P_{xy} \neq 0$ within the plate, as shown in Figs. 5.5–5.7. The plate length, width, and uniform thickness are denoted by a, b, and h. Although analytical solution for the in-plane stress is not available and difficult to obtain, the in-plane forces P_x, P_y, P_{xy} at all grid points are obtained by using the DQM, as presented in Chapter 5. Besides, it is more convenient to use the modified DQM, since the same grid points are used in both in-plane stress analysis and buckling analysis if formulation II is used in the in-plane stress analysis.

Table 7.7 shows the convergence study of the buckling coefficient for SSSS rectangular plates with three different aspect ratios. For comparisons, the buckling coefficient $\bar{P}_x$ is defined as

$$\bar{P}_x = \frac{\sigma_0 h b^2}{\pi^2 D} \tag{7.34}$$

where σ_0 is the maximum stress applied on the edge, shown in Fig. 5.1.

From Table 7.7, it is observed that converged results can be obtained by the modified DQM with Grid III and grid spacing of 13×13 for the three aspect ratios.

Table 7.8 shows the comparisons of results obtained by the modified DQM with grid spacing of 13×13 to available solutions cited from literatures [6–8].

From Table 7.8 it is observed that a large discrepancy exists. To check the data obtained by the modified DQM, finite element analysis is performed by using NASTRAN. Very fine meshes are used and the finite element data are also listed in Table 7.8 for comparisons. It can be seen that DQ results

Table 7.7 Convergence of the Buckling Coefficient $\bar{P}_x$ for SSSS Rectangular Plates Under Uniaxial Half-Cosine Distributed Compressive Load

a/b	N = 9	N = 11	N = 13	N = 15
0.5	7.452	7.452	7.452	7.452
1.0	5.419	5.419	5.419	5.419
3.0	5.930	5.845	5.849	5.849

Table 7.8 Comparisons of the Buckling Load $\bar{P}_x$ of SSSS Rectangular Plates Under Uniaxial Half-Cosine Distributed Compressive Load

a/b	DQM	FEM	Bert [6]	Benoy [8]	Van der Neut [7]
0.5	7.452	7.409	7.841	7.08	
1.0	5.419	5.383	5.146	4.59	4.68
3.0	5.849	5.818	5.748	4.53	

are the closest ones to the finite element data. Difference exists between DQ data and solutions obtained by Galerkin method in Ref. [6]. The possible reasons for this discrepancy are that a minor error exists in their derivations and stress boundary conditions are not satisfied to obtain in-plane stress solutions. Data in Refs. [7,8] are obviously too small and not accurate. More results on the buckling analysis by the modified DQM can be found in Refs. [2,9–13].

7.3.4 BUCKLING OF LAMINATED COMPOSITE RECTANGULAR PLATE UNDER COMBINED LOADS

For a laminated composite plate, the flexural rigidity of the plate in Eq. (7.1) is defined by

$$D_{ij} = \frac{1}{3}\sum_{k=1}^{n}(\bar{Q}_{ij})_k(z_k^3 - z_{k-1}^3) \quad (i,j=1,2,6) \tag{7.35}$$

where n is the number of lamina, z_k, and z_{k-1} are the coordinates of the top and bottom surfaces of the kth layer, and $\bar{Q}_{ij}(i,j=1,2,6)$ are computed by

$$\begin{aligned}
\bar{Q}_{11} &= Q_{11}\cos^4\theta + 2(Q_{12}+2Q_{66})\cos^2\theta\sin^2\theta + Q_{22}\sin^4\theta \\
\bar{Q}_{22} &= Q_{11}\sin^4\theta + 2(Q_{12}+2Q_{66})\cos^2\theta\sin^2\theta + Q_{22}\cos^4\theta \\
\bar{Q}_{12} &= \bar{Q}_{21} = (Q_{12}+Q_{22}-4Q_{66})\cos^2\theta\sin^2\theta + Q_{12}\left(\cos^4\theta+\sin^4\theta\right) \\
\bar{Q}_{66} &= (Q_{12}+Q_{22}-2Q_{12}-2Q_{66})\cos^2\theta\sin^2\theta + Q_{66}\left(\cos^4\theta+\sin^4\theta\right) \\
\bar{Q}_{16} &= (Q_{11}-Q_{22}-2Q_{66})\cos^3\theta\sin\theta + (Q_{12}-Q_{22}+2Q_{66})\cos\theta\sin^3\theta \\
\bar{Q}_{26} &= (Q_{11}-Q_{22}-2Q_{66})\cos\theta\sin^3\theta + (Q_{12}-Q_{22}+2Q_{66})\cos^3\theta\sin\theta
\end{aligned} \tag{7.36}$$

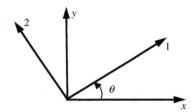

FIGURE 7.4

Principal Axes (1,2) and Reference Axes (x, y)

where the positive angle θ is measured counterclockwise from the x axis to the fiber direction, i.e., from the x axis to the principal material axis 1, shown in Fig. 7.4.

In Eq. (7.36), Q_{ij} $(i, j = 1, 2, 6)$ are defined by

$$
\begin{aligned}
Q_{11} &= \frac{E_{11}}{(1 - \mu_{12}\mu_{21})} \\
Q_{22} &= \frac{E_{22}}{(1 - \mu_{12}\mu_{21})} \\
Q_{12} &= Q_{21} = \mu_{21}Q_{11} = \mu_{12}Q_{22} \\
Q_{66} &= G_{12}
\end{aligned}
\tag{7.37}
$$

where E_{11} is the modulus of elasticity in the first direction and E_{22} is the modulus of elasticity in the second direction, G_{12} is the shear modulus, μ_{12} and μ_{21} are the major and minor Poisson's ratios satisfying the following equation:

$$
\frac{\mu_{12}}{E_{11}} = \frac{\mu_{21}}{E_{22}}
\tag{7.38}
$$

EXAMPLE 7.6

Consider a simply supported rectangular graphite/epoxy composite plate under combined uniaxial compression in the x direction and shear loading. The plate length, width, and uniform thickness are denoted by a, b, and h. For convenience in presentation of the results, the buckling load parameters are introduced and defined by

$$
R_x = P_x / P_{x0}, \quad R_{xy} = P_{xy} / P_{xy0}
\tag{7.39}
$$

where P_{x0}, P_{xy0} are the buckling loads under uniaxial compression and pure shear load, and P_x, P_{xy} are the buckling loads under combined loads, respectively.

The material properties are $Q_{11} / Q_{22} = 25$, $Q_{12} / Q_{22} = 0.25$, and $Q_{66} / Q_{22} = 0.5$. The geometric parameter is $a / b = 1.0$ and θ varies from $0°$ to $90°$.

If the composite rectangular plate contains only one layer, one has

$$
D_{ij} = \frac{h^3}{12} \overline{Q}_{ij} \quad (i, j = 1, 2, 6)
\tag{7.40}
$$

where $\overline{Q}_{ij}$ $(i, j = 1, 2, 6)$ are computed by Eq. (7.36).

Table 7.9 Buckling Load of Anisotropic SSSS Rectangular Plates

$\theta°$	$\bar{P}_{x0}$	$\bar{P}_{xy0}$	$-\bar{P}_{xy0}$
0	23.4403	39.7093	39.7093
45	17.3172	152.210	14.1442

Table 7.9 lists the DQ results for plates under uniaxial compression or pure shear alone. Grid III with a mesh of 15 × 15 is used. MMWC-3 is used for applying the multiple boundary conditions where the buckling load coefficients, $\bar{P}_{x0}$ and $\bar{P}_{xy0}$, are defined by

$$\bar{P}_{x0} = P_{x0}b^2 / (Q_{22}h^3), \ \bar{P}_{xy0} = P_{xy0}b^2 / (Q_{22}h^3) \tag{7.41}$$

From Table 7.9, it is seen that due to the anisotropy, the pure shear buckling loads for the rectangular plate with $\theta = 45°$ are quite different for positive shear and negative shear.

Figure 7.5 shows the buckling load parameters for the SSSS square plate($\theta = 0°$) under combined uniaxial compressive and shear loads.

For clarity, $-R_{xy}$ is used in Fig. 7.5 for negative shear, although $R_{xy} > 0$ according to the definition of Eq. (7.39). The dotted line represents the following equation:

$$R_{x0} + R_{xy}^2 = 1 \tag{7.42}$$

From Fig. 7.5, it is seen that Eq. (7.42) yields more conservative buckling load, which is close to the actual results. In other words, the buckling load calculated by Eq. (7.42) is a little smaller for the combined loading and thus is safer if it is used in practice.

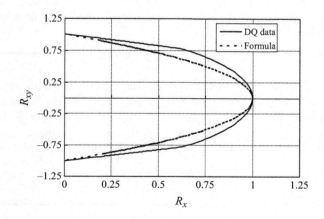

FIGURE 7.5

Buckling Load Parameters for the SSSS Plate ($\theta = 0°$)

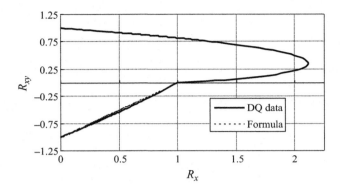

FIGURE 7.6

Buckling Load Parameters for SSSS Plates ($\theta = 45°$)

Figure 7.6 shows the buckling load parameters for the SSSS square plate($\theta = 45°$) under combined uniaxial compressive and shear loads. Again for clarity, $-R_{xy}$ is used for negative shear loading, although $R_{xy} > 0$ according to the definition of Eq. (7.39).

In Fig. 7.6, the dotted line for negative shear loading represents the following equation:

$$R_{x0} + R_{xy} = 1 \tag{7.43}$$

From Fig. 7.6, it is seen that Eq. (7.43) yields a little conservative buckling load for the negative shear loading and is very close to the actual results under combined loading. For positive shear, however, neither Eq. (7.42) nor Eq. (7.43) can predict the buckling load accurately. Otherwise, too conservative buckling load would be obtained. Therefore, direct computing of the buckling load by using various numerical methods may be required, such as the modified DQM or finite element method. More information on the buckling analysis of composite plates by the DQM may be found in Refs. [10,14]. Different ways to apply the multiple boundary conditions are used in Refs. [10,14]; MMWC-3 is used in Ref. [10] and the δ approach is used in Ref. [14]. Therefore, the results listed in Ref. [10] should be more accurate than the ones listed in Ref. [14] if the same number of grid points and the same nonuniform grid distributions are used.

EXAMPLE 7.7

Consider an SSSS square ($a/b = 1$) graphite/epoxy plate under biaxial uniformly distributed edge compressions. The edge loads are $\sigma_{x0} = -\sigma_0$, $\sigma_{y0} = -\sigma_0$, $\tau_{xy0} = 0$. For this loading case, the in-plane stress components within the entire plate are $P_{xy} = 0$ and $P_x = P_y = -\sigma_0 h$. The material properties are $Q_{11}/Q_{22} = 25$, $Q_{12}/Q_{22} = 0.25$, and $Q_{66}/Q_{22} = 0.5$ with principal material direction at 45° to the plate side (x-axis). For comparison purposes, the buckling coefficient $\bar{P}_x$ is defined as

$$\bar{P}_x = \frac{\sigma_0 b^2}{Q_{22} h^2} \tag{7.44}$$

Table 7.10 shows the convergence study of the buckling coefficient obtained by the DQM with different numbers of grid points. Grid III and MMWC-3 are used.

Table 7.10 Convergence of the Buckling Coefficient $\bar{P}_x$ of SSSS Square Plate Under Biaxial Uniformly Distributed Compressive Loads

N	7	9	11	15	19	21	25
DQM	8.49105	8.58390	8.65099	8.72514	8.76193	8.77380	8.79054
DQM [15]	8.740	8.574					
Ashton		11.565 ($m = n = 5$)			11.060 ($m = n = 7$)		
Whitney		8.418 ($m = n = 7$)			8.556 ($m = n = 9$)		

It is seen that the rate of convergence is low due to high anisotropy. Table 7.10 also shows the comparisons of DQ results with available solutions by Ashton and Whitney, which are cited from literature [15]. It is observed from Table 7.10 that large discrepancy exists.

The results given by Ashton are obtained by employing the Rayleigh–Ritz method, m and n are the terms of double sine series. Due to extra constraint exists in the assumed displacement, the results are too high and not correct. The results given by Whitney are obtained by employing the Fourier analysis; m and n are the terms in the Fourier series. It is expected that due to lower rate of convergence, the actual result should be higher than 8.556.

The results obtained by the DQM in Ref. [15] used MMWC-1 to apply the multiple boundary conditions. The bending moment condition is not satisfied exactly and extra constraint is introduced. Although the result is close to the one obtained by Fourier analysis, the trend in Ref. [15] seems incorrect because the larger the N, the smaller the buckling coefficient. The modified DQM with grid spacing of 25×25 should yield the most accurate results listed in Table 7.10. This will be further verified by the QEM in Chapter 11.

7.3.5 ISOTROPIC SKEW PLATE UNDER UNIAXIAL EDGE COMPRESSION

For an isotropic skew plate with uniform thickness of h, the governing equation for buckling analysis in the oblique coordinate system is given by

$$
D_{11}\frac{\partial^4 w}{\partial \xi^4} + 4D_{16}\frac{\partial^4 w}{\partial \xi^3 \partial \eta} + 2(D_{12}+2D_{66})\frac{\partial^4 w}{\partial \xi^2 \partial \eta^2} +
$$
$$
4D_{26}\frac{\partial^4 w}{\partial \xi \partial \eta^3} + D_{22}\frac{\partial^4 w}{\partial \eta^4} = \cos^4\theta\left(P_\xi \frac{\partial^2 w}{\partial \xi^2} + 2P_{\xi\eta}\frac{\partial^2 w}{\partial \xi \partial \eta} + P_\eta \frac{\partial^2 w}{\partial \eta^2}\right) \tag{7.45}
$$

where (ξ, η) is the oblique coordinate and θ is the skew angle shown in Fig. 7.7.

The flexural rigidity of the skew plate in Eq. (7.45) is defined by

$$
D_{11} = D_{22} = D = \frac{Eh^3}{12(1-\mu^2)}
$$
$$
D_{16} = D_{26} = -D\sin\theta
$$
$$
D_{12} = D(\mu\cos^2\theta + \sin^2\theta) \tag{7.46}
$$
$$
D_{66} = \frac{D(1+\sin^2\theta - \mu\cos^2\theta)}{2}
$$

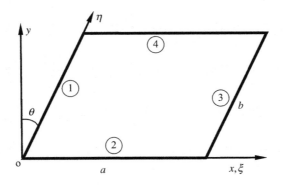

FIGURE 7.7

Sketch of a Thin Skew Plate

where E and μ are the modulus of elasticity and Poisson's ratio, respectively.

The boundary conditions are similar to Eqs. (7.3)–(7.5). Thus, the DQ equations for inner nodes and boundary nodes are similar to the ones of the anisotropic rectangular plate. In other words, the PROGRAM for analyzing buckling of anisotropic rectangular plate can be directly used to analyze the buckling of isotropic skew plates with Eq. (7.46).

Consider a rhombic plate ($a/b = 1$) under uniaxial compression in ξ direction, i.e., only $P_\xi \neq 0$ and $P_{\xi\eta} = P_\eta = 0$. Four boundary conditions, SSSS, SCSC (loaded edges are simply supported) and CCCC, CSCS (loaded edges are clamped), are investigated. The skew angles vary from 0° to 60°. The results obtained by DQM with grid spacing of 17 × 17 are listed in Table 7.11. For comparisons, the buckling load coefficient k is defined by

$$k = \frac{P_\xi b^2}{\pi^2 D} \tag{7.47}$$

where D is defined in Eq. (7.46) and b is the edge length of the plate.

From Table 7.11, it is seen that the DQ results are either the same or slightly lower than the results cited from Ref. [16] with type 1 corner. The published results by many researchers are quite different for the SSSS skew plates with larger skew angles [16]. It is believed that the modified DQ results with Grid III are accurate enough since the increase in the number of grid points does not affect the result

Table 7.11 Buckling Load Coefficients k of a Rhombic Plate Under Uniaxial Compression					
θ	0°	15°	30°	45°	60°
SSSS	4.0000 (4.00 [16])	4.39232 (4.39 [16])	5.86115 (5.98 [16])	9.71876 (9.87 [16])	20.8955
CCCC	10.0739 (10.08 [16])	10.8347 (10.89 [16])	13.5380 (13.75 [16])	20.1048 (20.69 [16])	38.5607
SCSC	7.69128	8.30193	10.2346	15.9140	30.9871
CSCS	6.74319	7.32982	9.47514	14.9101	30.5765

significantly. It is also seen that the buckling load for SCSC skew plate is slightly larger than the one for CSCS skew plate. In other words, the clamped unloaded edges make the plate stiffer than the plate with simply supported unloaded edges.

7.4 SUMMARY

In this chapter, the modified DQM is used to successfully perform the linear buckling analysis for plates under edge compressions. The detailed solution procedures are given and the formulation is general and suitable for rectangular plates with anisotropic materials and for skew plates with isotropic materials. If the in-plane stress distributions are unknown, such as for the case of a rectangular plate under nonuniformly distributed edge compressive load, the stress components at all grid points should be obtained by using the DQM first, which was presented in Chapter 5.

Care should be taken that the expression of shear force is slightly different from the one used in the static analysis. For demonstration, several examples are given. Numerical results show that the modified DQM can yield accurate results with relatively small number of grid points. Thus, the attractive features of rapid convergence, high accuracy, and computational efficiency of the DQM are demonstrated. It is shown that applying the multiple boundary conditions appropriately is very important, otherwise even the convergence tread may be incorrect if highly anisotropic material is involved.

REFERENCES

[1] S.K. Jang, C.W. Bert, A.G. Striz, Application of differential quadrature to deflection and buckling of structural components, Int. J. Numer. Meth. Eng. 28 (1989) 561–577.
[2] X. Wang, L. Gan, Y. Wang, A differential quadrature analysis for vibration and buckling of an SS-C-SS-C rectangular plate loaded by linearly varying in-plane stresses, J. Sound Vib. 298 (2006) 420–431.
[3] A.W. Leissa, J.H. Kang, Exact solutions for vibration and buckling of an SS-C-SS-C rectangular plate loaded by linearly varying inplane stresses, Int. J. Mech. Sci. 44 (2002) 1925–1945.
[4] S. Timoshenko, J. Gere, Theory of Elastic Stability, second ed., McGraw-Hill Book Company, Inc, New York, 1963.
[5] Handbook for Aircraft Design, vol. 3, The Defense Industry Publisher, 1981, (in Chinese).
[6] C.W. Bert, K.K. Devarakonda, Buckling of rectangular plates subjected to nonlinearly distributed in-plane loading, Int. J. Solids Struct. 40 (2003) 4097–4106.
[7] A. Van der Neut, Buckling caused by thermal stresses, in: High Temperature Effects in Aircraft Structures, AGARDograph, vol. 28, 1958, pp. 215–247.
[8] M.B. Benoy, An energy solution for the buckling of rectangular plates under nonuniform in-plane loading, Aeronaut. J. 73 (1969) 974–977.
[9] X. Wang, L. Gan, Y. Zhang, Differential quadrature analysis of the buckling of thin rectangular plates with cosine-distributed compressive loads on two opposite sides, Adv. Eng. Softw. 39 (2008) 497–504.
[10] X. Wang, M. Tan, Y. Zhou, Buckling analyses of anisotropic plates and isotropic skew plates by the new version differential quadrature method, Thin Wall. Struct. 41 (2003) 15–29.
[11] X. Wang, X. Wang, Shi. X Accurate buckling loads of thin rectangular plates under parabolic edge compressions by differential quadrature method, Int. J. Mech. Sci. 49 (2007) 447–453.
[12] X. Wang, X. Wang, X. Shi, Differential quadrature buckling analyses of rectangular plates subjected to nonuniform distributed in-plane loadings, Thin Wall. Struct. 44 (2006) 837–843.

[13] L. Jiang, Y. Wang, X. Wang, Buckling analysis of stiffened circular cylindrical panels using differential quadrature element method, Thin Wall. Struct. 46 (4) (2008) 390–398.

[14] X. Wang, Differential quadrature for buckling analysis of laminated plates, Comput. Struct. 57 (4) (1995) 715–719.

[15] C.W. Bert, X. Wang, A.G. Striz, Differential quadrature for static and free vibration analyses of anisotropic plates, Int. J. Solids Struct. 30 (1993) 1737–1744.

[16] C.M. Wang, K.M. Liew, W.A.M. Alwis, Buckling of skew plates and corner condition for simply supported edges, J. Eng. Mech. 118 (4) (1992) 651–662.

FREE VIBRATION ANALYSIS OF THIN PLATE

8.1 INTRODUCTION

This chapter demonstrates the free vibration analysis of isotropic and anisotropic rectangular plate by using the modified DQM. Since a fourth-order partial differential equation is solved, MMWC-3 or MMWC-4 should be used to apply the multiple boundary conditions. Formulations are worked out in detail and examples are given for illustrations. The importance of using reliable grid spacing in the application of the DQM is demonstrated via an example, free vibration of an isotropic skew plate with a large skew angle.

8.2 FREE VIBRATION OF RECTANGULAR THIN PLATE

Consider the problem of free vibration of the rectangular thin plate. The plate has length a, width b, and uniform thickness h. Anisotropic material is considered.

The governing equation for the out-of-plane vibration of a plate involves the position variables (x, y) and time variable (t).

On the assumption of a sinusoidal time response for free vibration analysis, the transverse deflection $w_o(x, y, t)$ is given by

$$w_o(x, y, t) = w(x, y)e^{i\omega t} \tag{8.1}$$

Thus the governing equation for free vibration analysis is given by

$$D_{11}\frac{\partial^4 w}{\partial x^4} + 4D_{16}\frac{\partial^4 w}{\partial x^3 \partial y} + 2(D_{12} + 2D_{66})\frac{\partial^4 w}{\partial x^2 \partial y^2} +$$
$$4D_{26}\frac{\partial^4 w}{\partial x \partial y^3} + D_{22}\frac{\partial^4 w}{\partial y^4} = \rho h\omega^2 w \tag{8.2}$$

where D_{ij} is the flexural rigidity of the plate, w is the deflection, ρ is the mass density of the plate material, and ω is the circular frequency.

The boundary conditions are summarized as follows,

1. Simply supported edge (S):

$$w = M_x = 0 \text{ at } x = 0, a \text{ or } w = M_y = 0 \text{ at } y = 0, b \tag{8.3}$$

2. Clamped edge (C):

$$w = \frac{\partial w}{\partial x} = 0 \text{ at } x = 0, \ a \text{ or } w = \frac{\partial w}{\partial y} = 0 \text{ at } y = 0, b \tag{8.4}$$

3. Free edge (F):

$$Q_x = M_x = 0 \text{ at } x = 0, a \text{ or } Q_y = M_y = 0 \text{ at } y = 0, b \qquad (8.5)$$

where

$$M_x = D_{11} \frac{\partial^2 w}{\partial x^2} + D_{12} \frac{\partial^2 w}{\partial y^2} + 2D_{16} \frac{\partial^2 w}{\partial x \partial y} \qquad (8.6)$$

$$M_y = D_{12} \frac{\partial^2 w}{\partial x^2} + D_{22} \frac{\partial^2 w}{\partial y^2} + 2D_{26} \frac{\partial^2 w}{\partial x \partial y} \qquad (8.7)$$

$$Q_x = D_{11} \frac{\partial^3 w}{\partial x^3} + 4D_{16} \frac{\partial^3 w}{\partial x^2 \partial y} + (D_{12} + 4D_{66}) \frac{\partial^3 w}{\partial x \partial y^2} + 2D_{26} \frac{\partial^3 w}{\partial y^3} \qquad (8.8)$$

$$Q_y = D_{22} \frac{\partial^3 w}{\partial y^3} + 4D_{26} \frac{\partial^3 w}{\partial x \partial y^2} + (D_{12} + 4D_{66}) \frac{\partial^3 w}{\partial x^2 \partial y} + 2D_{16} \frac{\partial^3 w}{\partial x^3} \qquad (8.9)$$

If a corner is free, three boundary conditions, that is, $M_x = M_y = R = 0$, should be applied, where the concentrated force R is given by

$$R = 2D_{16} \frac{\partial^2 w}{\partial x^2} + 2D_{26} \frac{\partial^2 w}{\partial y^2} + 4D_{66} \frac{\partial^2 w}{\partial x \partial y} \qquad (8.10)$$

Let N_x and N_y be the total number of grid points in the x- and y-directions. Usually, $N_x = N_y = N$. A 7×7 grid spacing is schematically shown in Fig. 8.1. Each inner point has one degree of freedom (DOF), that is, $w_{ij} (i = 2, 3, ..., N_y - 1, j = 2, 3, ..., N_x - 1)$, each corner node has three DOFs, that is, w_{ij}, w_{xij}, w_{yij} $(i = 1, N_y, j = 1, N_x)$, each boundary node at $x = 0$ and $x = a$ has two DOFs, that is, w_{ij}, w_{xij} $(i = 2, ..., N_y - 1, j = 1, N_x)$, and each boundary node at $y = 0$ and $y = b$ has two DOFs, that is, w_{ij}, w_{yij} $(i = 1, N_y, j = 2, ..., N_x - 1)$.

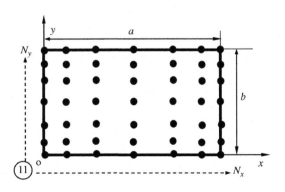

FIGURE 8.1

Sketch of a Rectangular Plate with 7×7 Grid Spacing

MMWC-3 or MMWC-4 is used to introduce the additional DOFs at the two end points in one dimension, that is, $w_1' = (\partial w / \partial x)_{x=0}$ and $w_N' = (\partial w / \partial x)_{x=a}$. The first- to the fourth-order derivatives of the solution function w at a grid point i can be symbolically written by

$$w_i^{(1)} = \sum_{j=1}^{N_x+2} A_{ij}^x \, \delta_j$$

$$w_i^{(2)} = \sum_{j=1}^{N_x+2} B_{ij}^x \, \delta_j$$

$$w_i^{(3)} = \sum_{j=1}^{N_x+2} C_{ij}^x \, \delta_j \tag{8.11}$$

$$w_i^{(4)} = \sum_{j=1}^{N_x+2} D_{ij}^x \, \delta_j \quad (i=1,2,...,N_x)$$

where $A_{ij}^x, B_{ij}^x, C_{ij}^x, D_{ij}^x$ are the weighting coefficients of the first- to the fourth-order derivatives with respect to x, N_x is the total number of grid points and δ_j is defined by

$$\delta_j = w_j \left(j=1,2,...,N_x\right), \delta_{N_x+1} = w_1', \delta_{N_x+2} = w_{N_x}' \tag{8.12}$$

The weighting coefficients of the first- to the fourth-order derivatives with respect to y can be formulated in a similar way. The total number of grid points in both x- and y-direction is assumed the same and denoted by N. In terms of the DQ, Eq. (8.1) at all inner grid points can be expressed as

$$D_{11} \sum_{k=1}^{N+2} D_{ik}^x \overline{w}_{kl} + 4D_{16} \sum_{j=1}^{N+2} \sum_{k=1}^{N+2} C_{ij}^x A_{lk}^y \overline{w}_{jk} + 2(D_{12}+2D_{66}) \sum_{j=1}^{N+2} \sum_{k=1}^{N+2} B_{ij}^x B_{lk}^y \overline{w}_{jk}$$

$$+ 4D_{26} \sum_{j=1}^{N+2} \sum_{k=1}^{N+2} A_{ij}^x C_{lk}^y \overline{w}_{jk} + D_{22} \sum_{k=1}^{N+2} D_{lk}^y \overline{w}_{ik} = \rho h \omega^2 w_{il} \tag{8.13}$$

$$(i=2,3,...,N_y-1; l=2,3,...,N_x-1)$$

For simplicity and clarity, N, N_x and N_y may appear in the same equation at the same time. Superscripts x and y denote the corresponding derivatives taken with x and y. The over bar on w means that the derivative DOFs are included.

To formulate a complete set of algebraic equations, the bending moment equation is placed at the position where the DOF of $\partial w / \partial x$ or $\partial w / \partial y$ is, and the shear force equation is placed at the position where the boundary deflection w is.

In detail, at the positions where $(\partial w / \partial x)_{x=0}$ and $(\partial w / \partial x)_{x=a}$ are, the following DQ equations are placed,

$$D_{11} \sum_{k=1}^{N+2} B_{ik}^x \overline{w}_{kl} + D_{12} \sum_{k=1}^{N+2} B_{lk}^y \overline{w}_{ik} + 2D_{16} \sum_{j=1}^{N+2} \sum_{k=1}^{N+2} A_{ij}^x A_{lk}^y \overline{w}_{jk} = (M_x)_{il}$$

$$(i=1,2,...,N_y; l=1,N_x) \tag{8.14}$$

At the positions $w_{il} (i=2,3,...,N_y-1; l=1,N_x)$, the following DQ equations are applied,

$$D_{11} \sum_{k=1}^{N+2} C_{ik}^x \overline{w}_{kl} + 4D_{16} \sum_{j=1}^{N+2} \sum_{k=1}^{N+2} B_{ij}^x A_{lk}^y \overline{w}_{jk} + (D_{12}+4D_{66}) \sum_{j=1}^{N+2} \sum_{k=1}^{N+2} A_{ij}^x B_{lk}^y \overline{w}_{jk}$$

$$+ 2D_{26} \sum_{k=1}^{N+2} C_{lk}^y \overline{w}_{ik} = (Q_x)_{il} \quad (i=2,3,...,N_y-1; l=1,N_x) \tag{8.15}$$

Note that the four corner points, w_{il} $(i = 1, N_y; l = 1, N_x)$, are not included in Eq. (8.15).

Similarly, at the positions where $(\partial w / \partial y)_{y=0}$ and $(\partial w / \partial y)_{y=b}$ are, the following DQ equations are placed,

$$D_{12} \sum_{k=1}^{N+2} B_{ik}^x \overline{w}_{kl} + D_{22} \sum_{k=1}^{N+2} B_{ik}^y \overline{w}_{ik} + 2D_{26} \sum_{j=1}^{N+2} \sum_{k=1}^{N+2} A_{ij}^x A_{ik}^y \overline{w}_{jk} = \left(M_y \right)_{il}$$
$$(i = 1, N_y; \; l = 1, 2, ..., N_x)$$

(8.16)

And at the positions w_{il} $(i = 1, N_y; l = 2, 3, ..., N_x - 1)$, the following DQ equations are applied,

$$D_{22} \sum_{k=1}^{N+2} C_{ik}^y \overline{w}_{ik} + 4D_{26} \sum_{j=1}^{N+2} \sum_{k=1}^{N+2} A_{ij}^x B_{ik}^y \overline{w}_{jk} + \left(D_{12} + 4D_{66} \right) \sum_{j=1}^{N+2} \sum_{k=1}^{N+2} B_{ij}^y A_{ik}^x \overline{w}_{jk}$$
$$+ 2D_{16} \sum_{k=1}^{N+2} C_{ik}^x \overline{w}_{kl} = \left(Q_y \right)_{il} \quad (i = 1, N_y; \; l = 2, 3, ..., N_x - 1)$$

(8.17)

Again the four corner points, w_{il} $(i = 1, N_y; l = 1, N_x)$, are not included in Eq. (8.17).

Finally, at the four corner points, w_{il} $(i = 1, N_y; l = 1, N_x)$, the following DQ equations are applied

$$2D_{16} \sum_{k=1}^{N+2} B_{ik}^x \overline{w}_{kl} + 2D_{26} \sum_{k=1}^{N+2} B_{ik} \overline{w}_{ik} + 4D_{66} \sum_{j=1}^{N+2} \sum_{k=1}^{N+2} A_{ij}^x A_{ik}^y \overline{w}_{jk} = R_{il}$$
$$(i = 1, N_y; l = 1, N_x)$$

(8.18)

Combining Eqs. (8.13)–(8.18) together results in the following partitioned matrix equation,

$$\begin{bmatrix} K_{bb} & K_{bi} \\ K_{ib} & K_{ii} \end{bmatrix} \left\{ \begin{array}{c} \{\Delta_b\} \\ \{w_i\} \end{array} \right\} = \left\{ \begin{array}{c} \{F_b\} \\ \lambda[I]\{W_i\} \end{array} \right\}$$

(8.19)

where subscripts b and i denote the quantities related to the DOFs at boundary points and the displacement at all inner points, $\{F_b\}$ is the generalized force vector, and $\lambda = \rho h \omega^2$. Equation (8.19) contains $(N+2) \times (N+2) - 4$ equations. The number of equations is the same as the number of unknowns.

Any combinations of boundary conditions can be easily applied by using Eqs. (8.14)–(8.18), since either the deflection or the shear force, as well as either the slope or the bending moment is specified. In other words, the element in $\{F_b\}$ is either zero if the corresponding DOF in $\{\Delta_b\}$ is unknown or unknown if the corresponding DOF in $\{\Delta_b\}$ is given. After imposing the regular boundary conditions, that is, the zero generalized displacement conditions, the DQ equation for free vibration analysis of an anisotropic rectangular plate becomes

$$\begin{bmatrix} \overline{K}_{bb} & \overline{K}_{bi} \\ \overline{K}_{ib} & K_{ii} \end{bmatrix} \left\{ \begin{array}{c} \{\overline{\Delta}_b\} \\ \{w_i\} \end{array} \right\} = \left\{ \begin{array}{c} \{0\} \\ \lambda[I]\{W_i\} \end{array} \right\}$$

(8.20)

where the over bar means that the quantities have been modified since the zero generalized displacements have been removed.

After eliminating the nonzero $\{\bar{\Delta}_b\}$, Eq. (8.20) can be rewritten as

$$\left[K_{ii} - \bar{K}_{ib}\bar{K}_{bb}^{-1}\bar{K}_{bi}\right]\{W_i\} = \lambda[I]\{W_i\} \tag{8.21}$$

Equation (8.21) can be further simplified to

$$\left[\bar{K}\right]\{W_i\} = \lambda[I]\{W_i\} \tag{8.22}$$

Equation (8.22) is a standard eigenvalue matrix equation and can be solved by a standard eigenvalue solver.

For free vibration analysis of isotropic or orthotropic rectangular plate without a free edge, the boundary conditions can be directly built in during formulation of weighting coefficients by using MMWC-1 together with MMWC-3 (or MMWC-4). Therefore, only DQ equations at all inner grid points are needed, namely,

$$D_{11}\sum_{k=1}^{N}D_{ik}^{x}w_{kl} + 4D_{16}\sum_{j=1}^{N}\sum_{k=1}^{N}C_{ij}^{x}A_{ik}^{y}w_{jk} + 2\left(D_{12}+2D_{66}\right)\sum_{j=1}^{N}\sum_{k=1}^{N}B_{ij}^{x}B_{lk}^{y}w_{jk}$$
$$+ 4D_{26}\sum_{j=1}^{N}\sum_{k=1}^{N}A_{ij}^{x}C_{lk}^{y}w_{jk} + D_{22}\sum_{k=1}^{N}D_{lk}^{y}w_{ik} = \rho h\omega^2 w_{il} \tag{8.23}$$
$$(i = 2,3,...,N_y - 1; l = 2,3,...,N_x - 1)$$

Equation (8.23) is different from Eq. (8.13) since the boundary conditions have been built in during formulations of $B_{ij}^{x}, B_{lk}^{y}, C_{ij}^{x}, C_{lk}^{y}, D_{ik}^{x}, D_{lk}^{y}$, the weighting coefficients of the second-, third- and fourth-order derivatives with respect to x or y. Besides, the DOFs do not contain the first-order derivatives with respect to x or y, thus the summation range is from 1 to N.

In matrix form, Eq. (8.23) can be rewritten as

$$\left[K_{ii}\right]\{W_i\} = \lambda[I]\{W_i\} \tag{8.24}$$

or

$$[K]\{W_i\} = \lambda[I]\{W_i\} \tag{8.25}$$

Note that deflections at all boundary points are not included in $\{W_i\}$ since they are zero. Therefore, the implementation is much simpler.

8.3 EXAMPLES

Two FORTRAN programs are provided for analyzing free vibration of isotropic, orthotropic, and anisotropic rectangular plate without and with free edges, which are included in Appendices III and V. The two programs have been converted to MATLAB files, too. Free vibration of rectangular plate with any combinations of boundary conditions and materials can be analyzed by the written programs with only a few modifications.

8.3.1 FREE VIBRATION OF ISOTROPIC RECTANGULAR PLATE

EXAMPLE 8.1

Free vibration of isotropic rectangular plate with three combinations of boundary conditions, that is, SSSS, CCCC, and FFFF, is considered. Poisson's ratio μ is 0.3. Grid III and grid spacing of 21 × 21 are used in the DQ analysis. The results obtained by the modified DQM are listed in Tables 8.1–8.3. The exact or accurate upper-bound solutions given by Leissa [1] are also cited for comparison. For comparison, the frequency parameter $\bar{\lambda}$ is defined by

$$\bar{\lambda} = \omega a^2 \sqrt{\rho h / D} \qquad (8.26)$$

where $D = Eh^3 / 12(1 - \mu^2)$ is the flexural rigidity and E is Young's modulus of the plate material.

Table 8.1 lists the frequency parameters for the free vibration of SSSS isotropic rectangular plates. Since the frequency parameters obtained by the modified DQM are exactly the same as the ones given by Leissa [1], only the DQ solutions are listed in Table 8.1 for most cases. For some cases in which the data are not exactly the same, the results cited from Ref. [1] are also included and put in parenthesis. It is believed that the difference is caused by a misprint and the DQ solution should be the accurate one. The ninth-mode frequency parameters given by Leissa for aspect ratios of 0.4 and 2.5 are incorrect. According to the DQ results, they are actually the tenth-mode frequency parameters. It is believed that this is also caused by the misprint.

For the free vibration of CCCC rectangular plates, the results are summarized in Table 8.2. Since the frequency parameters obtained by the DQM are slightly smaller than the ones given by Leissa [1], the upper-bound solution is also listed in Table 8.2 (in parenthesis) for comparisons. As is expected, the DQ data are all slightly lower and agree with the upper-bound solutions except for one case. It is seen that the

Table 8.1 Frequency Parameter $\bar{\lambda}$ for SSSS Isotropic Rectangular Plates ($\mu = 0.3$)

Mode Sequence	a/b				
	0.4	2/3	1.0	1.5	2.5
1	11.4487 (11.4487)	14.2561 (14.2561)	19.7392 (19.7392)	32.0762 (37.0762)	71.5546 (71.5564)
2	16.1862	27.4156	49.3480	61.6850	101.1634
3	24.0818	43.9649	49.3480	98.6960	150.5115
4	35.1358	49.3480	78.9568	111.0330	219.5987
5	41.0576	57.0244	98.6960	128.3049	256.6097
6	45.7950	78.9568	98.6960	177.6529	286.2185
7	49.3480	80.0535	128.3049	180.1203	308.4251
8	53.6906	93.2129	128.3049	209.7291	335.5665
9	64.7446 (**66.7185**)	106.3724 (106.3724)	167.7833 (167.7833)	239.3379 (239.3379)	404.6538 (**416.9908**)

Table 8.2 Frequency Parameter $\bar{\lambda}$ **for CCCC Isotropic Rectangular Plates**

Mode Sequence	*a/b*				
	0.4	**2/3**	**1.0**	**1.5**	**2.5**
1	23.644	27.005	35.985	60.761	147.77
	(23.468)	(27.010)	(35.992)	(60.772)	(147.80)
2	27.807	41.704	73.394	93.833	173.79
	(27.817)	(41.716)	(73.413)	(93.860)	(173.85)
3	35.417	66.124	73.394	148.78	221.36
	(35.446)	(66.143)	(73.413)	(148.82)	(221.54)
4	46.617	66.522	108.22	149.67	291.70
	(46.702)	(66.552)	(108.27)	(149.74)	(291.89)
5	61.495	79.805	131.58	179.56	384.34
	(61.554)	(79.850)	(131.64)	(179.66)	(384.71)
6	63.083	100.81	132.20	226.82	394.27
	(63.110)	(100.85)	(132.24)	(226.92)	(394.37)

fundamental frequency parameter given by Leissa for aspect ratio of 0.4 (the bold number) is slightly different from the DQ result and believed that this is also caused by a misprint. The DQ solution should be the accurate one. In other words, the upper-bound solution should be 23.648 and not 23.468. If the number of grid points is increased from 21 to 25, the results obtained by the modified DQM remains the same.

For free vibration of FFFF rectangular plates, the results are summarized in Table 8.3. Since the first six nonzero frequency parameters obtained by the modified DQM are all slightly smaller than the ones

Table 8.3 Frequency Parameter $\bar{\lambda}$ **for FFFF Isotropic Rectangular Plates** ($\mu = 0.3$)

Mode Sequence	*a/b*				
	0.4	**2/3**	**1.0**	**1.5**	**2.5**
1	3.4326	8.9313	13.468	20.096	21.454
	(3.4629)	(8.9459)	(13.489)	(20.128)	(21.643)
2	5.2782	9.5170	19.596	21.413	32.989
	(5.2881)	(9.6015)	(19.789)	(21.603)	(33.050)
3	9.5406	20.599	24.270	46.347	59.629
	(9.6220)	(20.735)	(24.432)	(46.654)	(60.137)
4	11.329	22.182	34.801	49.910	70.803
	(11.437)	(22.353)	(35.024)	(50.293)	(71.484)
5	18.627	25.650	34.801	57.713	116.42
	(18.793)	(25.867)	(35.024)	(58.201)	(117.45)
6	18.923	29.791	61.093	67.030	118.27
	(19.100)	(29.973)	(61.526)	(67.494)	(119.38)

given by Leissa [1], the upper-bound solutions are also listed in Table 8.3 (in parenthesis) for comparison. As is expected, the DQ data agree with the upper-bound solutions for all cases. It is believed that the DQ data are more accurate than the ones given by Leissa.

From Tables 8.1–8.3, it is seen that the DQM with grid spacing of 21×21 can yield very accurate lower-mode frequencies. Due to space limitations, DQ results for other combinations of boundary conditions are not listed in the book. With a few modifications, the provided FORTRAN programs or MATLAB files can be used to obtain the solutions for rectangular plates with any combinations of boundary conditions if they are needed.

8.3.2 FREE VIBRATION OF ISOTROPIC SKEW PLATE

EXAMPLE 8.2

Consider the free vibration of isotropic rhombic plates shown in Fig. 8.2. Three combinations of boundary conditions, that is, SSSS, CCCC, and FFFF, are considered. Poisson's ratio μ is 0.3. Define

$$D_{11} = D_{22} = D = \frac{Eh^3}{12\left(1-\mu^2\right)}$$
$$D_{16} = D_{26} = -D\sin\theta$$
$$D_{12} = D\left(\mu\cos^2\theta + \sin^2\theta\right) \tag{8.27}$$
$$D_{66} = \frac{D\left(1+\sin^2\theta - \mu\cos^2\theta\right)}{2}$$
$$\lambda = \cos^4\theta\rho h\omega^2$$

where E is the modulus of elasticity, and θ is the skew angle shown in Fig. 8.2. It is seen that skew angle $\theta = 0°$ corresponds to the isotropic square plate. With Eq. (8.27), a FORTRAN program or MATLAB file is written and included in Appendix V, which can be used to obtain the frequencies of isotropic skew plates.

The results obtained by the modified DQM with grid spacing of 21×21 are listed in Tables 8.4–8.6 for rhombic plates with SSSS, CCCC, and FFFF boundary conditions. Skew angle θ varies from $15°$ to $75°$. For SSSS and CCCC skew plates, Grid III is used in the DQ analysis. However, Grid V is used in the DQ analysis for the case of FFFF skew plate, since all other grid spacing listed in Chapter 1 cannot yield reliable solutions when the skew angle is relatively large.

In Table 8.4, the DQ results for the SSSS skew plate are summarized. The accurate upper-bound solutions cited from Ref. [2] are also included for comparisons. It is seen that except one datum ($\theta = 45°$), the DQ data are either exactly the same or slightly smaller than the accurate upper-bound solutions reported in Ref. [2] for $\theta \le 45°$. Even for large skew angle ($\theta = 75°$), the difference between the DQ results and upper-bound solutions is small. Therefore, the DQM with Grid III can yield reliable and accurate frequency parameters for SSSS isotropic rhombic plates.

In Table 8.5, the DQ results for the CCCC isotropic skew plate are summarized. The existing solutions cited from [4,5] are also included for comparisons. For $\theta = 75°$, the results obtained by NASTRAN with fine meshes (100×100) are included for comparisons, since data are not available in Refs. [4,5].

Table 8.4 Frequency Parameter $\bar{\lambda}$ for SSSS Isotropic Rhombic Plates [3]

$\theta°$	15		30		45		60		75	
Mode	DQM	[2]	DQM	[2]	DQM	[2]	DQM	[2]	DQM	[2]
1	20.868	20.868	24.899	24.899	34.755	34.749	62.331	62.409	196.64	199.28
2	48.205	48.205	52.638	52.638	66.277	66.277	104.95	104.95	283.05	283.06
3	56.107	56.107	71.711	71.711	100.25	100.25	147.65	147.67	357.66	358.45
4	79.043	79.043	83.829	83.829	107.01	107.04	196.29	196.29	438.78	438.83
5	104.00	104.00	122.82	122.82	140.80	140.80	205.35	205.86	521.03	524.41

From Table 8.5, it is seen that the DQ data are exactly the same as the accurate upper-bound solutions reported in Ref. [5] for $\theta = 15°$ and $\theta = 45°$. For other cases, the DQ data are close to the results reported in Ref. [4] or calculated by NASTRAN with fine meshes. Since data obtained by FEM are not upper-bound solutions for the case of thin plates, the DQ results may be either slightly larger or smaller than the FE data. From Table 8.5, one may conclude that the modified DQM with Grid III can yield reliable and accurate frequency parameters for CCCC isotropic rhombic plates. More results for skew plates without free edges and free corners can be found in Ref. [3].

Table 8.6 shows the comparison of the first five nonzero mode frequency parameters $\bar{\lambda}$ obtained by the modified DQM with Grid V for FFFF isotropic rhombic plates with θ varying from 15° to 75°. It is seen that the DQ data are fairly close to the FE data obtained by NASTRAN with fine meshes of 100×100. Although the rate of convergence of the modified DQM is decreased with the increase in skew angles, the difference is not large even for $\theta = 75°$.

Quite different from isotropic rhombic plates without free edges, the DQM with commonly used Grid III cannot yield accurate and reliable solutions. After testing various nonuniform grid distributions presented in Chapter 1, which have been successfully used in the applications of the DQM for solving a variety problems in the area of structural mechanics, it is found that only one grid spacing, that is, Grid V, can yield accurate and reliable solutions. Besides, correctly applying boundary conditions at free corners is also important for success.

Table 8.5 Frequency Parameter $\bar{\lambda}$ for CCCC Isotropic Rhombic Plates [3]

$\theta°$	15		30		45		60		75	
Mode	DQM	[5]	DQM	[4]	DQM	[5]	DQM	[4]	DQM	FEM
1	38.187	38.187	46.089	46.089	65.643	65.643	121.64	121.79	407.49	407.04
2	72.896	72.896	81.601	81.599	106.49	106.49	177.72	177.71	520.63	520.30
3	82.618	82.618	105.17	105.16	148.31	148.31	231.75	231.89	619.86	619.45
4	109.56	109.56	119.25	119.25	157.23	157.24	291.52	291.79	723.33	722.90
5	138.97	138.97	164.99	164.98	196.77	196.77	304.78	305.41	827.72	826.93

Table 8.6 Frequency Parameter $\bar{\lambda}$ for FFFF Isotropic Rhombic Plates ($\mu = 0.3$) [6]

$\theta°$	15		30		45		60		75	
Mode	DQM	FEM	DQM	FEM	DQM	FEM	DQM	FEM	DQM	FEM
1	12.761	12.749	11.531	11.521	10.499	10.491	9.7793	9.7730	9.3337	9.3505
2	20.292	20.283	22.646	22.633	24.034	24.010	22.268	22.248	21.207	21.200
3	27.108	27.096	26.659	26.630	27.763	27.737	39.316	39.239	38.099	38.068
4	30.298	30.262	35.364	35.341	41.771	41.722	39.859	39.812	59.733	59.655
5	39.304	39.248	43.933	43.866	51.537	51.443	62.686	62.429	76.075	75.629

When the skew angle is very large, the DQ solutions for skew plate with free edges are very sensitive to the grid spacing. Table 8.7 summarizes the DQ results with various grid points for the SFFS skew plate with $\theta = 75°$. The DQM with grid spacing of 21×21 is used. It is seen that only Grid V can yield accurate solutions. All others and even the widely and successfully used grid in literature, that is, Grid III, cannot yield reliable solutions for the case considered.

Note that the edge numbering sequence is shown in Fig. 8.2. From Table 8.7, it is obvious that some lower-order mode frequencies cannot be even caught by the DQM with grid spacing other than Grid V. This example tells us the importance of selecting a proper grid spacing in using the DQM for solving problems in the area of structural mechanics.

In Chapter 3, there are nine ways to apply the multiple boundary conditions. Existing results show that these methods yield similar accurate solutions for isotropic or orthotropic rectangular plates. For highly anisotropic rectangular plate or isotropic skew plate with large skew angles, however, not all the methods can yield reliable and accurate solutions, especially when the plate has free edges and free corners. Table 8.8 summarizes the DQ results with various ways to apply the multiple boundary conditions. Again the SFFS skew plate with skew angle of $\theta = 75°$ is considered. The DQM with grid spacing of 21×21 is used in the analysis. On the basis of the results listed in Table 8.7, only Grid V is used in the analysis to ensure the reliability of the DQ solutions.

From Table 8.8, it can be seen that both MMWC-3 and the mixed method can yield reliable and accurate solutions. The DQEM yields slightly higher frequencies as compared to the accurate upper-bound

Table 8.7 Comparison of Frequency Parameters $\bar{\lambda}$ for SFFS Isotropic Rhombic Plates by DQM With Various Grid Spacing ($\theta = 75°, \mu = 0.3$) [6]

Mode	II	III	IV	V	VI	VII	[7]	[8]
1	76.448	69.658	69.322	6.3145	62.960	68.413	6.2683	6.2578
2	78.598	204.38	199.62	27.321	200.67	201.54	27.137	27.096
3	114.93	265.61	354.40	66.075	354.22	352.89	65.635	65.546
4	188.00	273.40	462.35	83.456	455.65	449.68	82.837	82.705
5	223.27	352.78	500.42	120.15	501.49	507.48	119.46	119.29

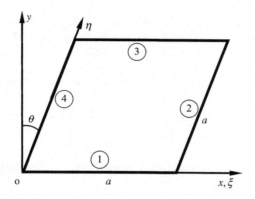

FIGURE 8.2

Sketch of a Rhombic Plate

solutions, since more constrains are used, namely, $Q_\xi(a,a)=0$ is applied at the free corner ($\xi=a$, $\eta=a$) and $w_{\xi\eta}=0$ is applied at other three corners. The extra constraints obviously affect the solution accuracy, since they are only approximately valid. If the governing equation at the free corners is used as the extra condition, the accuracy of the results obtained by the DQEM is improved, as shown in Table 3.4. The δ method is very sensitive to the value of δ and the fundamental frequency cannot be even caught when δ takes certain values. This example tells us the importance of selecting a proper method to apply the multiple boundary conditions in using the DQM for solving problems in the area of structural mechanics.

8.3.3 FREE VIBRATION OF ANISOTROPIC RECTANGULAR PLATE

Free vibration of thin anisotropic rectangular plate or laminated composite plate is investigated by using the modified DQM. For comparisons, the nondimensional frequency parameters are introduced, which are defined by

$$\bar{\lambda} = \omega a^2 \sqrt{\rho h / D_0} \tag{8.28}$$

Table 8.8 Comparison of Frequency Parameters $\bar{\lambda}$ for SFFS Isotropic Rhombic Plates by DQM with Various Methods to Apply the Boundary Conditions ($\theta = 75°, \mu = 0.3$, Grid V) [6]

Mode	Mixed	MMWC-3	δ Method (0.001a)	δ Method (0.0003a)	δ Method (0.003a)	DQEM	[7]	[8]
1	6.1705	6.3145	9.4606	–	–	6.6780	6.2683	6.2578
2	26.926	27.321	29.450	25.038	30.063	29.329	27.137	27.096
3	65.535	66.075	68.997	69.055	65.988	70.583	65.635	65.546
4	83.451	83.456	86.845	86.574	87.283	88.814	82.837	82.705
5	119.98	120.15	123.85	124.81	123.57	126.16	119.46	119.29

where $D_0 = E_{11}h^3/12(1 - \mu_{12}\mu_{21})$, h is the total thickness of the laminated plate, E_{11} is the modulus of elasticity in the principal direction 1, and μ_{12} and μ_{21} are the major and minor Poisson's ratios, respectively.

The material properties of a lamina are $E_{11}/E_{22} = 2.45$, $G_{12} = 0.48E_{22}$, $\mu_{12} = 0.23$. The geometric parameters are $h/a = 0.006$, and $a/b = 1.0$.

For the three-layer angle-ply $(\theta, -\theta, \theta)$ symmetric laminated rectangular plates, the flexural rigidity of the plate in Eq. (8.2) is defined by

$$D_{ij} = \frac{1}{3}\sum_{k=1}^{3}(\bar{Q}_{ij})_k(z_k^3 - z_{k-1}^3) \quad (i,j = 1,2,6) \tag{8.29}$$

where z_k and z_{k-1} are the coordinates of the top and bottom surface of the k-th layer, and $\bar{Q}_{ij}$ $(i,j = 1,2,6)$ are computed by

$$\begin{aligned}
\bar{Q}_{11} &= Q_{11}\cos^4\theta + 2(Q_{12} + 2Q_{66})\cos^2\theta\sin^2\theta + Q_{22}\sin^4\theta \\
\bar{Q}_{22} &= Q_{11}\sin^4\theta + 2(Q_{12} + 2Q_{66})\cos^2\theta\sin^2\theta + Q_{22}\cos^4\theta \\
\bar{Q}_{12} &= \bar{Q}_{21} = (Q_{12} + Q_{22} - 4Q_{66})\cos^2\theta\sin^2\theta + Q_{12}(\cos^4\theta + \sin^4\theta) \\
\bar{Q}_{66} &= (Q_{12} + Q_{22} - 2Q_{12} - 2Q_{66})\cos^2\theta\sin^2\theta + Q_{66}(\cos^4\theta + \sin^4\theta) \\
\bar{Q}_{16} &= (Q_{11} - Q_{22} - 2Q_{66})\cos^3\theta\sin\theta + (Q_{12} - Q_{22} + 2Q_{66})\cos\theta\sin^3\theta \\
\bar{Q}_{26} &= (Q_{11} - Q_{22} - 2Q_{66})\cos\theta\sin^3\theta + (Q_{12} - Q_{22} + 2Q_{66})\cos^3\theta\sin\theta
\end{aligned} \tag{8.30}$$

where the positive angle θ is measured counterclockwise from the x-axis to the fiber direction, that is, from the x-axis to the principal material 1 axis, shown in Fig. 7.4.

With a few modifications, Program 2 can be used to get the frequencies. Table 8.9 lists the first eight frequency parameters for the three-layer angle-ply $(\theta, -\theta, \theta)$ symmetric laminated rectangular plates with all edges simply supported (SSSS). The thickness of each layer is the same and the total thickness of the laminated plate is h. The results listed in Table 8.9 are obtained by the modified DQM with Grid III

Table 8.9 Frequency Parameters $\bar{\lambda}$ for Three-Layer Angle-Ply $(\theta, -\theta, \theta)$ Symmetric Laminated Square Plates (SSSS)

Ply Angle	Method	Mode Number							
		1	2	3	4	5	6	7	8
0°	DQM	15.171	33.248	44.387	60.682	64.457	90.145	93.631	108.46
	CLPT [9]	15.19	33.30	44.42	60.78	64.53	90.29	–	–
	Exact	15.171	33.248	44.387	60.682	64.457	90.145	–	–
15°	DQM	15.396	34.030	43.820	60.733	66.560	91.340	91.377	108.78
	CLPT [9]	15.43	34.09	43.80	60.85	66.67	91.40	–	–
30°	DQM	15.853	35.768	42.524	61.275	71.546	85.589	93.489	108.65
	CLPT [9]	15.90	35.86	42.62	61.45	71.71	85.72	–	–
45°	DQM	16.084	36.832	41.688	61.643	76.862	79.814	94.388	108.68
	CLPT [9]	16.14	36.93	41.81	61.85	77.04	80.00	–	–

Table 8.10 Frequency Parameters $\bar{\lambda}$ **for Three-Layer Angle-Ply** $(\theta,-\theta,\theta)$ **Symmetric Laminated Square Plates (FFFF)**

Ply Angle	Method	Mode Number							
		1	2	3	4	5	6	7	8
0°	DQM	9.9836	14.119	22.379	24.671	29.862	39.387	48.556	49.301
	FEM	9.9820	14.121	22.381	24.669	29.860	39.402	48.549	49.310
15°	DQM	10.250	14.160	21.731	24.851	30.210	39.930	47.865	51.398
	FEM	10.248	14.161	21.733	24.849	30.206	39.945	47.862	51.400
30°	DQM	10.793	14.350	20.299	25.287	30.604	42.287	48.047	53.550
	FEM	10.791	14.352	20.301	25.284	30.599	42.303	48.044	53.551
45°	DQM	11.067	14.550	19.465	25.547	30.643	46.116	48.293	48.368
	FEM	11.063	14.545	19.460	25.531	30.621	46.091	48.251	48.343

and the grid number N is 21. MMWC-3 is used for applying the multiple boundary conditions. Leissa's data [9] are also included for comparisons. It is seen that the results obtained by the modified DQM are very close to and consistently smaller than Leissa's upper-bound solutions. For $\theta = 0°$, the DQ results are exactly the same as the exact solutions.

Further, the free vibration of three-layer angle-ply laminated plates with all edges free (FFFF) is considered. The first eight nonzero mode frequency parameters are listed in Table 8.10. Since upper-bound solutions are not available, thus finite element results obtained by NASTRAN with fine meshes are included for comparisons.

The results listed in Table 8.10 are obtained by the modified DQM with Grid V and the grid number N is 21. MMWC-3 is used for applying the multiple boundary conditions. In the finite element analysis, a 100×100 mesh (30603 DOFs) is used to ensure the solution accuracy. The results listed in Table 8.10 are for the cases of the material principal direction θ varies from 0° to 45° with an increment of 15°. If Grid III is used, the results remain the same. In other words, DQM with either Grid III or Grid V can yield similar accurate results. This is quite different from the cases of isotropic skew plates with free corners. Strong stress singularity at the obtuse plate corners exists when the skew angle is large.

From Table 8.10, it can be seen that the results obtained by the modified DQM are very close to the finite element data and actually should be more accurate than the finite element results. Since finite element data are neither upper-bound solutions nor lower-bound solutions, thus the DQ data are either slightly smaller or larger than the finite element data.

8.4 SUMMARY

In this chapter, the modified DQM is used to successfully perform free vibration analysis for thin rectangular and skew plates. Detailed formulation and solution procedures are worked. Free vibration of plates with various combinations of boundary conditions is analyzed. Some results are obtained by using the programs attached in Appendices III and V.

Numerical results show that the modified DQM can yield accurate frequencies with relatively small number of grid points for all combinations of boundary conditions. Thus the attractive features of rapid convergence, high accuracy, and computational efficiency of the DQM are demonstrated. It is found that for isotropic or orthotropic rectangular plates with or without free corners, the results are insensitive to the grid spacing. For isotropic skew plates with large skew angles, however, the results are very sensitive to grid spacing. Only the modified DQM with Grid V can yield reliable frequencies if a free edge is involved. Even with the widely used grid spacing, the modified DQM cannot yield reliable and accurate frequencies.

REFERENCES

[1] A.W. Leissa, The free vibration of rectangular plates, J. Sound Vib. 31 (1973) 257–293.
[2] C.S. Huang, O.G. McGee, A.W. Leissa, et al. Accurate vibration analysis of simply supported rhombic plates by considering stress singularities, J. Vib. Acoust. 117 (1995) 245–251.
[3] X. Wang, Y. Wang, Z. Yuan, Accurate vibration analysis of skew plates by the new version of the differential quadrature method, Appl. Math. Model 38 (2014) 926–937.
[4] K.M. Liew, Y. Xiang, S. Kitipornchai, C.M. Wang, Vibration of thick skew plates based on Mindlin shear deformation plate theory, J. Sound Vib. 168 (1) (1993) 39–69.
[5] K.S. Woo, C.H. Hong, P.K. Basu, C.G. Seo, Free vibration of skew Mindlin plates by p-version of F, E. M. J. Sound Vib. 268 (2003) 637–656.
[6] X. Wang, Z. Wu, Differential quadrature analysis of free vibration of rhombic plates with free edges, Appl. Math. Comput. 225 (2013) 171–183.
[7] G. McGee, J.W. Kim, A.W. Leissa, The influence of corner stress singularities on the vibration characteristics of rhombic plates with combinations of simply supported and free edges, Int. J. Mech. Sci. 41 (1999) 17–41.
[8] L. Zhou, W.X. Zheng, Vibration of skew plates by the MLS-Ritz method, Int. J. Mech. Sci. 50 (2008) 1133–1141.
[9] A.W. Leissa, Y. Narita, Vibration studies for simply supported symmetrically laminated rectangular plates, Compos. Struct. 12 (1989) 113–132.

GEOMETRIC NONLINEAR ANALYSIS

9.1 INTRODUCTION

This chapter demonstrates the geometrical nonlinear buckling and postbuckling analysis by using the modified differential quadrature method (DQM), HDQM, and the DQ-based time integration scheme. Three typical structural members, thin doubly curved orthotropic shallow shell, rod, and beam, are considered. If fourth-order partial differential equations are solved, method of modifying the weighting coefficients (MMWC)-1, MMWC-3, or MMWC-4 should be used to apply the multiple boundary conditions. Formulations are worked out in detail and solution procedures are given. Newton–Raphson method together with load increment is used to obtain the load–displacement relations for the shallow shell under a uniformly distributed pressure and the buckling and postbuckling behavior of a rod sitting in a rigid cylinder. Small or large disturbances are used to obtain the lateral buckling mode and/ or helical buckling mode. The method of the DQ-based time integration is used to perform geometric nonlinear analysis for thin beam under an axial compressive load. Numerical examples are given for illustrations. The important finding is emphasized to cause readers' attention.

9.2 NONLINEAR STABILITY OF THIN DOUBLY CURVED ORTHOTROPIC SHALLOW SHELL

9.2.1 BASIC EQUATIONS OF THIN DOUBLY CURVED ORTHOTROPIC SHALLOW SHELL

Thin orthotropic doubly curved rectangular shallow shell shown in Fig. 9.1 is considered. The dimension of the rectangular base plane is $2a \times 2b$, and the thickness of the shell is h. Uniform thickness is assumed and the shell is subjected to a uniformly distributed load q. Since the shell thickness is assumed negligible in comparison to the least radius of curvature of the shell middle surface, Donnell-type thin shell theory is adopted. Cartesian coordinate system (xyz) is set at the middle plane of the shell. The geometric nonlinear strain displacement relationship is given by

$$\begin{cases} \varepsilon_x = u,_x + k_x w + 0.5\left(w,_x\right)^2 - zw,_{xx} = \varepsilon_x^\circ - zw,_{xx} \\ \varepsilon_y = v,_y + k_y w + 0.5\left(w,_y\right)^2 - zw,_{yy} = \varepsilon_y^\circ - zw,_{yy} \\ \gamma_{xy} = u,_y + v,_x + w,_x\, w,_y - 2zw,_{xy} = \gamma_{xy}^\circ - 2zw,_{xy} \end{cases} \tag{9.1}$$

where u, v, and w are the displacement components in the x, y, and z directions, and the comma represents the partial derivative, for example, $u,_x = \partial u / \partial x$; k_x and k_y are the curvature of the middle surface of the shallow shell in the x and y directions, respectively.

Differential Quadrature and Differential Quadrature Based Element Methods. 978-0-12-803081-3

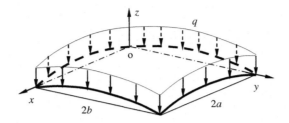

FIGURE 9.1

Thin Doubly Curved Shallow Shell Over a Rectangular Plane

For the thin shallow shell, plane stress condition is assumed. The stress–strain relationship is given by

$$
\begin{cases}
\sigma_x = \dfrac{E_x}{1-\mu_{xy}\mu_{yx}}\left(\varepsilon_x + \mu_{yx}\varepsilon_y\right) \\[2mm]
\sigma_y = \dfrac{E_y}{1-\mu_{xy}\mu_{yx}}\left(\mu_{xy}\varepsilon_x + \varepsilon_y\right) \\[4mm]
\tau_{xy} = G_{xy}\gamma_{xy}
\end{cases}
\tag{9.2}
$$

where $E_x, E_y, \mu_{xy}, \mu_{yx}$, and G_{xy} are the modulus of elasticity in the x and y directions, major and minor Poisson's ratio, and shear modulus, respectively.

The generalized in-plane forces N_x, N_y, N_{xy} are

$$
\begin{cases}
N_x = \dfrac{E_x h}{1-\mu_{xy}\mu_{yx}}\left(\varepsilon_x^\circ + \mu_{yx}\varepsilon_y^\circ\right) = A_{11}\varepsilon_x^\circ + A_{12}\varepsilon_y^\circ \\[2mm]
N_y = \dfrac{E_y h}{1-\mu_{xy}\mu_{yx}}\left(\mu_{xy}\varepsilon_x^\circ + \varepsilon_y^\circ\right) = A_{12}\varepsilon_x^\circ + A_{22}\varepsilon_y^\circ \\[4mm]
N_{xy} = G_{xy}h\gamma_{xy}^\circ = A_{66}\gamma_{xy}^\circ
\end{cases}
\tag{9.3}
$$

And the generalized bending moments M_x, M_y, M_{xy} are

$$
\begin{cases}
M_x = -\dfrac{E_x h^3}{12\left(1-\mu_{xy}\mu_{yx}\right)}\left(w,_{xx} + \mu_{yx}w,_{yy}\right) = -D_{11}w,_{xx} - D_{12}w,_{yy} \\[2mm]
M_y = -\dfrac{E_y h^3}{12\left(1-\mu_{xy}\mu_{yx}\right)}\left(\mu_{xy}w,_{xx} + w,_{yy}\right) = -D_{12}w,_{xx} - D_{22}w,_{yy} \\[4mm]
M_{xy} = -\dfrac{2G_{xy}h^3}{12}w,_{xy} = -2D_{66}w,_{xy}
\end{cases}
\tag{9.4}
$$

The governing equations are

$$
\begin{cases}
\dfrac{\partial N_x}{\partial x} + \dfrac{\partial N_{xy}}{\partial y} + X = 0 \\[3mm]
\dfrac{\partial N_{xy}}{\partial x} + \dfrac{\partial N_y}{\partial y} + Y = 0 \\[3mm]
\left[N_x(k_x - w,_{xx}) + N_y(k_y - w,_{yy}) - 2N_{xy}w,_{xy}\right] - \left(\dfrac{\partial^2 M_x}{\partial x^2} + 2\dfrac{\partial^2 M_{xy}}{\partial x \partial y} + \dfrac{\partial^2 M_y}{\partial y^2}\right) = Z
\end{cases}
\tag{9.5}
$$

where X, Y, and Z are body force in the x, y, and z directions integrated over the shell thickness. Assume k_x and k_y are constant for simplicity. Substituting Eqs. (9.1)–(9.4) into Eq. (9.5) results

$$
\begin{aligned}
&A_{11}u,_{xx} + A_{66}u,_{yy} + (A_{12} + A_{66})v,_{xy} + w,_x (A_{11}w,_{xx} + A_{66}w,_{yy}) \\
&+ (A_{12} + A_{66})w,_y\, w,_{xy} + w,_x (A_{11}k_x + A_{12}k_y) + X = 0
\end{aligned}
\tag{9.6}
$$

$$
\begin{aligned}
&A_{22}v,_{yy} + A_{66}v,_{xx} + (A_{12} + A_{66})u,_{xy} + w,_y (A_{22}w,_{yy} + A_{66}w,_{xx}) \\
&+ (A_{12} + A_{66})w,_x\, w,_{xy} + w,_y (A_{22}k_y + A_{12}k_x) + Y = 0
\end{aligned}
\tag{9.7}
$$

$$
\begin{aligned}
&D_{11}w,_{xxxx} + 2(D_{12} + 2D_{66})w,_{xxyy} + D_{22}w,_{yyyy} \\
&- (w,_{xx} - k_x)\left[A_{11}\left(u,_x + \frac{1}{2}(w,_x)^2 + k_x w\right) + A_{12}\left(v,_y + \frac{1}{2}(w,_y)^2 + k_y w\right) \right] \\
&- (w,_{yy} - k_y)\left[A_{12}\left(u,_x + \frac{1}{2}(w,_x)^2 + k_x w\right) + A_{22}\left(v,_y + \frac{1}{2}(w,_y)^2 + k_y w\right) \right] \\
&- 2A_{66}w,_{xy}(u,_y + v,_x + w,_x\, w,_y) = Z
\end{aligned}
\tag{9.8}
$$

For simplicity in presentation, only two kinds of boundary conditions are considered.

1. Clamped (C) boundary conditions,

$$
u = v = w = 0;\ w,_x = 0 \text{ at } x = 0 \text{ and } x = 2a
\tag{9.9}
$$

$$
u = v = w = 0;\ w,_y = 0 \text{ at } y = 0 \text{ and } y = 2b
\tag{9.10}
$$

2. Hinged (H) boundary conditions,

$$
u = v = w = 0;\ w,_{xx} + \mu_{yx}w,_{yy} = 0 \text{ at } x = 0 \text{ and } x = 2a
\tag{9.11}
$$

$$
u = v = w = 0;\ w,_{yy} + \mu_{xy}w,_{xx} = 0 \text{ at } y = 0 \text{ and } y = 2b
\tag{9.12}
$$

9.2.2 SOLUTION PROCEDURES BY THE HDQM

Since the shell material is orthotropic and only the clamped and hinged boundary conditions are considered, MMWC-1 is used for applying the hinged boundary conditions and MMWC-3 (or MMWC-4) is used for applying the clamped boundary conditions. In this way, the zero bending moment at hinged boundaries and the zero slope boundary condition at clamped boundaries have been built in during formulation of the weighting coefficients and only HDQ equations at all inner grid points are needed to be formulated.

Let the rectangular domain be discrete into $N \times N$ points; thus, the number of inner grid points is $(N-2) \times (N-2)$. Three degrees of freedom (DOFs) are assigned to each inner grid point, i.e., u, v, and w. In terms of the harmonic differential quadrature, the three governing equations, i.e., Eqs. (9.6)–(9.8), at all inner grid points can be expressed by

$$
\begin{aligned}
&A_{11}\sum_{k=2}^{N-1} B_{ik}^x u_{kl} + A_{66}\sum_{k=2}^{N-1} B_{lk}^y u_{ik} + (A_{12} + A_{66})\sum_{j=2}^{N-1}\sum_{k=2}^{N-1} A_{ij}^x A_{lk}^y v_{jk} \\
&+ \sum_{k=2}^{N-1} A_{ik}^x w_{kl}\left(A_{11}\sum_{k=2}^{N-1} B_{ik}^x w_{kl} + A_{66}\sum_{k=2}^{N-1} B_{lk}^y w_{ik} \right) \\
&+ (A_{12} + A_{66})\sum_{k=2}^{N-1} A_{lk}^y w_{ik}\left(\sum_{j=2}^{N-1}\sum_{k=2}^{N-1} A_{ij}^x A_{lk}^y w_{jk} \right) + \sum_{k=2}^{N-1} A_{ik}^x w_{kl}(A_{11}k_x + A_{12}k_y) = 0
\end{aligned}
\tag{9.13}
$$

$$A_{66}\sum_{k=2}^{N-1}B_{ik}^{x}v_{kl} + A_{22}\sum_{k=2}^{N-1}B_{lk}^{y}v_{ik} + \left(A_{12}+A_{66}\right)\sum_{j=2}^{N-1}\sum_{k=2}^{N-1}A_{ij}^{x}A_{lk}^{y}u_{jk}$$

$$+\sum_{k=2}^{N-1}A_{lk}^{y}w_{ik}\left(A_{66}\sum_{k=2}^{N-1}B_{ik}^{x}w_{kl}+A_{22}\sum_{k=2}^{N-1}B_{lk}^{y}w_{ik}\right) \tag{9.14}$$

$$+\left(A_{12}+A_{66}\right)\sum_{k=2}^{N-1}A_{ik}^{x}w_{kl}\left(\sum_{j=2}^{N-1}\sum_{k=2}^{N-1}A_{ij}^{x}A_{lk}^{y}w_{jk}\right)+\sum_{k=2}^{N-1}A_{lk}^{y}w_{ik}\left(A_{12}k_{x}+A_{22}k_{y}\right)=0$$

$$D_{11}\sum_{k=2}^{N-1}D_{ik}^{x}w_{kl}+2\left(D_{12}+2D_{66}\right)\sum_{j=2}^{N-1}\sum_{k=2}^{N-1}B_{ij}^{x}B_{lk}^{y}w_{jk}+D_{22}\sum_{k=2}^{N-1}D_{lk}^{y}w_{ik}$$

$$-\left(\sum_{k=2}^{N-1}B_{ik}^{x}w_{kl}-k_{x}\right)\times A_{11}\left(\sum_{k=2}^{N-1}A_{ik}^{x}u_{kl}+\frac{1}{2}\left[\sum_{k=2}^{N-1}A_{ik}^{x}w_{kl}\right]^{2}+k_{x}w_{il}\right)$$

$$-\left(\sum_{k=2}^{N-1}B_{ik}^{x}w_{kl}-k_{x}\right)\times A_{12}\left(\sum_{k=2}^{N-1}A_{lk}^{y}v_{ik}+\frac{1}{2}\left[\sum_{k=2}^{N-1}A_{lk}^{y}w_{ik}\right]^{2}+k_{y}w_{il}\right)$$

$$-\left(\sum_{k=2}^{N-1}B_{lk}^{y}w_{ik}-k_{y}\right)\times A_{12}\left(\sum_{k=2}^{N-1}A_{ik}^{x}u_{kl}+\frac{1}{2}\left[\sum_{k=2}^{N-1}A_{ik}^{x}w_{kl}\right]^{2}+k_{x}w_{il}\right)$$

$$-\left(\sum_{k=2}^{N-1}B_{lk}^{y}w_{ik}-k_{y}\right)\times A_{22}\left(\sum_{k=2}^{N-1}A_{lk}^{y}v_{ik}+\frac{1}{2}\left[\sum_{k=2}^{N-1}A_{lk}^{y}w_{ik}\right]^{2}+k_{y}w_{il}\right) \tag{9.15}$$

$$-2A_{66}\sum_{j=2}^{N-1}\sum_{k=2}^{N-1}A_{ij}^{x}A_{lk}^{y}w_{jk}\left(\sum_{j=2}^{N-1}A_{ik}^{x}u_{kl}+\sum_{k=2}^{N-1}A_{lk}^{y}v_{ik}+\sum_{k=2}^{N-1}A_{ik}^{x}w_{kl}\sum_{k=2}^{N-1}A_{lk}^{y}w_{ik}\right)=-q_{il}$$

$$\left(i=2,3,\ldots N-1;\ l=2,3,\ldots N-1\right)$$

where $q(x, y)$ is the distributed load in the opposite z direction (i.e., external pressure), $X = Y = 0$ for the problem to be considered, the weighting coefficient of the first-order derivative, A_{ij}, is calculated by using Eq. (1.33), and the weighting coefficients B_{ij}, C_{ij}, and D_{ij} are obtained by using MMWC-1, MMWC-3, or MMWC-4, depending on the boundary conditions.

Equations (9.13)–(9.15) can be written in the following matrix form,

$$[K]\{\bar{D}\}+\left[F\left(\{\bar{D}\}\right)\right]\{\bar{D}\}=\{P\} \tag{9.16}$$

where the size of both linear matrix $[K]$ and nonlinear matrix $[F(\{\bar{D}\})]$ is $3(N-2)^{2}\times3(N-2)^{2}$. Vectors $\{\bar{D}\}$ and $\{p\}$ are defined by

$$\{\bar{D}\}^{T}=\lfloor u_{22},u_{23},\ldots,u_{(N-1)(N-1)},v_{22},v_{23},\ldots,v_{(N-1)(N-1)},w_{22},w_{23},\ldots,w_{(N-1)(N-1)}\rfloor \tag{9.17}$$

$$\{P\}^{T}=\lfloor 0,0,\ldots,0,\quad 0,0,\ldots,0,\quad -q_{22},-q_{23},\ldots,-q_{(N-1)(N-1)}\rfloor \tag{9.18}$$

The nonlinear matrix equation, Eq. (9.16), is to be solved by Newton–Raphson method. Define matrix $[G]$ as

$$\left[G\left(\{\bar{D}\}\right)\right] = \frac{\partial\left(\left[F\left(\{\bar{D}\}\right)\right]\{\bar{D}\}\right)}{\partial\{\bar{D}\}}$$ (9.19)

For convenience, matrices $[K]$, $[F]$, $[G]$, are written in the following partitioned forms,

$$[K] = \begin{bmatrix} [K_{11}] & [K_{12}] & [K_{13}] \\ [K_{21}] & [K_{22}] & [K_{23}] \\ [K_{31}] & [K_{32}] & [K_{33}] \end{bmatrix}$$ (9.20)

$$[F] = \begin{bmatrix} [F_{11}] & [F_{12}] & [F_{13}] \\ [F_{21}] & [F_{22}] & [F_{23}] \\ [F_{31}] & [F_{32}] & [F_{33}] \end{bmatrix}$$ (9.21)

$$[G] = \begin{bmatrix} [G_{11}] & [G_{12}] & [G_{13}] \\ [G_{21}] & [G_{22}] & [G_{23}] \\ [G_{31}] & [G_{32}] & [G_{33}] \end{bmatrix}$$ (9.22)

where the dimension of each submatrix is $m \times m$, where $m = (N - 2) \times (N - 2)$. Detailed expressions for the submatrix are given as follows.

$$[K_{11}] = A_{11}[I] \otimes [B^x] + A_{66}[B^y] \otimes [I]$$ (9.23)

$$[K_{12}] = (A_{12} + A_{66})[A^y] \otimes [A^x]$$ (9.24)

$$[K_{13}] = (A_{11}k_x + A_{12}k_y)[I] \otimes [A^x]$$ (9.25)

$$[K_{21}] = (A_{12} + A_{66})[A^y] \otimes [A^x]$$ (9.26)

$$[K_{22}] = A_{22}[B^y] \otimes [I] + A_{66}[I] \otimes [B^x]$$ (9.27)

$$[K_{23}] = (A_{12}k_x + A_{22}k_y)[A^y] \otimes [I]$$ (9.28)

$$[K_{31}] = (A_{11}k_x + A_{12}k_y)[I] \otimes [A^x]$$ (9.29)

$$[K_{32}] = (A_{12}k_x + A_{22}k_y)[A^y] \otimes [I]$$ (9.30)

$$[K_{33}] = D_{11}[I] \otimes [D^x] + 2(D_{12} + 2D_{66})[B^y] \otimes [B^x]$$
$$+ D_{22}[D^y] \otimes [I] + (A_{11}k_x^2 + 2A_{12}k_xk_y + A_{22}k_y^2)[I] \otimes [I] \tag{9.31}$$

$$[F_{11}] = [0] \tag{9.32}$$

$$[F_{12}] = [0] \tag{9.33}$$

$$[F_{13}] = \frac{1}{2}[W_x]\left(A_{11}[I] \otimes [B^x] + A_{66}[B^y] \otimes [I]\right)$$
$$+ \frac{1}{2}\left(A_{11}[W_{xx}] + A_{66}[W_{yy}]\right)\left([I] \otimes [A^x]\right)$$
$$+ \frac{(A_{12} + A_{66})}{2}[W_y]\left([A^y] \otimes [A^x]\right)$$
$$+ \frac{(A_{12} + A_{66})}{2}[W_{xy}]\left([A^y] \otimes [I]\right) \tag{9.34}$$

$$[F_{21}] = [0] \tag{9.35}$$

$$[F_{22}] = [0] \tag{9.36}$$

$$[F_{23}] = \frac{1}{2}[W_y]\left(A_{22}[B^y] \otimes [I] + A_{66}[I] \otimes [B^x]\right)$$
$$+ \frac{1}{2}\left(A_{22}[W_{yy}] + A_{66}[W_{xx}]\right)\left([A^y] \otimes [I]\right)$$
$$+ \frac{(A_{12} + A_{66})}{2}[W_x]\left([A^y] \otimes [A^x]\right)$$
$$+ \frac{(A_{12} + A_{66})}{2}[W_{xy}]\left([I] \otimes [A^x]\right) \tag{9.37}$$

$$[F_{31}] = -\frac{1}{2}\left(A_{11}[W_{xx}] + A_{12}[W_{yy}]\right)\left([I] \otimes [A^x]\right)$$
$$- A_{66}[W_{xy}]\left([A^y] \otimes [I]\right) \tag{9.38}$$

$$[F_{32}] = -\frac{1}{2}\left(A_{12}[W_{xx}] + A_{22}[W_{yy}]\right)\left([A^y] \otimes [I]\right)$$
$$- A_{66}[W_{xy}]\left([I] \otimes [A^x]\right) \tag{9.39}$$

$$[F_{33}] = -\frac{1}{2}[U_x]\left(A_{11}[I]\otimes[B^x] + A_{12}[B^y]\otimes[I]\right)$$
$$-A_{66}[U_y]\left([A^y]\otimes[A^x]\right)$$
$$-\frac{1}{2}[V_y]\left(A_{12}[I]\otimes[B^x] + A_{22}[B^y]\otimes[I]\right)$$
$$-A_{66}[V_x]\left([A^y]\otimes[A^x]\right)$$
$$+\frac{(A_{11}k_x + A_{12}k_y)}{2}[W_x]\left([I]\otimes[A^x]\right)$$
$$+\frac{(A_{12}k_x + A_{22}k_y)}{2}[W_y]\left([A^y]\otimes[I]\right)$$
$$-\frac{(A_{11}k_x + A_{12}k_y)}{2}[W_{xx}]\left([I]\otimes[I]\right)$$
$$-\frac{(A_{11}k_x + A_{12}k_y)}{2}[W]\left([I]\otimes[B^x]\right)$$
$$-\frac{(A_{12}k_x + A_{22}k_y)}{2}[W_{yy}]\left([I]\otimes[I]\right)$$
$$-\frac{(A_{12}k_x + A_{22}k_y)}{2}[W]\left([B^y]\otimes[I]\right)$$
$$-\frac{1}{3}\left(A_{11}[W_{xx}] + A_{12}[W_{yy}]\right)[W_x]\left([I]\otimes[A^x]\right)$$
$$-\frac{1}{6}[W_x][W_x]\left(A_{11}[I]\otimes[B^x] + A_{12}[B^y]\otimes[I]\right)$$
$$-\frac{1}{3}\left(A_{12}[W_{xx}] + A_{22}[W_{yy}]\right)[W_y]\left([A^y]\otimes[I]\right)$$
$$-\frac{1}{6}[W_y][W_y]\left(A_{12}[I]\otimes[B^x] + A_{22}[B^y]\otimes[I]\right)$$
$$-\frac{2A_{66}}{3}[W_{xy}][W_y]\left([I]\otimes[A^x]\right) - \frac{2A_{66}}{3}[W_x][W_y]\left([A^y]\otimes[A^x]\right)$$
$$-\frac{2A_{66}}{3}[W_x][W_{xy}]\left([A^y]\otimes[I]\right)$$

(9.40)

$$[G_{11}] = [0] \tag{9.41}$$

$$[G_{12}] = [0] \tag{9.42}$$

$$[G_{13}] = [W_x]\left(A_{11}[I]\otimes[B^x] + A_{66}[B^y]\otimes[I]\right)$$
$$+\left(A_{11}[W_{xx}] + A_{66}[W_{yy}]\right)\left([I]\otimes[A^x]\right)$$
$$+(A_{12} + A_{66})[W_y]\left([A^y]\otimes[A^x]\right)$$
$$+(A_{12} + A_{66})[W_{xy}]\left([A^y]\otimes[I]\right)$$

(9.43)

$$[G_{21}] = [0] \tag{9.44}$$

$$[G_{22}] = [0] \tag{9.45}$$

$$\begin{aligned}
[G_{23}] = {} & [W_y]\left(A_{22}[B^y] \otimes [I] + A_{66}[I] \otimes [B^x]\right) \\
& + \left(A_{22}[W_{yy}] + A_{66}[W_{xx}]\right)\left([A^y] \otimes [I]\right) \\
& + (A_{12} + A_{66})[W_x]\left([A^y] \otimes [A^x]\right) \\
& + (A_{12} + A_{66})[W_{xy}]\left([I] \otimes [A^x]\right)
\end{aligned} \tag{9.46}$$

$$\begin{aligned}
[G_{31}] = {} & -\left(A_{11}[W_{xx}] + A_{12}[W_{yy}]\right)\left([I] \otimes [A^x]\right) \\
& - 2A_{66}[W_{xy}]\left([A^y] \otimes [I]\right)
\end{aligned} \tag{9.47}$$

$$\begin{aligned}
[G_{32}] = {} & -\left(A_{12}[W_{xx}] + A_{22}[W_{yy}]\right)\left([A^y] \otimes [I]\right) \\
& - 2A_{66}[W_{xy}]\left([I] \otimes [A^x]\right)
\end{aligned} \tag{9.48}$$

$$\begin{aligned}
[G_{33}] = {} & -[U_x]\left(A_{11}[I] \otimes [B^x] + A_{12}[B^y] \otimes [I]\right) - 2A_{66}[U_y]\left([A^y] \otimes [A^x]\right) \\
& - [V_y]\left(A_{12}[I] \otimes [B^x] + A_{22}[B^y] \otimes [I]\right) - 2A_{66}[V_x]\left([A^y] \otimes [A^x]\right) \\
& + (A_{11}k_x + A_{12}k_y)[W_x]\left([I] \otimes [A^x]\right) + (A_{12}k_x + A_{22}k_y)[W_y]\left([A^y] \otimes [I]\right) \\
& - (A_{11}k_x + A_{12}k_y)[W_{xx}]\left([I] \otimes [I]\right) - (A_{11}k_x + A_{12}k_y)[W]\left([I] \otimes [B^x]\right) \\
& - (A_{12}k_x + A_{22}k_y)[W_{yy}]\left([I] \otimes [I]\right) - (A_{12}k_x + A_{22}k_y)[W]\left([B^y] \otimes [I]\right) \\
& - \left(A_{11}[W_{xx}] + A_{12}[W_{yy}]\right)[W_x]\left([I] \otimes [A^x]\right) \\
& - \frac{1}{2}[W_x][W_x]\left(A_{11}[I] \otimes [B^x] + A_{12}[B^y] \otimes [I]\right) \\
& - \left(A_{12}[W_{xx}] + A_{22}[W_{yy}]\right)[W_y]\left([A^y] \otimes [I]\right) \\
& - \frac{1}{2}[W_y][W_y]\left(A_{12}[I] \otimes [B^x] + A_{22}[B^y] \otimes [I]\right) \\
& - 2A_{66}[W_{xy}][W_y]\left([I] \otimes [A^x]\right) - 2A_{66}[W_x][W_y]\left([A^y] \otimes [A^x]\right) \\
& - 2A_{66}[W_x][W_{xy}]\left([A^y] \otimes [I]\right)
\end{aligned} \tag{9.49}$$

where the subscripts or superscripts x and y denote the derivatives with respect to x and y, and the Kronecker product $\otimes$ for two matrices $[A]$ and $[B]$ is defined by

$$[A] \otimes [B] = \left[A_{ij}[B]\right] \tag{9.50}$$

In Eqs. (9.23)–(9.49), one subscript means the first-order derivative and two subscripts means the second-order derivative. For example, the diagonal matrix $[W_{xy}]$ is defined as the mixed second-order derivative of displacement w with respect to x and y at all inner grid points. Matrices $[U_x], [U_y], [V_x], [V_y], [W], [W_x], [W_y], [W_{xx}], [W_{xy}], [W_{yy}]$ are all diagonal matrix with dimension of $m \times m$. Matrices $[A^x], [B^x], [D^x]$ are the weighting coefficient matrices of the first-, second-, and fourth-order derivatives with respect to x, and $\left[A^y\right], \left[B^y\right], \left[D^y\right]$ are the weighting coefficient matrices of the first-, second-, and fourth-order derivatives with respect to variable y. These equations are symbolically the same as the DQ equations presented in [1]. Note that the dimension of weighting coefficient matrices and unit matrix $[I]$ is $m \times m$, since one of the two boundary conditions has been built in and the other, the zero displacement at the boundary point, has not been included in Eq. (9.17).

Equation (9.16) contains $3m$ nonlinear algebraic equations. In order to obtain the solutions, Newton–Raphson method is used. Applying Newton–Raphson method to Eq. (9.16) results [1]:

$$[K]\{\Delta\bar{D}\}_k + \left[G\left(\{\bar{D}\}_{k-1}\right)\right]\{\Delta\bar{D}\}_k = \{\Delta P\}_k$$
$$\{\bar{D}\}_k = \{\bar{D}\}_{k-1} + \{\Delta\bar{D}\}_k \qquad k=1,2,\dots \tag{9.51}$$

where matrix $[G]$ is given by Eq. (9.19) and vector $\{\Delta P\}_k$ is defined by

$$\{\Delta P\}_k = \{P\} - \left([K] + \left[F\left(\{\bar{D}\}_{k-1}\right)\right]\right)\{\bar{D}\}_{k-1} \tag{9.52}$$

Iterations end when $\left\|\{\Delta\bar{D}\}_k\right\| \le \varepsilon$, where $\|\bullet\|$ is a norm and ε is a prescribed tolerance.

Once the displacements are obtained, in-plane forces and bending moments can be computed by Eq. (9.3) and Eq. (9.4) together with Eq. (9.1). In terms of the harmonic differential quadrature, the in-plane forces are computed by

$$(N_x)_{il} = A_{11}\left(\sum_{k=2}^{N-1} A_{ik}^x u_{kl} + \frac{1}{2}\left[\sum_{k=2}^{N-1} A_{ik}^x w_{kl}\right]^2 + k_x w_{il}\right)$$
$$+ A_{12}\left(\sum_{k=2}^{N-1} A_{lk}^y v_{ik} + \frac{1}{2}\left[\sum_{k=2}^{N-1} A_{lk}^y w_{ik}\right]^2 + k_y w_{il}\right) \tag{9.53}$$

$$(N_y)_{il} = A_{12}\left(\sum_{k=2}^{N-1} A_{ik}^x u_{kl} + \frac{1}{2}\left[\sum_{k=2}^{N-1} A_{ik}^x w_{kl}\right]^2 + k_x w_{il}\right)$$
$$+ A_{22}\left(\sum_{k=2}^{N-1} A_{lk}^y v_{ik} + \frac{1}{2}\left[\sum_{k=2}^{N-1} A_{lk}^y w_{ik}\right]^2 + k_y w_{il}\right) \tag{9.54}$$

$$(N_{xy})_{il} = A_{66}\left(\sum_{j=2}^{N-1} A_{ik}^x u_{kl} + \sum_{k=2}^{N-1} A_{lk}^y v_{ik} + \sum_{k=2}^{N-1} A_{ik}^x w_{kl} \sum_{k=2}^{N-1} A_{lk}^y w_{ik}\right) \tag{9.55}$$

$$(i, l = 1, 2,\dots,N)$$

And the bending moments are computed by

$$(M_x)_{il} = -D_{11}\sum_{k=2}^{N-1} B_{ik}^x w_{kl} - D_{12}\sum_{k=2}^{N-1} B_{lk}^y w_{ik} \tag{9.56}$$

$$(M_y)_{il} = -D_{12}\sum_{k=2}^{N-1} B_{ik}^x w_{kl} - D_{22}\sum_{k=2}^{N-1} B_{lk}^y w_{ik} \tag{9.57}$$

$$\left(M_{xy}\right)_{il} = -2D_{66}\sum_{j=2}^{N-1}\sum_{k=2}^{N-1}A_{ij}^{x}A_{lk}^{y}w_{jk} \tag{9.58}$$

$$(i, l = 1, 2,...,N)$$

To obtain the postbuckling load–deflection curves, the external pressure is applied incrementally. Equation (9.18) is rewritten by

$$\{P\} = -q\lfloor 0,0,...,0,\quad 0,0,...,0,\quad 1,1,...,1\rfloor^{T} = -q\{R\} \tag{9.59}$$

Define

$$[S] = c\{R\}\{R\}^{T} \tag{9.60}$$

where c is a constant and defined by $c = \bar{k}_{jj} \times p\ (j = m/2), \bar{k}_{jj}$ is the corresponding element of the submatrix $[K_{33}]$, and usually $p = 10^{-3} \sim 10^{-2}$.

Replace $[K]$ by the sum of $[K]$ and $[S]$ during the computation. The actual applied pressure at each end of the incremental load q_{ck} is computed by

$$q_{cj} = k \times \Delta q + c\{R\}^{T}\{\bar{D}\}\quad (k = 1, 2, 3,..., M) \tag{9.61}$$

where M is an integer number representing the total number of the load increment, and Δq is computed by

$$\Delta q = \frac{q_{max}}{M} \tag{9.62}$$

In Eq. (9.62), q_{max} is the maximum load to be applied during the analysis and equal load increment is adopted.

9.2.3 **APPLICATIONS AND DISCUSSION**

Two examples are given. In the HDQ analysis, Grid III is used and the grid spacing is 15 × 15. The tolerance in the present analysis is $\varepsilon = 10^{-8}$.

EXAMPLE 9.1

Consider first an isotropic rectangular plate with all edges clamped (CCCC). For this case, the curvatures k_x, k_y are zero. Three aspect ratios, namely, $a/b = 1,2/3,0.5$, are analyzed by the modified HDQM. The nondimensional load-center deflection curves are shown in Fig. 9.2, where lines are HDQ results and symbols are theoretical values obtained by using the energy method [2]. It is seen that the HDQ results compare quite well to the theoretical values, similar to the DQ results presented in Ref. [1].

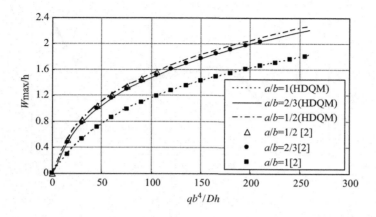

FIGURE 9.2

Nondimensional Center Deflection-Load Curves for CCCC Rectangular Plate

EXAMPLE 9.2

The nonlinear stability of thin shallow shells with all edges hinged (HHHH) is investigated by using the HDQM. The shell is under uniformly distributed load. Material parameters and dimensions are summarized in Table 9.1. It is seen that the shell curvature in x and y directions is unequal and the material behaves orthotropic.

In the analysis, M was determined by the method of trial and error such that only three or four iterations were required for each load increment. Fig. 9.3 shows the load–deflection curves at center as well as at the point where maximum deflection occurs. The critical load is defined when the deflection at the point where maximum deflection occurs reaches its first maximum value.

The critical load obtained by the HDQM with grid spacing of 15 × 15 is $q_{cr} = 50.77$ Pa. The corresponding buckling mode shape is shown in Fig. 9.4. It is clearly seen that the mode number is (1,3). The results are almost the same as the ones obtained by the modified DQM, since there is not much difference between the DQM and the HDQM when the number of grid points is large enough. Thus, more results can be found in Ref. [1].

Table 9.1 Material Parameters and Dimensions

E_x (N/m²)	E_y (N/m²)	G_{xy} (N/m²)	μ_{yx}	k_x (1/m)	k_y (1/m)	a (m)	b (m)	h (mm)
2.0×10^{10}	4.0×10^{10}	1.0×10^{10}	0.1	0.2	0.3	0.1	0.1	0.22

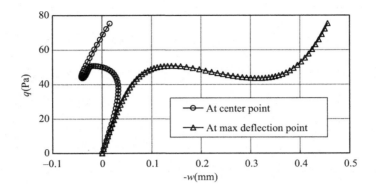

FIGURE 9.3

Load–Deflection Curves of HHHH Shallow Shell

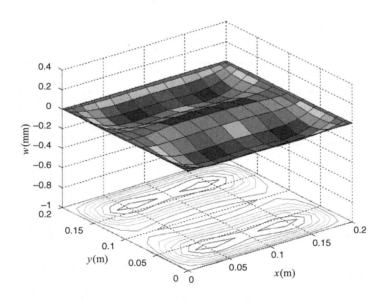

FIGURE 9.4

Buckling Mode Shape of HHHH Shallow Shell

9.3 BUCKLING OF INCLINED CIRCULAR CYLINDER-IN-CYLINDER

9.3.1 BASIC EQUATION AND SOLUTION PROCEDURES

Consider a circular cylinder constrained by an inclined rigid circular cylinder, shown in Fig. 9.5. L, q, and EI are the length, weight per unit length, and bending rigidity of the inner cylinder. The inner cylinder is subjected to a compressive force P at its upper end and a resulting compressive force F_b at its lower end. α is the inclined angle, W_n is the contact force per unit length, and θ is the deviation angle

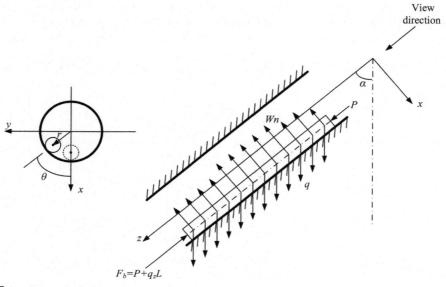

FIGURE 9.5

Schematic Diagram of a Circular Cylinder Sitting in an Inclined Rigid Cylinder

in the xy plane. The radial clearance between inner and outer cylinders is r. Assumed that the inner cylinder contacts initially with the outer cylinder and frictions between the inner and outer cylinders are neglected. For simplicity, the material of the inner cylinder is isotropic. The same problem is analyzed by using the DSC algorithm in Ref. [3].

Assume that $r(\theta')_{\max}$ is small. Then the governing equation in θ direction is given by [3]

$$EI\theta^{iv} + (P + q_z s)\theta'' + q_z\theta' - 6EI\left(\theta'\right)^2\theta'' + q_x \sin\theta / r = 0 \tag{9.63}$$

where

$$\begin{cases} q_x = q\sin\alpha \\ q_z = q\cos\alpha \end{cases} \tag{9.64}$$

At $s = 0$ and $s = L$, the boundary conditions are

$$\text{either } \theta \text{ is prescribed or } EI(\theta''' - 2\theta'^3) + (P + q_z s)\theta' = 0 \tag{9.65}$$

$$\text{either } \theta \text{ is prescribed or } EI\theta'' = 0 \tag{9.66}$$

Neglecting the higher-order derivatives, the contact force per unit length, W_n, can be computed by

$$W_n = (P + q_z s)r\left(\theta'\right)^2 + q_x \cos\theta - EIr\left(\theta'\right)^4 \tag{9.67}$$

It should be mentioned that Eq. (9.63) and Eq. (9.67) are derived based on the deformed configuration and all quantities are in terms of the curvilinear coordinate s, rather than the global coordinate z. In other words, the derivative is taken with respect to s and not to z. Detailed derivations may be found in Ref. [3].

In terms of the differential quadrature, the nonlinear governing equation (9.63) at all inner points is expressed by

$$EI\left(\sum_{j=1}^{N+2} D_{kj}\bar{\theta}_j\right) + (P + q_z s_k)\left(\sum_{j=1}^{N+2} B_{kj}\bar{\theta}_j\right) + q_z\left(\sum_{j=1}^{N+2} A_{kj}\bar{\theta}_j\right)$$
$$- 6EI\left(\sum_{j=1}^{N+2} A_{kj}\bar{\theta}_j\right)^2\left(\sum_{j=1}^{N+2} B_{kj}\bar{\theta}_j\right) + q_x \sin\theta_k / r = 0 \quad (9.68)$$
$$(k = 2, 3, \cdots, N-1)$$

where $\bar{\theta}_j$ is defined by

$$\bar{\theta}_j = \theta_j \quad (j = 1, 2, ..., N), \quad \bar{\theta}_{N+1} = \theta_1', \quad \bar{\theta}_{N+2} = \theta_N' \quad (9.69)$$

For this problem, either MMWC-3 (MMWC-4) or the DQEM can be used to introduce the additional DOF, since the problem is essentially a one-dimensional problem.

The four boundary equations are also expressed in terms of differential quadrature. The first and the Nth DQ equations are

$$EI\left(\sum_{j=1}^{N+2} C_{kj}\bar{\theta}_j\right) - 2EI\left(\sum_{j=1}^{N+2} A_{kj}\bar{\theta}_j\right)^3 + (P + q_z s_k)\left(\sum_{j=1}^{N+2} A_{kj}\bar{\theta}_j\right) = Q_k \quad (9.70)$$
$$(k = 1, N)$$

The $(N + 1)$th DQ equation is

$$EI\left(\sum_{j=1}^{N+2} B_{1j}\bar{\theta}_j\right) = M_1 \quad (9.71)$$

and the $(N + 2)$th DQ equation is

$$EI\left(\sum_{j=1}^{N+2} B_{Nj}\bar{\theta}_j\right) = M_N \quad (9.72)$$

Let

$$f_k = \left(\sum_{j=1}^{N+2} A_{kj}\bar{\theta}_j\right), \quad g_k = \left(\sum_{j=1}^{N+2} B_{kj}\bar{\theta}_j\right) \quad (k = 2, 3, ..., N-1)$$
$$f_k = \left(\sum_{j=1}^{N+2} A_{kj}\bar{\theta}_j\right), \quad g_k = \left(\sum_{j=1}^{N+2} A_{kj}\bar{\theta}_j\right) / 3 \quad (k = 1, N) \quad (9.73)$$

With the definition given by Eq. (9.73), the nonlinear term in Eq. (9.68) and Eq. (9.70) can be expressed as

$$6EI \left(\sum_{j=1}^{N+2} A_{kj}\bar{\theta}_j \right)^2 \left(\sum_{j=1}^{N+2} B_{kj}\bar{\theta}_j \right) = 6EI\, f_k^2 g_k \qquad (k=1,2,...,N) \tag{9.74}$$

After applying the appropriate boundary conditions, Eqs. (9.68), (9.70)–(9.72) can be rewritten in the following matrix form, namely,

$$\left\{ F\left(\bar{\theta}\right) \right\} = \left\{ F^* \right\} \tag{9.75}$$

where $F_i^* = 0$ at all inner nodes ($i = 2, 3,...,N − 1$). The other four terms, i.e., F_i^* ($i = 1, N, N + 1, N + 2$), depend on the boundary conditions. If θ is prescribed at $s = 0$ or $s = L$, then F_i^* is unknown for $i = 1$ or N, otherwise $F_i^* = 0$ for $i = 1$ or N. Similarly, if θ' is prescribed at $s = 0$ or $s = L$, then F_i^* is unknown for $i = N + 1$ or $N + 2$, otherwise $F_i^* = 0$ for $i = N + 1$ or $N + 2$. After applying the appropriate boundary conditions and eliminating the zero generalized displacements, $\{F^*\} = \{0\}$.

The nonlinear equations can be solved iteratively together with incrementally for a given load. The detailed solution procedures are given as follows:

1. Give an initial guess $\{\bar{\theta}\}_0$, then $\{\bar{\theta}\}_1$ can be obtained by using the Newton–Raphson method as follows,

$$\{\bar{\theta}\}_1 = \{\bar{\theta}\}_0 - \left[H\left(\{\bar{\theta}\}_0\right) \right]^{-1} \left\{ F\left(\{\bar{\theta}\}_0\right) \right\} \tag{9.76}$$

where

$$\left[H\left(\{\bar{\theta}\}_0\right) \right] = \left[\frac{\partial \left\{ F\left(\{\bar{\theta}\}\right) \right\}}{\partial \{\bar{\theta}\}} \right]_{\{\bar{\theta}\}=\{\bar{\theta}\}_0} \tag{9.77}$$

Usually $\{\bar{\theta}\}_1 \neq \{\bar{\theta}\}_0$ and more iterations are required.

2. Repeat the same procedures until the results are within the prescribed tolerance (ε), i.e.,

$$\{\bar{\theta}\}_{i+1} = \{\bar{\theta}\}_i - \left[H\left(\{\bar{\theta}\}_i\right) \right]^{-1} \left\{ F\left(\{\bar{\theta}\}_i\right) \right\} \tag{9.78}$$

and

$$Error = \sqrt{ \left(\{\bar{\theta}\}_{i+1} - \{\bar{\theta}\}_i \right)^T \left(\{\bar{\theta}\}_{i+1} - \{\bar{\theta}\}_i \right) } < \varepsilon \tag{9.79}$$

If the given load is smaller than the linear buckling load, a trivial solution will be obtained. Once the given load is greater than the linear buckling load, a nontrivial solution will be obtained. For such a case, the nontrivial solution can serve as the initial guess for the higher applied load to speed up the rate of convergence.

Once $\{\bar{\theta}\}$ is obtained for a given load, the contact force per unit length can be computed by Eq. (9.67). In terms of differential quadrature, W_n is computed by

$$(W_n)_k = r(P + q_z s_k) \left(\sum_{j=1}^{N+2} A_{kj} \bar{\theta}_j \right)^2 + q_x \cos \theta_k - EIr \left(\sum_{j=1}^{N+2} A_{kj} \bar{\theta}_j \right)^4 \tag{9.80}$$

$$(k = 1, 2, ..., N)$$

The assumptions made in deriving the governing equation, i.e., $r(\theta'_k)_{\max}$ is small and $(W_n)_k \geq 0$, should be checked. Otherwise the solution $\{\bar{\theta}\}$ may not be a valid solution.

To initiate a lateral buckling mode, $\{\bar{\theta}\}_0$ should be small enough. To initiate a helical buckling mode, however, $\{\bar{\theta}\}_0$ should be large enough. This is one of the main findings in Ref. [3], otherwise the helical buckling mode would not be obtained numerically.

If only linear buckling is considered, the nonlinear governing equation, i.e., Eq. (9.63), is reduced to the following linear equation,

$$EI\theta^{iv} + (P + q_z s)\theta'' + q_z \theta' + q_x \theta / r = 0 \tag{9.81}$$

In terms of differential quadrature, the linear governing equation (9.81) at all inner nodes becomes

$$EI \left(\sum_{j=1}^{N+2} D_{kj} \bar{\theta}_j \right) + (P + q_z s_k) \left(\sum_{j=1}^{N+2} B_{kj} \bar{\theta}_j \right) + q_z \left(\sum_{j=1}^{N+2} A_{kj} \bar{\theta}_j \right)$$
$$+ q_x \theta_k / r = 0 \tag{9.82}$$

$$(k = 2, 3, \cdots, N-1)$$

The two nonlinear DQ equations at the boundary points, i.e., Eq. (9.70), should be also linearized, namely,

$$EI \left(\sum_{j=1}^{N+2} C_{kj} \bar{\theta}_j \right) + (P + q_z s_k) \left(\sum_{j=1}^{N+2} A_{kj} \bar{\theta}_j \right) = Q_k \tag{9.83}$$

$$(k = 1, N)$$

Equations (9.82), (9.83), (9.71), and (9.72) form a completed set of algebraic equations for the $(N + 2)$ unknowns. After applying the boundary conditions, the set of DQ equations forms an eigenvalue equation. The linear buckling load can be obtained by using an eigensolver.

In practice, such as the buckling of drill-string in a wellbore, the rod is usually very long. In such cases, either the DQEM or the LaDQM should be used. Otherwise numerical instability by using the DQM may occur if the number of grid points N is too large.

To formulate a DQ circular cylinder element used in the buckling analysis of inclined circular cylinder-in-cylinder, Eq. (9.71) and Eq. (9.72) remain unchanged, but Eq. (9.68) and Eq. (9.70) should be slightly modified. Equation (9.68) should be modified as,

$$EI \left(\sum_{j=1}^{N+2} D_{kj} \bar{\theta}_j \right) + (P + q_z s_k^*) \left(\sum_{j=1}^{N+2} B_{kj} \bar{\theta}_j \right) + q_z \left(\sum_{j=1}^{N+2} A_{kj} \bar{\theta}_j \right)$$
$$- 6EI \left(\sum_{j=1}^{N+2} A_{kj} \bar{\theta}_j \right)^2 \left(\sum_{j=1}^{N+2} B_{kj} \bar{\theta}_j \right) + q_x \sin \theta_k / r = 0 \tag{9.84}$$

$$(k = 2, 3, \cdots, N-1)$$

where s_k^* is measured from the top end of the inner cylinder and varies element by element. Note that $s_k^* = s_k$ is only valid for the first element with node 1 at the top end of the inner cylinder shown in Fig. 9.5.

Equation (9.70) should be modified as

$$EI\left(\sum_{j=1}^{N+2} C_{kj}\bar{\theta}_j\right) - 2EI\left(\sum_{j=1}^{N+2} A_{kj}\bar{\theta}_j\right)^3 + (P + q_z s_k^*)\left(\sum_{j=1}^{N+2} A_{kj}\bar{\theta}_j\right) = Q_k \tag{9.85}$$
$$(k = 1, N)$$

where s_k^* is measured from the top end of the inner cylinder and varies element by element.

For convenient in the assemblage, $\bar{\theta}_j$ should be rearranged as follows,

$$\tilde{\theta}_1 = \theta_1, \tilde{\theta}_2 = \theta_1', \tilde{\theta}_j = \theta_{j-1} \quad (j = 3, 4, ..., N+1), \quad \tilde{\theta}_{N+2} = \theta_N' \tag{9.86}$$

The corresponding equations should be rearranged accordingly, namely, Eq. (9.85) is the first and the $(N+1)$th equation, Eq. (9.84) is the third to the Nth equations, Eq. (9.71) is the second equation, and Eq. (9.72) is the last equation. After assemblage and applying the appropriate boundary conditions, the following nonlinear matrix equation is obtained,

$$\{F(\theta^*)\} = \{F^*\} \tag{9.87}$$

The form of Eq. (9.87) is similar to Eq. (9.75). After eliminating the zero generalized displacements, $\{F^*\} = \{0\}$; thus, Eq. (9.87) can be solved iteratively together with incrementally for a given load, as described previously.

In order to draw the postbuckling configuration, the displacement components in the x, y, and z directions, i.e., u, v, w, should be obtained. With the known θ, u, v, w can be computed by [3]

$$\left\{\begin{matrix} u \\ v \\ w \end{matrix}\right\} = \left\{\begin{matrix} r\cos\theta \\ r\sin\theta \\ 0 \end{matrix}\right\}_{s^*=0} + \int_0^{s^*}\left\{\begin{matrix} -r\dfrac{d\theta}{ds}\sin\theta \\ r\dfrac{d\theta}{ds}\cos\theta \\ 1 - \dfrac{1}{2}r^2\left(\dfrac{d\theta}{ds}\right)^2 \end{matrix}\right\} ds \tag{9.88}$$

where r is the radial clearance between inner and outer cylinders.

9.3.2 APPLICATIONS AND DISCUSSION

Two examples are given. For simplicity, only the simply supported boundary condition at both ends is considered, i.e., $\theta = \theta'' = 0$ at both ends. Then MMWC-1 can be used to formulating the weighting coefficients of the third- and fourth-order derivatives if the DQM is used. If the differential quadrature element method is used, either MMWC-3 or the DQEM can be used to introduce the first order derivative DOFs at two ends. If the LaDQM is used, MMWC-4 can be used to introduce the first order derivative DOFs at two ends.

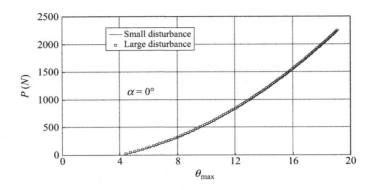

FIGURE 9.6

$P-\theta_{max}$ Curves with Small and Large Initial Disturbances

---

EXAMPLE 9.3

A simply supported circular cylinder is within a vertical rigid cylinder ($\alpha = 0°$) and subjected to an axial force P at $s = 0$. The geometrical and material parameters are $L = 21.34$ m, $r = 0.01727$ m, $q = 13.2$ N/m, $EI = 343.3498$ Nm^2, and the load $P = 2240.0$ N. The upper end of the inner cylinder can rotate freely with respect to its center axis.

The DQM is used in the analysis, since the length is relatively short. In the DQ analysis, $N = 101$ and $\varepsilon = 10^{-8}$. P is applied incrementally in variable steps depending on the required resolution of the load. The lateral buckling load, obtained by directly solving the linear eigenvalue problem, is -204.5 N [3]. In other words, the rod is buckled under its own weight even without a compressive force applied at its top end. Both small ($\theta_i = 0.095$; $i = 2,3,\ldots,N - 1$) and large ($\theta_i = 0.95$; $i = 2,3,\ldots,N - 1$) disturbances are tried for the initial guess $\{\bar{\theta}\}_0$.

Figure 9.6 shows the applied load P vs. θ_{max} curves obtained by the DQM with both small and larges disturbances. The load increment is 20 N. It is seen that the DQM with small or large initial guesses yields the same results for the problem considered. The reason is that only helical buckling will occur for this problem [4] and the half wave number of the postbuckling mode shape does not change with the increase in applied load; therefore, whether the initial guess $\{\bar{\theta}\}_0$ is large or small does not affect the final results. The DQ results are almost exactly the same as the DSC results presented in Ref. [3]; the formulations validate each other.

Figure 9.7 shows the postbuckling configuration at $P = 2240$ N. The maximum rotation is about 19.133 radians, i.e., $\theta_{max} = 6.09\pi$. The two plots represent the same postbuckling configuration but viewing at different directions. Note that different units are used in Fig. 9.7 for clarity. The unit in the x and y direction is centimeter and the unit in the z direction is meter.

To verify whether the results are valid data or not, $\beta (= rd\theta / ds)$ and the contact force per unit length (W_n) are also computed and plotted. Figure 9.8 shows the variations of β at various applied axial loads. From Fig. 9.8, it is seen that β is small and the absolute values are less than 0.04; therefore, the assumption of $\sin(\beta) \approx \beta$ is held. It is also observed that the pitch of the helix is almost constant for most portions of the cylinder but varies greatly in the middle portion of the cylinder. The deformation is no

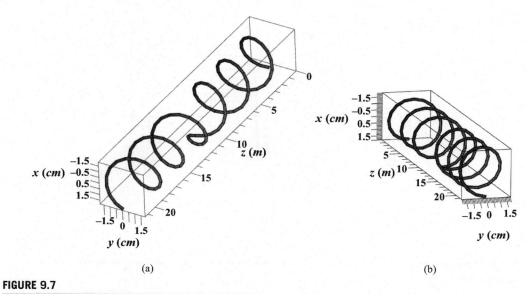

(a)   (b)

FIGURE 9.7

Postbuckling Configurations at $P = 2240$ N

longer symmetric about the middle portion of the cylinder due to the effect of the self-weight of the inner cylinder.

Figure 9.9 shows the distribution of the contact force per unit length (W_n) at various applied axial loads. It is seen that $(W_n) \geq 0$ everywhere. From Figs. 9.8 and 9.9, the assumptions in deriving the governing equation, namely, β is small and the inner cylinder always contacts with the outer cylinder, are valid. Therefore, the results shown in Figs. 9.6 and 9.7 are valid solutions.

To check, exactly same ranges of axes as the ones used in Ref. [3] are used in Figs. 9.8–9.9. It is easy to see that the results are almost the same as the data obtained by the discrete singular convolution (DSC) [3]; thus, the formulations are verified mutually.

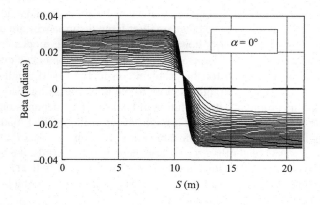

FIGURE 9.8

Variations of β Obtained by the DQM

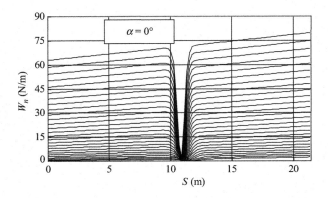

FIGURE 9.9

Variations of W_n Obtained at Various Applied Axial Loads

EXAMPLE 9.4

A simply supported circular cylinder is within an inclined rigid cylinder ($\alpha = 45°$) and subjected to an axial force P at $s = 0$. The geometrical and material parameters are exactly the same as the ones given in Example 9.3.

In the DQ analysis, $N = 101$ and $\varepsilon = 10^{-8}$. P is applied incrementally in variable steps depending on the required resolution of the load. The lateral buckling load, obtained by directly solving the linear eigenvalue problem, is 712 N [3]. In other words, the rod is not buckled under its own weight for $\alpha = 45°$ and a compressive force should be applied at its top end to cause the lateral buckling. Both small disturbance ($\theta_i = 0.095$; $i = 2, 3, ..., N-1$) and large disturbance ($\theta_i = 0.95$; $i = 2, 3, ..., N-1$) have been tried for the initial guess $\{\bar{\theta}\}_0$.

Figure 9.10 shows the applied load P vs. θ_{max} curves obtained by the DQM with small and larges disturbances. The load increment is 20 N. It is seen that when the applied axial load P is small, the inner cylinder retains its straight configuration whether small or large disturbances are used, since the linear buckling load is 712 N. Not alike Fig. 9.6, however, different buckling modes are obtained by using small and large disturbances at relatively large applied loads. The reason is that either lateral or helical buckling may occur under a given load for the problem considered [4]. If small disturbance is used, the load at which the rod is no longer straight is close to the linear buckling load obtained by directly solving the linear eigenvalue problem. The resolution of buckling load depends on the load increment. When large disturbance is used, helical buckling might occur at the applied load even slightly lower than the linear buckling load. However, the helical buckling solutions for load $P < 940$ N are not valid solution, since W_n is not always greater than zero everywhere. In other words, the helical buckling solutions at load range of 520 N − 940 N are invalid solutions. Obviously the helical buckling load is higher than the linear buckling load. The DQ results are almost exactly the same as the DSC results presented in Ref. [3], the formulations are validated mutually.

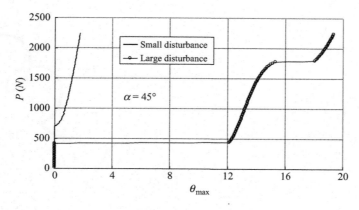

FIGURE 9.10

$P-\theta_{max}$ Curves with Small and Large Initial Disturbances

The postbuckling configurations, obtained by the DQM with small disturbance, are shown in Fig. 9.11. There are two plots in the figure representing the same postbuckling configuration but viewing at different directions. It is seen that the buckling mode is lateral and not helical. At $P = 1200$ N, the maximum rotation is only 0.9479 radians, i.e., $\theta_{max} = 0.30\pi$ for the lateral buckling shown in Fig. 9.11.

The postbuckling configurations, obtained by the DQM with large disturbance, are shown in Fig. 9.12. There are also two plots in the figure representing the same postbuckling configuration but viewing at different directions. It is seen that the buckling mode is helical and not lateral. At $P = 1200$ N, the maximum rotation is 13.532 radians, i.e., $\theta_{max} = 4.31\pi$ for helical buckling shown in Fig. 9.12.

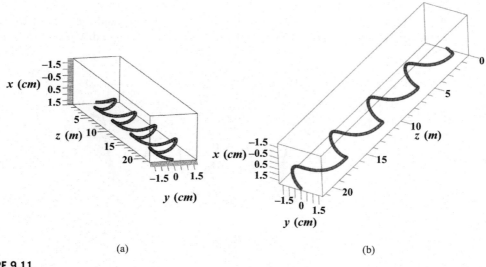

(a)   (b)

FIGURE 9.11

Postbuckling Configuration (Small Disturbance, $P = 1200$ N, $\alpha = 45°$)

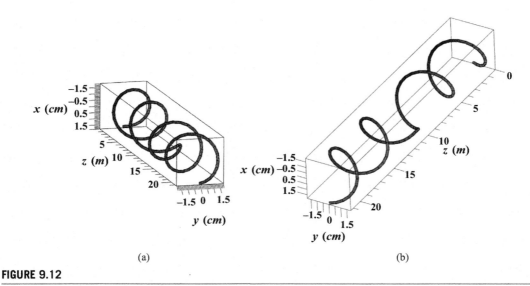

(a)   (b)

FIGURE 9.12

Postbuckling Configuration (Large Disturbance, $P = 1200$ N, $\alpha = 45°$)

The results obtained by the DQM, i.e., Figs. 9.10–9.12, are similar to the data obtained by the discrete singular convolution (DSC) [3]; thus, the formulations are verified each other. It should be mentioned that large disturbance should be used in order to catch the helical buckling mode [3], otherwise only lateral buckling mode can be captured by various numerical method, such as the DQM, DQEM, FEM, LaDQM, and DSC. This is an important finding and contribution to the postbuckling analysis of thin rods, plates, and shells.

To verify the results shown in Figs. 9.10–9.12 further, $\beta\,(= r d\theta\,/\,ds)$ and the contact force per unit length (W_n) are also plotted. Figure 9.13 shows the variations of β at three different P values. The

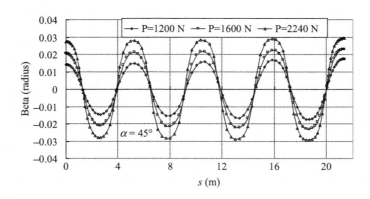

FIGURE 9.13

Variations of β Obtained by the DQM with Small Disturbance

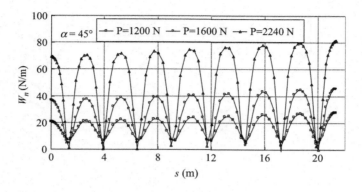

FIGURE 9.14

Variations of W_n at Various Applied Axial Loads (Small Disturbance)

results are obtained by the DQM with small disturbance. It is seen that β is small and the maximum value is less than 0.03; thus, the assumption of $\sin(\beta) \approx \beta$ is held.

Figure 9.14 shows the distribution of the contact force per unit length (W_n) at three different applied loads P. The results are obtained by the DQM with small disturbance. It is seen that $(W_n) \geq 0$ everywhere. From Figs. 9.13 and 9.14, the assumptions in deriving the governing equation are valid. Therefore, the results shown in Figs. 9.10 and 9.11 are valid solutions.

Figure 9.15 shows the variations of β at three different loads P. The results are obtained by the DQM with large disturbance. It is seen that β is also small and the maximum value is less than 0.04; thus, the assumption of $\sin(\beta) \approx \beta$ is held. Comparison of Fig. 9.15 with Fig. 9.13 shows that the variation of β is quite different due to different buckling configurations shown in Figs. 9.11 and 9.12.

Figure 9.16 shows the distribution of the contact force per unit length (W_n) at three applied loads. The results are obtained by the DQM with large disturbance. It is seen that $(W_n) \geq 0$ everywhere. From Figs. 9.15 and 9.16, the assumptions in deriving the governing equation are valid. Therefore, the results shown in Fig. 9.10 ($P \geq 1200\ N$) and Fig. 9.12 are also valid solutions.

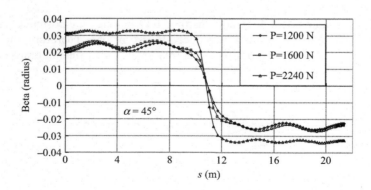

FIGURE 9.15

Variations of β Obtained by the DQM with Large Disturbance

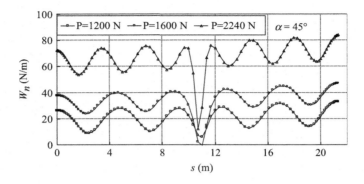

FIGURE 9.16

Variations of W_n at Three Different Applied Axial Loads (Large Disturbance)

The results obtained by the DQM are more or less similar to the data obtained by the DSC; thus, Figs. 9.6–9.16 are almost the same as the corresponding figures presented in [3,5,6]. For helical buckling analysis, especially the rod is long, the DQEM or LaDQM is recommended since a large number of grid points are needed. Otherwise a very large number of grid points may result in numerical instability and cause divergence problems. More DQ elements with small number of grid points or the LaDQM with small computational bandwidth should be used in the analysis to avoid possible numerical instability.

9.4 BUCKLING AND POSTBUCKLING ANALYSIS OF EXTENSIBLE BEAM

9.4.1 BASIC EQUATION

For completeness, the nonlinear equilibrium formulation of extensible beam is briefly described. More details on the derivations may be found in references [7,8].

Consider an isotropic beam–column with length l, cross-section A, and moment of inertia I. The elasticity modulus is E. The beam–column is under an axial compression force P, shown in Fig. 9.17.

To use the DQ-based numerical integration method for obtaining the postbuckling solutions, the order of the equilibrium equation should be reduced to two or one. Therefore, the formulation of the beam–column is in terms of the cross-sectional rotation φ, rather than in terms of the deflection w [7,8].

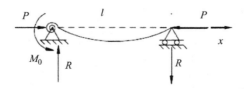

FIGURE 9.17

Sketch of the Beam–Column

Using the geometric equation, constitutive equation, and equilibrium equation yields

$$EI\frac{d^2\varphi}{dx^2}+\left[1-\left(\frac{P\cos\varphi}{EA}-\frac{R\sin\varphi}{EA}\right)\right](R\cos\varphi+P\sin\varphi)=0 \tag{9.89}$$

where R is the reaction force shown in Fig. 9.17.

Equation (9.89) can be rewritten in dimensionless form, namely,

$$\frac{d^2\varphi}{d\xi^2}+\frac{P}{\eta}[1-p(\cos\varphi-\alpha\sin\varphi)](\alpha\cos\varphi+\sin\varphi)=0 \tag{9.90}$$

where

$$p=\frac{P}{EA},\quad \alpha=\frac{R}{P},\quad \xi=\frac{x}{l},\quad \eta=\frac{I}{Al^2}. \tag{9.91}$$

The axial displacement can be computed by

$$\frac{\bar{u}(\xi)}{l}=\int_0^\xi[(1-p\cos\varphi+p\alpha\sin\varphi)\cos\varphi-1]d\xi \tag{9.92}$$

where $\bar{u}$ is the axial displacement of the cross-section center, and $\bar{u}(0)=0$ for the case shown in Fig. 9.17.

The transverse displacement is given by

$$\frac{\bar{w}(\xi)}{l}=\frac{\bar{w}(1)}{l}+\frac{\eta}{p}\left(\frac{d\varphi(1)}{d\xi}-\frac{d\varphi}{d\xi}\right)+\alpha\left[\left(1+\frac{\bar{u}(1)}{l}\right)-\left(\xi+\frac{\bar{u}}{l}\right)\right]$$

$$=\frac{\eta}{p}\left(\frac{d\varphi(0)}{d\xi}-\frac{d\varphi}{d\xi}\right)-\alpha\left(\xi+\frac{\bar{u}}{l}\right) \tag{9.93}$$

where $\bar{w}$ is the transverse displacement of the cross-section center, and $d\varphi(0)/d\xi$ and $d\varphi(1)/d\xi$ denote $d\varphi/d\xi$ at $\xi=0$ and $\xi=1$. Similar symbols are used in the remaining part of Section 9.4.

In terms of cross-sectional rotation φ, the boundary conditions shown in Fig. 9.17 are

$$\begin{cases}\dfrac{d\varphi(0)}{d\xi}=\dfrac{lM_0}{EI}=\dfrac{lK\varphi(0)}{EI}\\[3mm]\dfrac{d\varphi(1)}{d\xi}=0\end{cases} \tag{9.94}$$

where M_0 is bending moment applied at the left end of the beam shown in Fig. 9.17, K is the elastic constant of the rotational spring. It is seen that when K is zero, the left end is hinged; when $K=\infty$, the left end is clamped. It is worth noting that the φ in Eq. (9.94) is the cross-sectional rotation.

For the case considered, Eq. (9.90) and Eq. (9.92) are coupled with each other and solved simultaneously.

9.4.2 DQ-BASED NUMERICAL INTEGRATION SCHEME

Equation (9.90) is rewritten as

$$\frac{d^2\varphi}{d\xi^2} = -\frac{p}{\eta}[1 - p(\cos\varphi - \alpha\sin\varphi)](\alpha\cos\varphi + \sin\varphi) \tag{9.95}$$

or

$$\frac{d^2\varphi}{d\xi^2} = f_1(\varphi) \tag{9.96}$$

The second-order differential equation, Eq. (9.95) or Eq. (9.96), is solved by the DQ-based numerical integration presented in Chapter 1. Two initial conditions are needed, namely, $\varphi(0) = \bar{\varphi}_0, d\varphi(0)/d\xi = \bar{\varphi}_0'$, where $\bar{\varphi}_0$ and $\bar{\varphi}_0'$ should be given and $\xi \in [0,1]$. According to Eq. (9.94), only one of them, i.e., $\bar{\varphi}_0'$, is given and the other is missing and to be found iteratively by satisfying the second boundary condition in Eq. (9.94), namely, by satisfying $d\varphi(1)/d\xi = 0$.

Similar to the numerical integration in time domain, the entire domain is divided into M integration intervals. Uniformly distributed integral interval, namely, $\Delta\xi = 1/M$, is used. Each integral interval $[\xi_i, \xi_{i+1}]$ is regularized to [0, 1] by $\tau = (\xi - \xi_i)/\Delta\xi$. Thus, one has

$$\varphi(\tau = 0) \equiv \varphi_0 = \varphi(\xi_i) \tag{9.97}$$

$$\frac{d\varphi(\tau = 0)}{d\tau} = \dot{\varphi}(\tau = 0) \equiv \dot{\varphi}_0 = \Delta\xi \frac{d\varphi(\xi)}{d\xi}|_{\xi = \xi_i} \equiv \Delta\xi\varphi'(\xi_i) \tag{9.98}$$

since

$$\frac{d\varphi}{d\xi} = \frac{d\varphi}{d\tau}\frac{d\tau}{d\xi} = \frac{1}{\Delta\xi}\frac{d\varphi}{d\tau} \equiv \frac{1}{\Delta\xi}\dot{\varphi} \tag{9.99}$$

$$\frac{d^2\varphi}{d\xi^2} = \frac{d^2\varphi}{d\tau^2}\frac{d\tau^2}{d\xi^2} = \frac{1}{\Delta\xi^2}\frac{d^2\varphi}{d\tau^2} \equiv \frac{1}{\Delta\xi^2}\ddot{\varphi} \tag{9.100}$$

Take the typical integral interval [0, 1] as an example. The values of φ_0 and $\dot{\varphi}_0$ at $\tau = 0$, i.e., at the nodal point ξ_i, are initial conditions.

Let ξ_i be the first nodal point, i.e., τ_0 ($\tau = 0$). The remaining nodal points are

$$0 < \tau_1 < \tau_2 < \cdots < \tau_n < 1 \tag{9.101}$$

For unconditionally stable, higher-order accurate and computationally efficient, τ_i is computed by [9]

$$\tau_i = (1 + x_i)/2, \qquad i = 1, 2, ..., n \tag{9.102}$$

where x_i are the abscissas of Gaussian quadrature in domain $[-1, 1]$. For $n = 3$, x_i takes the values of $-\sqrt{0.6}$, 0, $\sqrt{0.6}$.

In terms of the DQ-based numerical integration method, Eq. (9.96) at all inner points can be expressed as

$$\begin{Bmatrix} \ddot{\varphi}(\tau_1) \\ \ddot{\varphi}(\tau_2) \\ \vdots \\ \ddot{\varphi}(\tau_n) \end{Bmatrix} = \Delta \xi^2 \begin{Bmatrix} f_1(\varphi(\tau_1)) \\ f_1(\varphi(\tau_2)) \\ \vdots \\ f_1(\varphi(\tau_n)) \end{Bmatrix} \tag{9.103}$$

or in short,

$$\{\ddot{\varphi}\} = \Delta \xi^2 \{f_1(\varphi)\} \tag{9.104}$$

Regarding τ as the time, the first-order derivative is given by Eq. (1.61), i.e.,

$$\{\dot{\varphi}\} = \{G_0\} \varphi_0 + [G]\{\varphi\} \tag{9.105}$$

And the second-order derivative is given by Eq. (1.63), namely,

$$\{\ddot{\varphi}\} = \{G_0\} \dot{\varphi}_0 + [G]\{G_0\} \varphi_0 + [G][G]\{\varphi\} \tag{9.106}$$

where

$$\{\dot{\varphi}\} = \begin{pmatrix} \dot{\varphi}_1 \\ \vdots \\ \dot{\varphi}_n \end{pmatrix}, \quad \{\varphi\} = \begin{pmatrix} \varphi_1 \\ \vdots \\ \varphi_n \end{pmatrix}, \quad \{G_0\} = \begin{pmatrix} A_{10}^{(1)} \\ \vdots \\ A_{n0}^{(1)} \end{pmatrix}, \quad [G] = \begin{pmatrix} A_{11}^{(1)} & \cdots & A_{1n}^{(1)} \\ \vdots & \ddots & \vdots \\ A_{n1}^{(1)} & \cdots & A_{nn}^{(1)} \end{pmatrix} \tag{9.107}$$

and A_{ij} are the weighting coefficient of the first-order derivative with respect to τ, computed by Eq. (1.30).

Substituting Eq. (9.106) into Eq. (9.103) results

$$[G][G]\{\varphi\} = \Delta \xi^2 \{f_1(\varphi)\} - \varphi_0 [G]\{G_0\} - \dot{\varphi}_0 \{G_0\} \tag{9.108}$$

Equation (9.108) can be rewritten as

$$\{\varphi\} = ([G][G])^{-1} \{\bar{f}_1(\varphi)\} \tag{9.109}$$

where

$$\{\bar{f}_1(\varphi)\} = \left[\Delta \xi^2 \{f_1(\varphi)\} - \varphi_0 [G]\{G_0\} - \dot{\varphi}_0 \{G_0\} \right] \tag{9.110}$$

Since $\bar{f}_1(\varphi)$ is a nonlinear function, the direct iteration method is used to obtain the solutions, namely,

$$\{\varphi^{k+1}\} = ([G][G])^{-1} \{\bar{f}_1(\varphi^k)\} \tag{9.111}$$

Note that $([G][G])$ is a constant matrix; thus, its inverse, $([G][G])^{-1}$, is only performed one time. Besides, the dimension of matrix $[G]$ is very small.

To start with the iteration, set the value of $\varphi_1, \varphi_2 \ldots \ldots \varphi_n$ as φ_0 and the value of $\dot{\varphi}_1, \dot{\varphi}_2 \ldots \ldots \dot{\varphi}_n$ as $\dot{\varphi}_0$. Note that φ_0 is unknown for the problem considered; thus, a way to determine it iteratively is proposed in Ref. [7] and will be described in Section 9.4.3.

Let $u = \bar{u} / l$, which can be obtained by Eq. (9.92). Several ways can be used to perform the integration numerically. Since Eq. (9.92) and Eq. (9.90) are coupled with each other, u is also integrated by the same DQ-based numerical integration scheme.

In order to use the DQ-based numerical integration scheme, change the integration equation to a differential equation by taking the first-order derivative with respect to ξ on both sides of Eq. (9.92). After doing so and using Eq. (9.99), one has

$$\frac{du}{d\tau} = \Delta\xi\left[(1 - p\cos\varphi + p\alpha\sin\varphi)\cos\varphi - 1\right] \tag{9.112}$$

In terms of the differential quadrature and using Eq. (9.105), Eq. (9.112) can be written by

$$[G]\{u\} = \Delta\xi\{f_2(\varphi)\} - \{G_0\}u_0 \tag{9.113}$$

where $\{f_2(\varphi)\}$ is defined by

$$\{f_2(\varphi)\} = \{(1 - p\cos\varphi + p\alpha\sin\varphi)\cos\varphi - 1\} \tag{9.114}$$

For the case considered, as shown in Fig. 9.17, $\mu_0 = 0$ for the first integral interval. The displacement u can be obtained by Eq. (9.113) step by step. Iteration may be necessary, since Eq. (9.113) and Eq. (9.109) are coupled with each other.

Once $\varphi_1, \varphi_2, \ldots \ldots \varphi_n, \dot{\varphi}_1, \dot{\varphi}_2, \ldots, \dot{\varphi}_n$ and $\mu_1, \mu_2 \ldots \ldots \mu_n$ for the integral interval are obtained, the value of $\varphi, \dot{\varphi}, u$ at $\tau = 1$, i.e., at ξ_{k+1}, can be computed by using the method of extrapolation, namely,

$$\varphi(1) = \sum_{j=0}^{N} l_j(1)\varphi_j$$

$$\dot{\varphi}(1) = \sum_{j=0}^{N} l_j(1)\dot{\varphi}_j \tag{9.115}$$

$$u(1) = \sum_{j=0}^{N} l_j(1)u_j$$

where $l_j(\tau)$ is the Lagrange interpolation functions defined by

$$l_j(\tau) = \frac{(\tau - \tau_0)(\tau - \tau_1)\cdots(\tau - \tau_{j-1})(\tau - \tau_{j+1})\cdots(\tau - \tau_{N-1})(\tau - \tau_N)}{(\tau_j - \tau_0)(\tau_j - \tau_1)\cdots(\tau_j - \tau_{j-1})(\tau_j - \tau_{j+1})\cdots(\tau_j - \tau_{N-1})(\tau_j - \tau_N)} \tag{9.116}$$

in which $\tau_0 = 0$ and τ_1 $(i = 1, 2, \ldots, N)$ are the $(N + 1)$ grid points within interval $(0,1)$.

It is shown [9] that the order of accuracy at $\tau_0 = 1$ is even higher than the order of accuracy at other locations within the integral interval. The formulation for obtaining u in terms of differential quadrature is first reported in Ref. [7].

The deflection $\bar{w}(\xi)/l$ can be directly computed by Eq. (9.93).

9.4.3 **SOLUTION PROCEDURES**

Since the problem considered is a boundary value problem, the DQ-based time integration scheme is not strictly applicable. Two initial conditions at $\xi = 0$ are required; however, from Eq. (9.94) only one initial condition at $\xi = 0$, i.e., $d\varphi(0)/d\xi$, is given. Therefore, a way to find the other initial condition is proposed in Ref. [7].

To start with, assign the missing initial condition φ_0 a certain value. Usually the assigned value is not the real one; therefore, the missing initial condition is determined iteratively by satisfying the other boundary condition at $\xi = 1$.

Since $\dot{\varphi}_0$ is known and φ_0 at $\xi = 0$ is assumed, φ can be obtained iteratively by Eq. (9.111) for the first integral interval. Then, φ_0 and $\dot{\varphi}_0$ for the next integration interval can be obtained by Eq. (9.115). The integration proceeds for increasing the value of ξ until it reaches 1. Check if the other boundary condition in Eq. (9.94), i.e., $d\varphi(1)/d\xi = 0$, is satisfied. If the relative error between the known value and calculated one is within a prescribed tolerance, the solution is obtained. Otherwise, modify the value of the missing initial condition at $\xi = 0$ and repeat the process until $d\varphi(1)/d\xi = 0$ is approximately satisfied.

For example, to start with the numerical integration one can simply assign 0 to the missing initial condition, i.e., $\varphi_0 = \varphi(0) = 0$. Equation (9.96) is then integrated step by step until $\xi = 1$. Check if $|d\varphi(1)/d\xi| \leq \mathrm{err}$, where err is a prescribed tolerance. If $|d\varphi(1)/d\xi| > \mathrm{err}$, then φ_0 is increased by Δ, that is,

$$\varphi(0)_{n+1} = \varphi(0)_n + \Delta \tag{9.117}$$

where Δ is a small positive value, since the direction of the buckling mode has no significant influence on the buckling load and on the postbuckling behavior.

It is seen from Eqs. (9.90) and (9.91) that $\bar{u}(1)/l$, $d\varphi(0)/d\xi$, and α are coupled with each other. It was experienced [8] that divergence of the iteration may be encountered, especially under high applied loads. Therefore, a way to ensure the convergence of the iteration has been proposed in Ref. [7] and briefly described below.

Through careful study, it is found that the key to ensure the convergence is to find an appropriate value of $\alpha = R/P$. Once an appropriate α is obtained, Eq. (9.113) and Eq. (9.109) are decoupled and can be solved separately.

The way to find an appropriate value of α is given as follows. Note that the relationship between transverse force R and the bending moment at $\xi = 0$ is

$$R\left(l + \bar{u}(1)\right) = M_0 = EI\frac{d\varphi(0)}{dx} \tag{9.118}$$

Using Eq. (9.91) one obtains

$$\alpha = \frac{R}{P} = \frac{d\varphi(0)/dx}{\dfrac{Pl}{EI}\left(1 + \dfrac{\bar{u}(1)}{l}\right)} = \frac{d\varphi(0)/d\xi}{\dfrac{p}{\eta}\left(1 + \dfrac{\bar{u}(1)}{l}\right)} = \frac{Kl\varphi(0)/EI}{\dfrac{p}{\eta}\left(1 + \dfrac{\bar{u}(1)}{l}\right)} \tag{9.119}$$

From Eq. (9.119), it is seen that α is related to the initial value of $d\varphi(0)/d\xi$ and $\bar{u}(1)/l$. Therefore, direct iteration can be used to find α, i.e.,

$$\alpha_{n+1} = \frac{d\varphi(0)/d\xi}{\dfrac{P}{\eta}\left(1 + \dfrac{\bar{u}_{\alpha_n}}{l}\right)} = \frac{Kl\varphi(0)/EI}{\dfrac{P}{\eta}\left(1 + \dfrac{\bar{u}_{\alpha_n}}{l}\right)} \tag{9.120}$$

where $\bar{u}_{\alpha_n}/l$ stands for the value of $\bar{u}(1)/l$ with $\alpha = \alpha_n$.

To increase the rate of convergence, α_{n+1} can be modified by

$$\alpha_{n+1} = \lambda\alpha_n + (1-\lambda)\alpha_{n+1}^{\text{temp}} \tag{9.121}$$

where $\alpha_{n+1}^{\text{temp}}$ stands for the value of α_{n+1} computed by using Eq. (9.120) with $\alpha = \alpha_n$, and $\lambda \in [0,1)$.

To start with, an appropriate initial iterative value for α is needed. At the beginning, small deformation can be assumed, since the applied load is small. From Fig. 9.17 one has

$$\frac{\bar{u}(1)}{l} \approx -\frac{P}{EA} = -p \tag{9.122}$$

Thus, Eq. (9.119) can be approximated by

$$\alpha \approx \frac{Kl\varphi(0)/EI}{(p/\eta)(1-p)} \tag{9.123}$$

Once p and $\varphi(0)$ are determined, the approximate value of α can be directly calculated by Eq. (9.123). Since p is given and $\varphi(0)$ can be determined by using the procedures described previously, an approximate value of α is always available at each step of integration.

Numerical results show that a satisfactory result can be obtained with such an approximate value of α as the initial iterative value when the applied load is relatively small. When the applied load is large, large deflection is encountered. In such cases, the convergence value of α under $\varphi(0)_n$ should be used as the initial iterative value of α to compute $\varphi(0)_{n+1}$. This is based on the fact that the increment Δ in Eq. (9.117) is small; thus, α would not change much. More details on this matter can be found in Ref. [7].

9.4.4 RESULTS AND DISCUSSION

EXAMPLE 9.5

Consider first the beam is clamped at $x = 0$ and simply supported at $x = l$. This case corresponds to $K = \infty$, where K is the stiffness of the rotational spring shown in Fig. 9.17.

The post-buckling deformations at $p/p_{cr} = 1.05$ is shown in Fig. 9.18. Symbol p_{cr} is the linear buckling load of the beam with one end fixed and the other simply supported. In Fig. 9.18, the results obtained by the multiple-scale (MS) method [8] are also included for comparison. As is expected, the results are very close to each other.

It was reported earlier [8] that convergent results could not be obtained by using the numerical integration at higher loads ($p/p_{cr} > 1.0672$). However, such difficulty is not encountered by using the DQ-based numerical integration scheme. Figure 9.19 shows the DQ results at two higher loads, i.e., $p/p_{cr} = 2.0$ and $p/p_{cr} = 4.0$. Since the rotations are quite large, the multiple-scale solutions presented in Ref. [8] are not valid any more, and thus are not included in Fig. 9.19. Table 9.2 lists the values of $(\bar{u}/l)_{max}$ and $(\bar{w}/l)_{max}$. For the same reason, only the multiple-scale solutions at $p/p_{cr} = 1.05$ are cited for comparison.

The buckling behavior for the beam-column shown in Fig. 9.17 is obtained for various Kl/EI varying from 10^{-3} to 10^4. The variations of the critical load with Kl/EI are shown in Fig. 9.20 for three different η. It is seen that the critical loads change with the increase in the rotational spring constant (Kl/EI).

For clarity, the critical loads at ten different values of Kl/EI are tabulated in Table 9.3 for the case of $\eta = 0015$. Detailed comparisons are made in Ref. [7] and revealed that when Kl/EI approaches zero, the left end is reduced to the hinged end, and when Kl/EI approaches infinity, the left end is reduced to the clamped end.

The postbuckling deformations at $p/p_{cr} = 1.05$ and 1.15 for $\eta = 0.001$ and $Kl/EI = 5.0$ are shown in Fig. 9.21. More information can be found in Ref. [7].

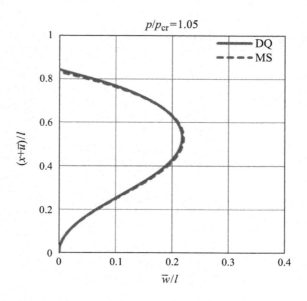

FIGURE 9.18

Postbuckling Configuration for CS Beam

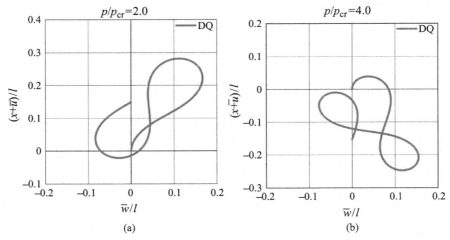

FIGURE 9.19

Postbuckling Configurations Under Different Loads for CS Beam

Table 9.2	Postbuckling Maximum Displacements of CS Beam			
p/p_{cr}		1.05	2.0	4.0
$(\bar{\mu}/l)_{max}$	DQ	−0.157761	−0.850249	−1.152352
	MS	−0.167727	–	–
	MS/DQ	1.063	–	–
$(\bar{w}/l)_{max}$	DQ	0.217760	0.165283	0.149866
	MS	0.220183	–	–
	MS/DQ	1.011	–	–

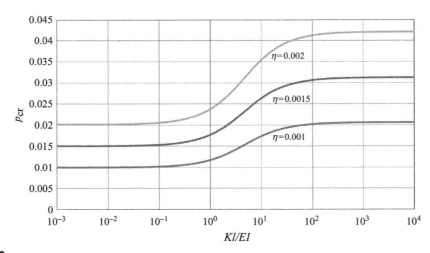

FIGURE 9.20

Critical Load Kl/EI Curves ($\eta = 0.001, 0.0015, 0.002$)

Table 9.3 Critical Loads Under Different Kl/EI ($\eta = 0.0015$)				
Kl/EI	0.001	0.01	1	5	10
p_{cr}	0.015033	0.015061	0.017711	0.023466	0.026306
Kl/EI	20	50	100	1000	10000
p_{cr}	0.028434	0.030037	0.030634	0.031199	0.031257

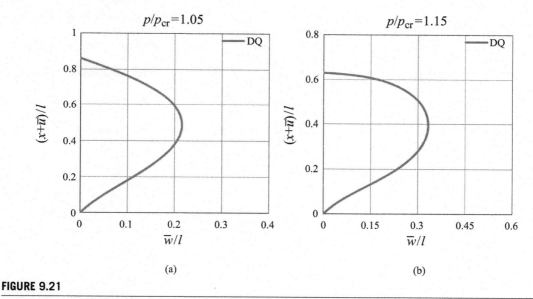

(a)   (b)

FIGURE 9.21

Postbuckling Configurations Under Different Applied Loads ($\eta = 0.001$, $Kl/EI = 5.0$)

9.5 SUMMARY

In this chapter, the modified DQM (or HDQM) is used to successfully perform geometric nonlinear analysis for thin rod, beam, and shallow shell. Different solution strategies are used to solve the nonlinear algebraic equations. Newton–Raphson method together with load increment is used to obtain the load–displacement relations for the shallow shell under a uniformly distributed pressure and the buckling and postbuckling behavior of a rod sitting in a rigid cylinder. For postbuckling analysis, small or large disturbances are used to obtain the lateral buckling mode and/or helical buckling mode. The method of the DQ-based time integration is used to perform geometric nonlinear analysis for thin beam under an axial compressive load. Numerical results are given and discussed.

The results show that the modified DQM (or HDQM) can yield accurate nonlinear solutions with relatively small number of grid points. Thus, the attractive features of rapid convergence, high accuracy, and computational efficiency of the DQM (or HDQM) are demonstrated. One important finding is that a large disturbance should be used in order to get helical postbuckling mode. If the rod is long, either the DQEM or the LaDQM should be used to ensure numerical stability.

REFERENCES

[1] X. Wang, Nonlinear stability analysis of thin doubly curved orthotropic shallow shells by the differential quadrature method, Comput. Meth. Appl. Mech. Eng. 196 (2007) 2242–2251.

[2] S. Timoshenko, S. Woinowsky-Krieger, Theory of Plates and Shells, McGraw-Hill Book Company, New York, 1959.

[3] Z. Yuan, X. Wang, Nonlinear buckling analysis of inclined circular cylinder-in-cylinder by the discrete singular convolution, Int. J. Nonlin. Mech. 47 (2012) 699–711.

[4] R.F. Mitchell, Effects of well deviation on helical buckling, SPE Drill. Completion SPE 29462 (1997) 63–69.

[5] X. Wang, Z. Yuan, Investigation of frictional effects on helical buckling of circular rods laterally constrained in horizontal cylinders, J. Petrol. Sci. Eng. 90-91 (2012) 70–78.

[6] Z. Yuan, Nonlinear buckling analysis of thin rods in circular cylinder by the DSC algorithm. Master Thesis. Nanjing University of Aeronautics and Astronautics, China, 2012 (in Chinese).

[7] Z. Yuan, X. Wang, Buckling and post-buckling analysis of extensible beams by using the differential quadrature method, Comput. Math. Appl. 62 (2011) 4499–4513.

[8] C.E.N. Mazzilli, Buckling and post-buckling of extensible rods revisited: a multiple-scale solution, Int. J. Nonlin. Mech. 44 (2009) 199–207.

[9] T.C. Fung, Solving initial value problems by differential quadrature method – Part 2: second- and higher-order equations, Int. J. Numer. Meth. Eng. 50 (2001) 1429–1454.

ELASTOPLASTIC BUCKLING ANALYSIS OF PLATE

10.1 INTRODUCTION

This chapter demonstrates the elastoplastic buckling analysis by using the modified differential quadrature method (DQM). Both thin and thick rectangular plates are considered. For comparisons, Ramberg–Osgood material is used. In the formulations, two plasticity theories are used: one is the incremental plasticity (IT) with Prandtl–Reuss constitutive equation and the other is the deformation plasticity (DT) with Hencky constitutive equation. Formulations are worked out in detail and examples are given for illustrations. Various combinations of boundary conditions and loadings are investigated. The importance to check the assumption of small deformation is emphasized to cause readers' attention.

10.2 ELASTOPLASTIC BUCKLING OF THIN RECTANGULAR PLATE

10.2.1 BASIC EQUATIONS OF THIN RECTANGULAR PLATE

Consider elastoplastic buckling of a thin rectangular plate under general loadings, shown in Fig. 10.1. The Cartesian coordinate system (xyz) is set at the middle plane of the plate. Two plasticity theories, namely, the J_2' flow theory of plasticity with Prandtl–Reuss constitutive equation and the deformation plasticity theory with Hencky constitutive equation (DT), are considered. The J_2' flow theory of plasticity is also called the incremental theory of plasticity (IT) in this chapter.

The J_2' flow theory of plasticity with Prandtl–Reuss constitutive equation is given by

$$\dot{\sigma}_{ij} = 2G\dot{e}_{ij} + \lambda\delta_{ij}\dot{e}_{kk} - 3(G - G_t)\frac{S_{ij}S_{kl}\dot{e}_{kl}}{\sigma_e^2} \tag{10.1}$$

where the over dot denotes the derivative with respect to a time-like parameter, G and λ are Lame constants since the material is assumed elastically isotropic, S_{ij} is the deviatoric stress tensor, G_t is the tangential shear modulus, and σ_e is the von-Mises equivalent stress given by

$$\sigma_e = \sqrt{1.5S_{ij}S_{ij}}, \frac{1}{G_t} = \frac{1}{G} + 3\left(\frac{1}{E_t} - \frac{1}{E}\right) \tag{10.2}$$

where E and $E_t = d\sigma_e/d\varepsilon_e$ are elastic and tangential Young's moduli determined by the uniaxial stress–strain curve.

The rate form of the deformation plasticity theory with Hencky constitutive equation is given by

$$\dot{\sigma}_{ij} = 2G_s\dot{e}_{ij} + \lambda_s\delta_{ij}\dot{e}_{kk} - 3(G_s - G_t)\frac{S_{ij}S_{kl}\dot{e}_{kl}}{\sigma_e^2} \tag{10.3}$$

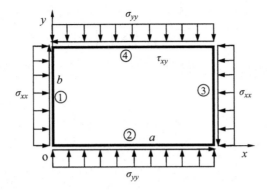

FIGURE 10.1

Isotropic Rectangular Plate Under General In-Plane Loadings

where λ_s and G_s are the secant Lame constants defined by

$$\begin{cases} \lambda_s = \dfrac{\mu_s E_s}{(1+\mu_s)(1-2\mu_s)} \\ G_s = \dfrac{E_s}{2(1+\mu_s)} \end{cases} \tag{10.4}$$

in which $E_s = \sigma_e/\varepsilon_e$ is the secant Young's modulus determined by the uniaxial stress–strain curve, and μ_s is the secant Poisson's ratio computed by

$$\mu_s = \frac{1}{2} - \left(\frac{1}{2} - \mu\right)\frac{E_s}{E} \tag{10.5}$$

where μ is Poisson's ratio of the material.

Since the material is assumed time independent,

$$\dot{\sigma}_{ij} = d\sigma_{ij}, \ \dot{\varepsilon}_{ij} = d\varepsilon_{ij} \tag{10.6}$$

For buckling of thin rectangular plate under general in-plane loadings, plane stress condition is assumed. The incremental constitutive equation of IT in matrix form is given by

$$\begin{Bmatrix} d\sigma_{xx} \\ d\sigma_{yy} \\ d\tau_{xy} \end{Bmatrix} = \begin{bmatrix} D_{11} & D_{12} & D_{16} \\ & D_{22} & D_{26} \\ \text{sym} & & D_{66} \end{bmatrix} \begin{Bmatrix} d\varepsilon_{xx} \\ d\varepsilon_{yy} \\ d\gamma_{xy} \end{Bmatrix} \tag{10.7}$$

where D_{ij} changes with the stress components $\sigma_{xx}, \sigma_{yy}, \tau_{xy}$.

The incremental constitutive equation of DT in matrix form is given by

$$\begin{Bmatrix} d\sigma_{xx} \\ d\sigma_{yy} \\ d\tau_{xy} \end{Bmatrix} = \begin{bmatrix} D'_{11} & D'_{12} & D'_{16} \\ & D'_{22} & D'_{26} \\ \text{sym} & & D'_{66} \end{bmatrix} \begin{Bmatrix} d\varepsilon_{xx} \\ d\varepsilon_{yy} \\ d\gamma_{xy} \end{Bmatrix} \tag{10.8}$$

where D'_{ij} also changes with the stress components $\sigma_{xx}, \sigma_{yy}, \tau_{xy}$.

Denote w the rate of deflection of the plate. The rate form of geometrical equation is

$$
\begin{cases}
\dot{\varepsilon}_{xx} = -z\dfrac{\partial^2 w}{\partial x^2} \\[2mm]
\dot{\varepsilon}_{yy} = -z\dfrac{\partial^2 w}{\partial y^2} \\[2mm]
\dot{\gamma}_{xy} = -2z\dfrac{\partial^2 w}{\partial x\partial y}
\end{cases}
\tag{10.9}
$$

Assume that no unloading occurs at the instant of the plastic buckling. The rate form of the governing equation for the elastoplastic buckling analysis of thin rectangular plates under general loadings is given by

$$
\frac{h^2}{12}\left[D_{11}\frac{\partial^4 w}{\partial x^4} + 2\left(D_{12}+2D_{66}\right)\frac{\partial^4 w}{\partial x^2 \partial y^2} + D_{22}\frac{\partial^4 w}{\partial y^4} + 4D_{16}\frac{\partial^4 w}{\partial x^3 \partial y} + 4D_{26}\frac{\partial^4 w}{\partial x\partial y^3}\right]
$$
$$
+\sigma_{xx}\frac{\partial^2 w}{\partial x^2} + 2\tau_{xy}\frac{\partial^2 w}{\partial x\partial y} + \sigma_{yy}\frac{\partial^2 w}{\partial y^2} = 0
\tag{10.10}
$$

where h is the thickness of the plate. The positive in-plane stress components, i.e., $\sigma_{xx}, \sigma_{yy}, \tau_{xy}$, are shown in Fig. 10.1. For simplicity, assumption is made that $\sigma_{xx}, \sigma_{yy}, \tau_{xy}$ are constant in the entire plate before buckling.

Boundary conditions based on the J'_2 flow theory of plasticity with Prandtl–Reuss constitutive equation are as follows.

1. Simply supported boundary conditions (S):

$$
w = 0 \text{ and } M_{xx} = 0 \text{ at } x = 0 \text{ and/or } x = a
\tag{10.11}
$$

or

$$
w = 0 \text{ and } M_{yy} = 0 \text{ at } y = 0 \text{ and/or } y = b
\tag{10.12}
$$

where

$$
M_{xx} = \frac{h^3}{12}\left(-D_{11}\frac{\partial^2 w}{\partial x^2} - D_{12}\frac{\partial^2 w}{\partial y^2} - 2D_{16}\frac{\partial^2 w}{\partial x\partial y}\right)
\tag{10.13}
$$

$$
M_{yy} = \frac{h^3}{12}\left(-D_{12}\frac{\partial^2 w}{\partial x^2} - D_{22}\frac{\partial^2 w}{\partial y^2} - 2D_{26}\frac{\partial^2 w}{\partial x\partial y}\right)
\tag{10.14}
$$

2. Clamped boundary conditions (C):

$$
w = 0 \text{ and } w_x = 0 \text{ at } x = 0 \text{ and/or } x = a
\tag{10.15}
$$

or

$$
w = 0 \text{ and } w_y = 0 \text{ at } y = 0 \text{ and/or } y = b
\tag{10.16}
$$

where subscripts x and y denote the first-order derivative with respect to x and y.

3. Free boundary conditions (F):

$$\begin{cases} M_{xx} = 0 \\ \bar{Q}_x = Q_x - h\sigma_{xx}\dfrac{\partial w}{\partial x} - h\tau_{xy}\dfrac{\partial w}{\partial y} + \dfrac{\partial M_{xy}}{\partial y} = 0 \text{ at } x = 0 \text{ and/or } x = a \end{cases} \tag{10.17}$$

or

$$\begin{cases} M_{yy} = 0 \\ \bar{Q}_y = Q_y - h\sigma_{yy}\dfrac{\partial w}{\partial y} - h\tau_{xy}\dfrac{\partial w}{\partial x} + \dfrac{\partial M_{yx}}{\partial x} = 0 \text{ at } y = 0 \text{ and/or } y = b \end{cases} \tag{10.18}$$

where

$$M_{xy} = M_{yx} = \frac{h^3}{12}\left(-D_{16}\frac{\partial^2 w}{\partial x^2} - D_{26}\frac{\partial^2 w}{\partial y^2} - 2D_{66}\frac{\partial^2 w}{\partial x \partial y}\right) \tag{10.19}$$

$$Q_x = \frac{h^3}{12}\left[-D_{11}\frac{\partial^3 w}{\partial x^3} - 3D_{16}\frac{\partial^3 w}{\partial x^2 \partial y} - (D_{12} + 2D_{66})\frac{\partial^3 w}{\partial x \partial y^2} - D_{26}\frac{\partial^3 w}{\partial y^3}\right] \tag{10.20}$$

$$Q_y = \frac{h^3}{12}\left[-D_{16}\frac{\partial^3 w}{\partial x^3} - (D_{12} + 2D_{66})\frac{\partial^3 w}{\partial x^2 \partial y} - 3D_{26}\frac{\partial^3 w}{\partial x \partial y^2} - D_{22}\frac{\partial^3 w}{\partial y^3}\right] \tag{10.21}$$

With Eqs. (10.19) and (10.20), the second equation of Eq. (10.17) becomes

$$\frac{\bar{Q}_x}{h} = \frac{h^2}{12}\left[-D_{11}\frac{\partial^3 w}{\partial x^3} - 4D_{16}\frac{\partial^3 w}{\partial x^2 \partial y} - (D_{12} + 4D_{66})\frac{\partial^3 w}{\partial x \partial y^2} - 2D_{26}\frac{\partial^3 w}{\partial y^3}\right]$$
$$-\sigma_{xx}\frac{\partial w}{\partial x} - \tau_{xy}\frac{\partial w}{\partial y} = 0 \tag{10.22}$$

With Eqs. (10.19) and (10.21), the second equation of Eq. (10.18) becomes

$$\frac{\bar{Q}_y}{h} = \frac{h^2}{12}\left[-2D_{16}\frac{\partial^3 w}{\partial x^3} - (D_{12} + 4D_{66})\frac{\partial^3 w}{\partial x^2 \partial y} - 4D_{26}\frac{\partial^3 w}{\partial x \partial y^2} - D_{22}\frac{\partial^3 w}{\partial y^3}\right]$$
$$-\sigma_{yy}\frac{\partial w}{\partial y} - \tau_{xy}\frac{\partial w}{\partial x} = 0 \tag{10.23}$$

As mentioned before, Eqs. (10.10)–(10.23) are based on the J_2' flow theory of plasticity with Prandtl–Reuss constitutive equation. For the deformation plasticity theory with Hencky constitutive equation, simply replace all D_{ij} in Eqs. (10.10)–(10.23) by D_{ij}'. Therefore, the governing equation and expressions of boundary conditions based on the deformation plasticity theory with Hencky constitutive equation are omitted.

10.2.2 SOLUTION PROCEDURES BY THE DQM

Equation (10.10) is similar to the differential equation of an anisotropic rectangular plate in form, i.e., Eq. (7.1). Thus, the DQM with MMWC-3, also called the modified DQM, is used in the analysis.

Let the rectangular domain be discrete into $N \times N$ points; thus, the number of inner grid points is $(N-2) \times (N-2)$. In terms of the differential quadrature, Eq. (10.10) at all inner grid points can be written as

$$
\begin{aligned}
\Bigg[D_{11} \sum_{k=1}^{N+2} D_{ik}^x \delta_{kj} &+ 2(D_{12} + 2D_{66}) \sum_{l=1}^{N+2} \sum_{k=1}^{N+2} B_{jl}^y B_{ik}^x \delta_{kl} + 4D_{16} \sum_{l=1}^{N+2} \sum_{k=1}^{N+2} A_{jl}^y C_{ik}^x \delta_{kl} \\
&+ D_{22} \sum_{l=1}^{N+2} D_{jl}^y \delta_{il} + 4D_{26} \sum_{l=1}^{N+2} \sum_{k=1}^{N+2} C_{jl}^y A_{ik}^x \delta_{kl} \Bigg] \frac{h^2}{12} = \\
&- \sigma \Bigg[\xi_1 \sum_{k=1}^{N+2} B_{ik}^x \delta_{kj} + 2\xi_3 \sum_{l=1}^{N+2} \sum_{k=1}^{N+2} A_{jl}^y A_{ik}^x \delta_{kl} + \xi_2 \sum_{l=1}^{N+2} B_{jl}^y \delta_{il} \Bigg] \\
&\qquad\qquad\qquad (i = 2, 3, ..., N-1;\, j = 2, 3, ..., N-1)
\end{aligned}
\tag{10.24}
$$

where $A_{ij}^x, B_{ij}^x, C_{ij}^x, D_{ij}^x$ are the weighting coefficients of the first- to the fourth-order derivatives with respect to x, $A_{ij}^y, B_{ij}^y, C_{ij}^y, D_{ij}^y$ are the weighting coefficients of the first- to the fourth-order derivatives with respect to y, the generalized displacement δ_{ij} in Eq. (10.24) contains w_{ij} as well as $(w_x)_{ij}$, $(w_y)_{ij}$, $\sigma_{xx} = \xi_1 \sigma$, $\sigma_{yy} = \xi_2 \sigma$, $\tau_{xy} = \xi_3 \sigma$, where $\sigma(> 0)$ is the buckling coefficient.

At the positions where $(\partial w / \partial x)_{x=0}$ and $(\partial w / \partial x)_{x=a}$ are, following DQ equations are placed,

$$
\frac{h^3}{12} \Bigg(-D_{11} \sum_{k=1}^{N+2} B_{ik}^x \delta_{kj} - D_{12} \sum_{l=1}^{N+2} B_{jl}^y \delta_{il} - 2D_{16} \sum_{l=1}^{N+2} \sum_{k=1}^{N+2} A_{jl}^y A_{ik}^x \delta_{kl} \Bigg) = (M_{xx})_{ij}
\tag{10.25}
$$
$$
(i = 1, 2, ..., N;\, j = 1, N)
$$

At the positions where $(\partial w / \partial y)_{y=0}$ and $(\partial w / \partial y)_{y=b}$ are, following DQ equations are placed,

$$
\frac{h^3}{12} \Bigg(-D_{12} \sum_{k=1}^{N+2} B_{ik}^x \delta_{kj} - D_{22} \sum_{l=1}^{N+2} B_{jl}^y \delta_{il} - 2D_{26} \sum_{l=1}^{N+2} \sum_{k=1}^{N+2} A_{jl}^y A_{ik}^x \delta_{kl} \Bigg) = (M_{yy})_{ij}
\tag{10.26}
$$
$$
(i = 1, N;\, j = 1, 2, ..., N)
$$

At the positions where w_{ij} $(i = 2, 3, ..., N-1;\, j = 1, N)$ are, following DQ equations are applied,

$$
\begin{aligned}
\frac{h^2}{12} \Bigg[-D_{11} \sum_{k=1}^{N+2} C_{ik}^x \delta_{kj} &- 4D_{16} \sum_{l=1}^{N+2} \sum_{k=1}^{N+2} A_{jl}^y B_{ik}^x \delta_{kl} - (D_{12} + 4D_{66}) \sum_{l=1}^{N+2} \sum_{k=1}^{N+2} B_{jl}^y A_{ik}^x \delta_{kl} \\
&- D_{26} \sum_{l=1}^{N+2} C_{jl}^y \delta_{il} \Bigg] - \sigma \xi_1 \sum_{k=1}^{N+2} A_{ik}^x \delta_{kj} - \sigma \xi_3 \sum_{l=1}^{N+2} A_{jl}^y \delta_{il} = \left(\frac{\bar{Q}_x}{h} \right)_{ij}
\end{aligned}
\tag{10.27}
$$
$$
(i = 2, 3, ..., N-1;\, j = 1, N)
$$

At the positions where w_{ij} $(i = 1, N;\, j = 2, 3, ..., N-1)$ are, following DQ equations are applied,

$$
\begin{aligned}
\frac{h^2}{12} \Bigg[-2D_{16} \sum_{k=1}^{N+2} C_{ik}^x \delta_{kj} &- 4D_{26} \sum_{l=1}^{N+2} \sum_{k=1}^{N+2} B_{jl}^y A_{ik}^x \delta_{kl} - (D_{12} + 4D_{66}) \sum_{l=1}^{N+2} \sum_{k=1}^{N+2} A_{jl}^y B_{ik}^x \delta_{kl} \\
&- D_{22} \sum_{l=1}^{N+2} C_{jl}^y \delta_{il} \Bigg] - \sigma \xi_3 \sum_{k=1}^{N+2} A_{ik}^x \delta_{kj} - \sigma \xi_2 \sum_{l=1}^{N+2} A_{jl}^y \delta_{il} = \left(\frac{\bar{Q}_y}{h} \right)_{ij}
\end{aligned}
\tag{10.28}
$$
$$
(i = 1, N;\, j = 2, 3, ..., N-1)
$$

And at the four corner points, w_{ij} $(i = 1, N;\, j = 1, N)$, following DQ equations are applied,

$$\frac{h^3}{12}\left(-2D_{16}\sum_{k=1}^{N+2}B_{ik}^{x}\delta_{kj}-2D_{26}\sum_{l=1}^{N+2}B_{jl}^{y}\delta_{il}-4D_{66}\sum_{l=1}^{N+2}\sum_{k=1}^{N+2}A_{jl}^{y}A_{ik}^{x}\delta_{kl}\right)=(2M_{xy})_{ij} \tag{10.29}$$
$$(i=1,N;\ j=1,N)$$

Combining Eqs. (10.24)–(10.29) forms the following matrix equation,

$$[F]\{\Delta\}=[G]\{\Delta\} \tag{10.30}$$

After rearrangement, Eq. (10.30) can be partitioned as

$$\begin{bmatrix} F_a & F_b \\ F_c & F_d \end{bmatrix}\begin{Bmatrix} \Delta^a \\ \Delta^b \end{Bmatrix}=\begin{Bmatrix} \sigma([G_a]\{\Delta^a\}+[G_b]\{\Delta^b\}) \\ f^b \end{Bmatrix} \tag{10.31}$$

The size of $\{\Delta^a\}$ and $\{\Delta^b\}$ varies with the different combinations of boundary conditions. If no free edge is involved, the size of $\{\Delta^a\}$ and $\{\Delta^b\}$ is $(8N-4)$ and (N^2-4N+4), respectively.

Applying boundary conditions and eliminating the nonzero $\{\bar{\Delta}^b\}$ in Eq. (10.31) result

$$([F_a]-[\bar{F}_b][\bar{F}_d]^{-1}[\bar{F}_c])\{\Delta^a\}=\sigma([G_a]-[\bar{G}_b][\bar{F}_d]^{-1}[\bar{F}_c])\{\Delta^a\} \tag{10.32}$$

where the over bar denotes that the submatrix is modified by eliminating the rows and columns that correspond to the zero generalized displacements. For an SSSS rectangular plate, the size of $\{\bar{\Delta}^b\}$ is $4N$.

Equation (10.32) can be simplified as

$$[FF]\{\Delta^a\}=\sigma[GG]\{\Delta^a\} \tag{10.33}$$

For the modified DQM, the simplest case is the CCCC rectangular plate. The size of $\{\bar{\Delta}^b\}$ in Eq. (10.32) is 0. In other words, $[FF]=[F_a]$ and $[GG]=[G_a]$.

With given material parameters, Eq. (10.33) can be solved by a generalized eigensolver. The lowest eigenvalue is the buckling load coefficient.

Alternatively, Eq. (10.33) can be deduced to a standard eigenvalue problem, namely,

$$\frac{1}{\sigma}\{\Delta^a\}=[FF]^{-1}[GG]\{\Delta^a\}\ \text{or}\ [FG]\{\Delta^a\}=P\{\Delta^a\} \tag{10.34}$$

The largest eigenvalue of Eq. (10.34), obtained by a standard eigensolver, is the required solution.

Since the material parameters depend on the applied in-plane stresses, Eqs. (10.33) and (10.34) are nonlinear equations. Direct iteration method may be used for solutions.

To solve Eq. (10.34) by the direct iteration method, set $P=P_0$ first. Note that P_0 should be small enough not to cause any plastic deformation. Next, compute D_{ij} or D_{ij}' depending on the plasticity theory to be used. Once D_{ij} or D_{ij}' is given, Eq. (10.34) becomes a standard eigenvalue equation that can be solved by any standard eigensolver. The largest eigenvalue corresponding to a positive P is denoted by P_1. Check if $|(P_1-P_0)/P_1|<$ err, where err is the prescribed error bound. If $|(P_1-P_0)/P_1|>$ err, replace P_0 by $[(1-\eta)P_0+\eta P_1]$ $(0.5\leq\eta\leq1)$. Repeat the iteration processes until $|(P_i-P_{i-1})/P_i|<$ err. Then P_i is the solution to be found.

10.2.3 APPLICATIONS AND DISCUSSION

For comparisons, the well-known Ramberg–Osgood elastoplastic stress–strain relationship is considered, namely,

$$\varepsilon_e = \frac{\sigma_e}{E} + k_2 \left(\frac{\sigma_e}{E} \right)^c \tag{10.35}$$

where the subscript e denotes the equivalent quantity, E, k_2, and c are the material constants.

Figure 10.2 shows the stress–strain curve for a typical aerospace aluminum alloy (Al 7075-T6); the material constants E, k_2, and c are [1] 72.4 Gpa, 3.94×10^{21}, and 10.9, respectively.

For comparisons, the buckling coefficient K and thickness parameter α are introduced, which are defined by [1]

$$K = \frac{\sigma(1-\mu^2)}{\alpha E}; \quad \alpha = \frac{\pi^2 h^2}{12 b^2} \tag{10.36}$$

where E is the modulus of elasticity.

EXAMPLE 10.1

Consider first an isotropic square plate ($a = b$) with all edges simply supported (SSSS). For comparisons, the thickness parameter α, defined by Eq. (10.36), takes the values of 0.0001, 0.001, and 0.002, which correspond to h/a of 0.0110, 0.0349, and 0.0493. All are within the range of classical thin plate theory ($h/a < 0.05$). $\mu = 0.32$. The plate is under constant biaxial edge loadings, i.e., $\sigma_{yy}(=\sigma)$, $\sigma_{xx}(=-\xi\sigma)$, and $\tau_{xy}(=0)$. In other words, $\xi_1 = -\xi$, $\xi_2 = 1$, $\xi_3 = 0$, and $\xi > 0$ means tensile loading.

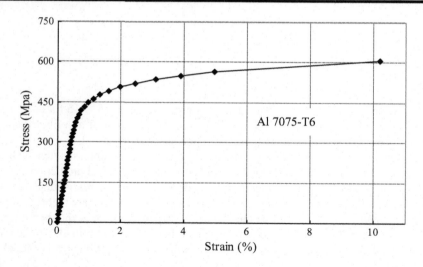

FIGURE 10.2

Stress–Strain Curve of Aluminum Alloy (Al 7075-T6)

For the constant biaxial edge loading, the instantaneous moduli D_{ij} in Eq. (10.7) is given by

$$D_{11} = \frac{4G(G+\lambda)-(G-G_t)[(2G+3\lambda)R_xR_x+4G/3]}{2G+\lambda-(G-G_t)(R_x+R_y)^2/3}$$

$$D_{16} = 0$$

$$D_{12}+2D_{66} = \frac{4G(G+\lambda)-(G-G_t)[(2G+3\lambda)R_xR_y-2G(R_x-R_y)^2+4G/3]}{2G+\lambda-(G-G_t)(R_x+R_y)^2/3}$$

$$D_{26} = 0$$

$$D_{22} = \frac{4G(G+\lambda)-(G-G_t)[(2G+3\lambda)R_yR_y+4G/3]}{2G+\lambda-(G-G_t)(R_x+R_y)^2/3}$$

(10.37)

And the instantaneous moduli D'_{ij} in Eq. (10.8) is given by

$$D'_{11} = \frac{4G_s(G_s+\lambda_s)-(G_s-G_t)[(2G_s+3\lambda_s)R_xR_y+4G_s/3]}{2G_s+\lambda_s-(G_s-G_t)(R_x+R_y)^2/3}$$

$$D'_{16} = 0$$

$$D'_{12}+2D'_{66} = \frac{4G_s(G_s+\lambda_s)-(G_s-G_t)[(2G_s+3\lambda_s)R_xR_y-2G_s(R_x-R_y)^2+4G_s/3]}{2G_s+\lambda_s-(G_s-G_t)(R_x+R_y)^2/3}$$

$$D'_{26} = 0$$

$$D'_{22} = \frac{4G_s(G_s+\lambda_s)-(G_s-G_t)[(2G_s+3\lambda_s)R_yR_y+4G_s/3]}{2G_s+\lambda_s-(G_s-G_t)(R_x+R_y)^2/3}$$

(10.38)

where G, G_s, and G_t are shear modulus, secant shear modulus, and tangential shear modulus; R_x and R_y are defined by

$$R_x = \frac{\xi}{\sqrt{1+\xi+\xi^2}}$$

$$R_y = \frac{-1}{\sqrt{1+\xi+\xi^2}}$$

(10.39)

In the DQ analysis, Grid III is used. The number of grid spacing is 15×15 for accuracy considerations. err $=1.0\times10^{-4}$. Since the problem is equivalent to the buckling of an orthotropic rectangular plate, MMWC-1 is used to apply the simply support boundary conditions.

Figure 10.3 shows the buckling coefficients under various biaxial compressive/tensile loadings with flow theory of plasticity [2]. Symbols are DQ data, and lines are theoretical predictions.

Figure 10.4 shows the buckling coefficients under various biaxial compressive/tensile loadings with deformation theory of plasticity [2]. Again symbols are DQ data, and lines are theoretical predictions.

From Figs. 10.3 and 10.4, it is seen that the DQ results are as accurate as the theoretical solutions. It was first pointed out by Durban and Zuckerman [1] that there exists an optimal loading path with the deformation theory model. Besides, the buckling coefficients with the J'_2 flow theory model are consistently higher than the ones with deformation theory.

For sufficiently thin plates ($\alpha = 0.0001$), the buckling coefficients obtained by the DQM are very close to the data obtained by the elastic buckling theory. Therefore, the difference of buckling coefficient

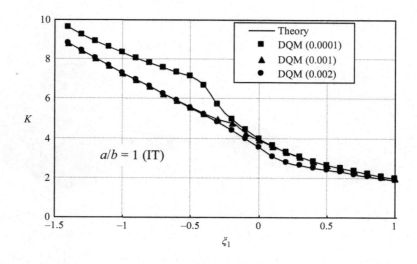

FIGURE 10.3

Influence of Thickness Ratio on K of an SSSS Square Plate

K obtained by the two theories is small over the entire range of loading parameter ξ_1. From Figs. 10.3 and 10.4, it is observed that the difference of buckling coefficient K obtained by the two theories is also small when the loading parameter $\xi_1 > -0.2$ for $\alpha = 0.001$ and $\xi_1 > 0.25$ for $\alpha = 0.002$, but increases dramatically with the decrease in loading parameter ξ_1.

The buckling modes of the SSSS rectangular plate at $\xi_1 = 1$ and $\xi_1 = -1$ are shown in Figs. 10.5 and 10.6. It is seen that under the same loading, the buckling mode is the same for the two plasticity theories. For different loading parameter ξ_1, the buckling mode is different.

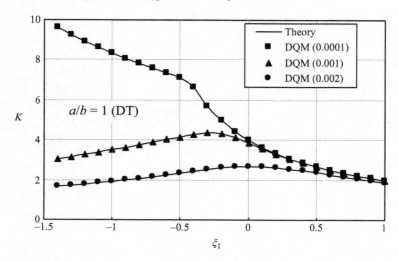

FIGURE 10.4

Influence of Thickness Ratio on K of an SSSS Square Plate

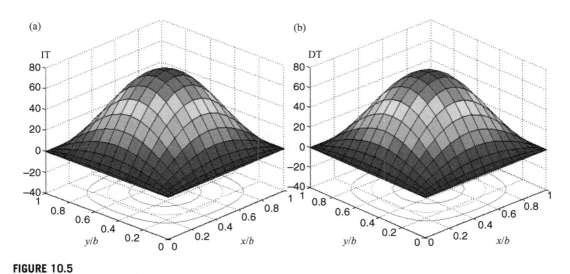

FIGURE 10.5

Buckling Mode of an SSSS Square Plate ($\xi_1 = 1$)

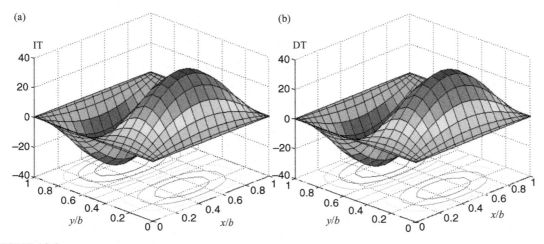

FIGURE 10.6

Buckling Mode of an SSSS Square Plate ($\xi_1 = -1$)

EXAMPLE 10.2

Consider next an isotropic square plate ($a = b$) with all edges clamped (CCCC). The thickness parameter α takes the same values as the ones in Example 10.1. The plate is under constant biaxial edge loading, and Poisson's ratio is 0.32.

In the DQ analysis, Grid III is used. The number of grid spacing is 15×15 for accuracy considerations. err $= 1.0 \times 10^{-4}$. Since the problem is equivalent to the buckling of an orthotropic rectangular plate, MMWC-3 is used to apply the clamped boundary conditions.

Figures 10.7 and 10.8 show the buckling coefficients under various biaxial compressive/tensile loadings with flow theory of plasticity and with deformation plasticity theory [2]. From Figs. 10.7 and 10.8, it can be seen that similar trends to the SSSS square plate exist. An optimal loading path is also shown in Fig. 10.8 for the deformation theory model. Since no analytical solutions are available and the DQ results are accurate enough, the DQ data may be useful for reference purposes.

Similar to the case of SSSS square plate, the buckling coefficients obtained by the DQM are very close to the results obtained by the elastic buckling theory for the sufficiently thin plate ($\alpha = 0.0001$). Therefore, the difference of buckling coefficient K obtained by the two theories is small over the entire range of loading parameter ξ_1.

The buckling modes of the CCCC rectangular plate at $\xi_1 = 1$ and $\xi_1 = -1$ are shown in Figs. 10.9 and 10.10. It is seen that under the same loading, the buckling mode is the same for the two different plasticity theories. For different loading parameter ξ_1, the buckling mode is different.

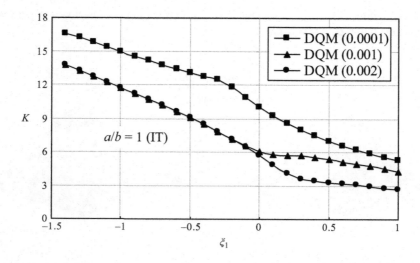

FIGURE 10.7

Influence of Thickness Ratio on K of a CCCC Square Plate

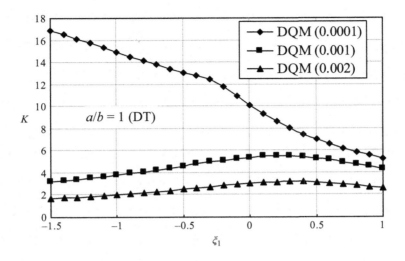

FIGURE 10.8

Influence of Thickness Ratio on K of a CCCC Square Plate

Results for square plates under biaxial edge loading with other combinations of boundary conditions, i.e., SCSC, CSCS, SSCC (or CCSS), and SCCC, can be found in Ref. [2].

Questions may arise as to why the difference is so large for positive loading in x direction ($\xi > 0$ or $\xi_1 < 0$) when $\alpha = 0.001$ and 0.002. Which one is the more reliable result?

To give a possible reason, one should check the assumptions made in deriving the differential equation. For small deformation, say $\varepsilon_{max} = 10.2\%$. Using Eq. (10.35) yields $\sigma_e = 608$ Mpa. Figure 10.11 shows the corresponding buckling coefficient K in the entire range of the loading parameter ξ_1 for $\alpha = 0.001$ and 0.002 when $\sigma_e = 608$ Mpa [2].

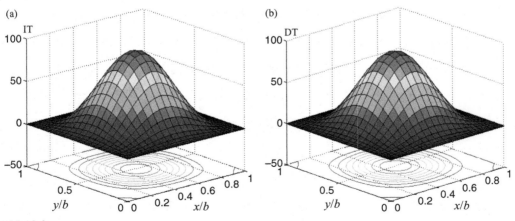

FIGURE 10.9

Buckling Mode of a CCCC Square Plate ($\xi_1 = 1$)

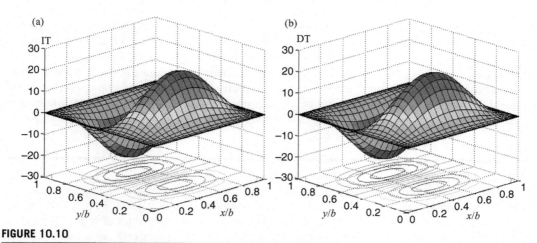

FIGURE 10.10

Buckling Mode of a CCCC Square Plate ($\xi_1 = -1$)

The corresponding buckling coefficient K of $\alpha = 0.0001$, not shown in Fig. 10.11, is 10 times that of $\alpha = 0.001$ [2]. Checking all results shown in Figs. 10.3, 10.4, 10.7, and 10.8 carefully reveals that all results based on the deformation theory model are smaller than the corresponding buckling coefficient K shown in Fig. 10.11. In other words, the small deformation assumption is always satisfied when deformation theory model is employed.

Only when $\alpha = 0.0001$, however, the results based on the J_2' flow theory model are smaller than the corresponding buckling coefficient K at $\sigma_e = 608$ Mpa in the entire range of the loading parameter ξ_1 and very close to the results obtained by the elastic buckling theory. When $\alpha = 0.001$ and

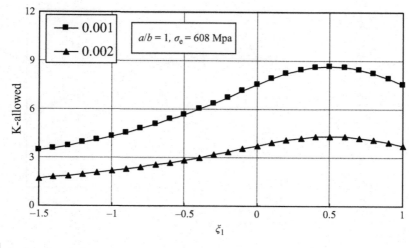

FIGURE 10.11

Maximum K Allowed When $\varepsilon_{max} = 10.2\%$

0.002, the results based on the J'_2 flow theory model are smaller than the corresponding buckling coefficient K at $\sigma_e = 608$ Mpa only in certain ranges of the loading parameter ξ_1, varied with the different combinations of boundary conditions. Within the valid range of the loading parameter ξ_1, the difference of buckling coefficients obtained by the two plasticity theories is small. If the discrepancy of buckling coefficients obtained by the two plasticity theories is large, the results obtained based on J'_2 flow theory model should not be trusted, since the assumption of small deformation is violated.

EXAMPLE 10.3

To investigate if free boundary condition will affect the results obtained by using the incremental theory or not, consider an isotropic square plate ($a = b$) with edges free at $x = 0$ and a, and simply supported at $y = 0$ and b (FSFS). The thickness parameter α takes the values of 0.0001, 0.001, and 0.002, which correspond to h/a of 0.0110, 0.0349, and 0.0493, respectively. All are within the range of classical thin plate theory ($h/a < 0.05$). The plate is under constant biaxial edge loadings, i.e., $\sigma_{yy}(=\sigma)$, $\sigma_{xx}(=-\xi\sigma)$, and $\tau_{xy}(=0)$, and Poisson's ratio is 0.32. In other words, $\xi_1 = -\xi$, $\xi_2 = 1$, and $\xi_3 = 0$.

Figures 10.12 and 10.13 show the buckling coefficients under various biaxial compressive/tensile loadings with using the flow theory of plasticity and the deformation plasticity theory [3]. Note that $\xi_1 = -\xi$ and $\xi_1 < 0$ means tensile loading in x direction. From Figs. 10.12 and 10.13, it can be seen that a different trend to the SSSS and CCCC square plates exists. The difference between IT and DT is

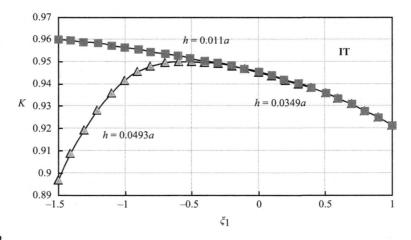

FIGURE 10.12

Influence of Thickness Ratio on K of a FSFS Square Plate

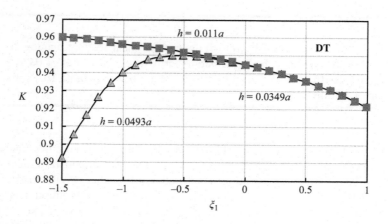

FIGURE 10.13

Influence of Thickness Ratio on K of a FSFS Square Plate

rather small. An optimal loading path is also shown in Figs. 10.12 and 10.13. For $\xi_1 > -0.4$ the buckling coefficient K for three thickness ratios is almost the same. The combinations of boundary conditions obviously affect the solutions for the IT model. Since no analytical solutions are available and the DQ results are accurate enough, the DQ data may be useful for reference purposes.

The buckling mode of the FSFS rectangular plate at $\xi_1 = 1$ is shown in Fig. 10.14 [3]. It is seen that under the same loading, the buckling mode is the same for the two different plasticity theories. This is similar to the cases of SSSS and CCCC square plates.

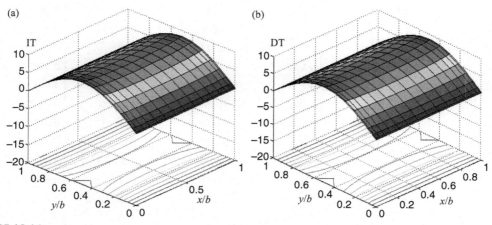

FIGURE 10.14

Buckling Mode of an FSFS Square Plate ($\xi_1 = 1$)

EXAMPLE 10.4

Consider last an isotropic square plate ($a = b$) under pure shear loading, namely, $\sigma_{xx} = \sigma_{yy} = 0$, $\tau_{xy} = \sigma$ ($\xi_1 = \xi_2 = 0$, $\xi_3 = 1$). The thickness parameter α takes the values of 0.0001, 0.001, and 0.002. The buckling coefficients obtained by the DQM are listed in Table 10.1 for six combinations of boundary conditions.

From Table 10.1, it is seen that for $h = 0.011a$, the results obtained by both flow theory of plasticity (IT) and deformation plasticity theory (DT) are almost the same. The effective strain at the buckling is small; thus, the small deformation assumption is satisfied. For the other two thicknesses, the buckling coefficients obtained by two theories are similar if the small deformation assumption is satisfied. When the small deformation assumption is violated, the IT predicts too high buckling loads; thus, the results should not be trusted. On the other hand, the results obtained by the deformation theory could be trusted since the small deformation assumption is always satisfied. When the small deformation assumption is satisfied and difference between results obtained by the two theories exists, verification may be necessary by experiments.

Table 10.1 Buckling Coefficients of Rectangular Plates Under Pure Shear Loading

Boundary Conditions	Thickness	K (IT)	ε_e (%)	K (DT)	ε_e (%)
SSSS	$h = 0.011a$	9.324	0.1789	9.324	0.1789
	$h = 0.0349a$	6.302	535.9	3.480	1.498
	$h = 0.0493a$	6.302	9.967E+05	1.923	3.138
CCCC	$h = 0.011a$	14.63	0.2809	14.63	0.2809
	$h = 0.0349a$	11.33	3.198 E+05	3.764	2.669
	$h = 0.0493a$	11.33	5.960E+08	2.046	5.500
CSCS	$h = 0.011a$	12.56	0.2411	12.56	0.2411
	$h = 0.0349a$	9.496	4.666 E+04	3.680	2.230
	$h = 0.0493a$	9.496	8.698 E+07	2.008	4.614
FSFS	$h = 0.011a$	4.190	0.08043	4.190	0.08043
	$h = 0.0349a$	2.868	0.6545	2.794	0.6153
	$h = 0.0493a$	2.045	5.474	1.629	1.020
CSFS	$h = 0.011a$	4.910	0.09425	4.910	0.09425
	$h = 0.0349a$	3.068	0.80202	2.903	0.6754
	$h = 0.0493a$	2.572	58.03	1.680	1.197
FCCC	$h = 0.011a$	8.440	0.1620	8.440	0.1620
	$h = 0.0349a$	5.533	130.5	3.375	1.243
	$h = 0.0493a$	5.530	2.398 E+05	1.872	2.509

10.3 ELASTOPLASTIC BUCKLING OF THICK RECTANGULAR PLATE
10.3.1 BASIC EQUATIONS OF THICK RECTANGULAR PLATE

For the elastoplastic buckling of thick plate, Mindlin plate theory is employed. Cartesian coordinate system (xyz) is set in the middle plane of the plate.

Let w denote the rate of deflection, and ϕ_x, ϕ_y the rate of rotation about the y and x axes. The geometric relation in rate form is given by

$$\dot{\varepsilon}_{xx} = z\frac{\partial \phi_x}{\partial x}$$

$$\dot{\varepsilon}_{yy} = z\frac{\partial \phi_y}{\partial y}$$

$$\dot{\gamma}_{xy} = z\left(\frac{\partial \phi_x}{\partial y} + \frac{\partial \phi_y}{\partial x}\right) \tag{10.40}$$

$$\dot{\gamma}_{xz} = \phi_x + \frac{\partial w}{\partial x}$$

$$\dot{\gamma}_{yz} = \phi_y + \frac{\partial w}{\partial y}$$

where $(\dot{\varepsilon}_{xx}, \dot{\varepsilon}_{yy}, \dot{\gamma}_{xy}, \dot{\gamma}_{xz}, \dot{\gamma}_{yz})$ are the strain rate components, and the over dot denotes differentiation with respect to a time-like parameter.

The rate of deflection w is assumed constant through the thickness. In order to avoid the thickness locking phenomenon, $\sigma_{zz} = 0$ is assumed.

Similar to the case of thin plate, two plasticity theories, namely, the J_2' flow theory of plasticity or the incremental theory of plasticity (IT) with Prandtl–Reuss constitutive equation and the deformation plasticity theory with Hencky constitutive equation (DT), are considered. For three dimensions, the rate forms of the two theories are shown in Eqs. (10.1) and (10.3), respectively.

The well-known three-parameter Ramberg–Osgood stress–strain relationship, slightly different from Eq. (10.35), is considered for comparisons with existing results in literature, namely,

$$\varepsilon_e = \frac{\sigma_e}{E} + \frac{k_1 \sigma_0}{E}\left(\frac{\sigma_e}{\sigma_0}\right)^c \tag{10.41}$$

where σ_0 is the nominal yield stress; $c\ (>1)$ is a nondimensional material parameter. $c = \infty$ for elastic perfect plastic materials; the meaning of k_1 is shown in Fig. 10.15 [3].

With Eq. (10.41), E_t and E_s can be computed by

$$\begin{cases} \dfrac{E}{E_t} = 1 + ck_1\left(\dfrac{\sigma_e}{\sigma_0}\right)^{c-1} \\[4mm] \dfrac{E}{E_s} = 1 + k_1\left(\dfrac{\sigma_e}{\sigma_0}\right)^{c-1} \end{cases} \tag{10.42}$$

where E is the Young's modulus of the isotropic material.

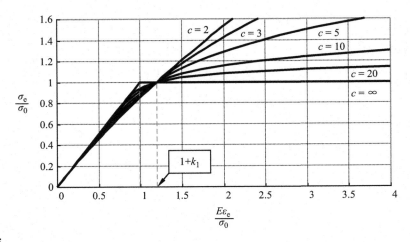

FIGURE 10.15

Three-Parameter Ramberg–Osgood Stress–Strain Curve

Comparing Eq. (10.42) to Eq. (10.35) reveals

$$k_2 = k_1 \left(\frac{E}{\sigma_0} \right)^{c-1} \tag{10.43}$$

Thus,

$$\frac{E}{E_t} = 1 + ck_2 \left(\frac{\sigma_e}{E} \right)^{c-1}, \quad \frac{E}{E_s} = 1 + k_2 \left(\frac{\sigma_e}{E} \right)^{c-1} \tag{10.44}$$

Since $\sigma_{zz} = 0$ is assumed and the general in-plane loading is considered, the incremental constitutive equation of IT in matrix form is given by

$$
\begin{Bmatrix} d\sigma_{xx} \\ d\sigma_{yy} \\ d\tau_{xy} \\ d\tau_{xz} \\ d\tau_{yz} \end{Bmatrix} =
\begin{bmatrix}
D_{11} & D_{12} & D_{16} & 0 & 0 \\
 & D_{22} & D_{26} & 0 & 0 \\
 & & D_{66} & 0 & 0 \\
 & & & \kappa^2 \cdot D_{44} & 0 \\
\text{sym} & & & & \kappa^2 \cdot D_{44}
\end{bmatrix}
\begin{Bmatrix} d\varepsilon_{xx} \\ d\varepsilon_{yy} \\ d\gamma_{xy} \\ d\gamma_{xz} \\ d\gamma_{yz} \end{Bmatrix} \tag{10.45}
$$

where κ^2 is the shear correction factor and is taken as 5/6. The stiffness matrix in Eq. (10.45) is obtained by the inverse of its corresponding flexural matrix, whose elements can be found in Refs. [3–5]. Thus, expressions for D_{ij} are not given.

Under the general in-plane loading, the incremental constitutive equation of DT in matrix form is given by

$$
\begin{Bmatrix} d\sigma_{xx} \\ d\sigma_{yy} \\ d\tau_{xy} \\ d\tau_{xz} \\ d\tau_{yz} \end{Bmatrix} =
\begin{bmatrix}
D'_{11} & D'_{12} & D'_{16} & 0 & 0 \\
 & D'_{22} & D'_{26} & 0 & 0 \\
 & & D'_{66} & 0 & 0 \\
 & & & \kappa^2 \cdot D'_{44} & 0 \\
\text{sym} & & & & \kappa^2 \cdot D'_{44}
\end{bmatrix}
\begin{Bmatrix} d\varepsilon_{xx} \\ d\varepsilon_{yy} \\ d\gamma_{xy} \\ d\gamma_{xz} \\ d\gamma_{yz} \end{Bmatrix} \tag{10.46}
$$

Note that D_{ij} and D'_{ij} depend on the applied in-plane loads. For general in-plane loading, D_{16}, D_{26}, D'_{16}, and D'_{26} are not zero. Thus, the material behaves like an anisotropic material.

Similar to the case of thin plate, assume that no unloading occurs at the instant of the plastic buckling. Take the IT as an example. The rate form of the governing equations for the elastoplastic buckling analysis of thick rectangular plate under general in-plane loading is given by

$$\kappa^2 D_{44}\left(\frac{\partial \phi_x}{\partial x} + \frac{\partial \phi_y}{\partial y} + \frac{\partial^2 w}{\partial x^2} + \frac{\partial^2 w}{\partial y^2}\right) = \sigma_{xx}\frac{\partial^2 w}{\partial x^2} + \sigma_{yy}\frac{\partial^2 w}{\partial y^2} + 2\tau_{xy}\frac{\partial^2 w}{\partial x \partial y} \tag{10.47}$$

$$\frac{h^2}{12}\left[D_{11}\frac{\partial^2 \phi_x}{\partial x^2} + D_{66}\frac{\partial^2 \phi_x}{\partial y^2} + 2D_{16}\frac{\partial^2 \phi_x}{\partial x \partial y} + D_{16}\frac{\partial^2 \phi_y}{\partial x^2} + D_{26}\frac{\partial^2 \phi_y}{\partial y^2}\right.$$
$$\left. + (D_{12} + D_{66})\frac{\partial^2 \phi_y}{\partial x \partial y}\right] - \kappa^2 D_{44}\left(\phi_x + \frac{\partial w}{\partial x}\right) = 0 \tag{10.48}$$

$$\frac{h^2}{12}\left[D_{16}\frac{\partial^2 \phi_x}{\partial x^2} + D_{26}\frac{\partial^2 \phi_x}{\partial y^2} + (D_{12} + D_{66})\frac{\partial^2 \phi_x}{\partial x \partial y} + D_{66}\frac{\partial^2 \phi_y}{\partial x^2}\right.$$
$$\left. + D_{22}\frac{\partial^2 \phi_y}{\partial y^2} + 2D_{26}\frac{\partial^2 \phi_y}{\partial x \partial y}\right] - \kappa^2 D_{44}\left(\phi_y + \frac{\partial w}{\partial y}\right) = 0 \tag{10.49}$$

Equations for generalized forces are defined as

$$Q_x = h\kappa^2 D_{44}\left(\phi_x + \frac{\partial w}{\partial x}\right) \tag{10.50}$$

$$Q_y = h\kappa^2 D_{44}\left(\phi_y + \frac{\partial w}{\partial y}\right) \tag{10.51}$$

$$M_{xx} = \frac{h^3}{12}\left(D_{11}\frac{\partial \phi_x}{\partial x} + D_{12}\frac{\partial \phi_y}{\partial y} + D_{16}\frac{\partial \phi_x}{\partial y} + D_{16}\frac{\partial \phi_y}{\partial x}\right) \tag{10.52}$$

$$M_{yy} = \frac{h^3}{12}\left(D_{12}\frac{\partial \phi_x}{\partial x} + D_{22}\frac{\partial \phi_y}{\partial y} + D_{26}\frac{\partial \phi_x}{\partial y} + D_{26}\frac{\partial \phi_y}{\partial x}\right) \tag{10.53}$$

$$M_{xy} = M_{yx} = \frac{h^3}{12}\left(D_{16}\frac{\partial \phi_x}{\partial x} + D_{26}\frac{\partial \phi_y}{\partial y} + D_{66}\frac{\partial \phi_x}{\partial y} + D_{66}\frac{\partial \phi_y}{\partial x}\right) \tag{10.54}$$

Boundary conditions are:

1. Simply supported boundary conditions (S):

$$w = 0, M_{xx} = 0, \phi_y = 0 \text{ at } x = 0 \text{ and/or } x = a \tag{10.55}$$

or

$$w = 0, M_{yy} = 0, \phi_x = 0 \text{ at } y = 0 \text{ and/or } y = b \tag{10.56}$$

2. Clamped boundary conditions (C):

$$w = 0, \ \phi_x = 0, \ \phi_y = 0 \text{ at } x = 0 \text{ and/or } x = a \tag{10.57}$$

or

$$w = 0, \ \phi_x = 0, \ \phi_y = 0 \text{ at } y = 0 \text{ and/or } y = b \tag{10.58}$$

3. Free boundary conditions (F):

$$M_{xx} = 0, \ M_{yx} = 0, \ Q_x - h\sigma_{xx}\frac{\partial w}{\partial x} - h\tau_{xy}\frac{\partial w}{\partial y} = 0 \text{ at } x = 0 \text{ and/or } x = a \tag{10.59}$$

or

$$M_{yy} = 0, \ M_{xy} = 0, \ Q_y - h\sigma_{yy}\frac{\partial w}{\partial y} - h\tau_{xy}\frac{\partial w}{\partial x} = 0 \text{ at } y = 0 \text{ and/or } y = b \tag{10.60}$$

10.3.2 SOLUTION PROCEDURES BY THE DQM

Since analytical solutions are available only for a few special combinations of boundary conditions, the differential quadrature method (DQM) can be used for solutions of rectangular plate with general combinations of boundary conditions. Different from the thin plate case, no additional degree is required for applying the boundary conditions. Thus, the formulations and implementation of the DQM are much easier.

Let the rectangular domain be discrete into $N \times N$ grid points. Three degrees of freedom (DOFs), namely, w, ϕ_x, ϕ_y, are assigned to each grid point. In terms of the differential quadrature, the governing equations (10.47)–(10.49) at all inner grid points can be expressed as

$$\kappa^2 D_{44}\left(\sum_{k=1}^{N} A_{ik}^x \phi_{xkj} + \sum_{l=1}^{N} A_{jl}^y \phi_{yil} + \sum_{k=1}^{N} B_{ik}^x w_{kj} + \sum_{l=1}^{N} B_{jl}^y w_{il}\right)$$
$$= \sigma\left(\xi_1 \sum_{k=1}^{N} B_{ik}^x w_{kj} + \xi_2 \sum_{l=1}^{N} B_{jl}^y w_{il} + 2\xi_3 \sum_{k=1}^{N}\sum_{l=1}^{N} A_{ik}^x A_{jl}^y w_{kl}\right) \tag{10.61}$$

$$\frac{h^2}{12}\left[D_{11}\sum_{k=1}^{N} B_{ik}^x \phi_{xkj} + D_{66}\sum_{l=1}^{N} B_{jl}^y \phi_{xil} + 2D_{16}\sum_{k=1}^{N}\sum_{l=1}^{N} A_{ik}^x A_{jl}^y \phi_{xkl} + D_{16}\sum_{k=1}^{N} B_{ik}^x \phi_{ykj} \right.$$
$$\left. + D_{26}\sum_{l=1}^{N} B_{jl}^y \phi_{yil} + (D_{12} + D_{66})\sum_{k=1}^{N}\sum_{l=1}^{N} A_{ik}^x A_{jl}^y \phi_{ykl}\right] - \kappa^2 D_{44}\left(\phi_{xij} + \sum_{k=1}^{N} A_{ik}^x w_{kj}\right) = 0 \tag{10.62}$$

$$\frac{h^2}{12}\left[D_{16}\sum_{k=1}^{N} B_{ik}^x \phi_{xkj} + D_{26}\sum_{l=1}^{N} B_{jl}^y \phi_{xil} + (D_{12} + D_{66})\sum_{k=1}^{N}\sum_{l=1}^{N} A_{ik}^x A_{jl}^y \phi_{xkl} + D_{66}\sum_{k=1}^{N} B_{ik}^x \phi_{ykj} \right.$$
$$\left. + D_{22}\sum_{l=1}^{N} B_{jl}^y \phi_{yil} + 2D_{26}\sum_{k=1}^{N}\sum_{l=1}^{N} A_{ik}^x A_{jl}^y \phi_{ykl}\right] - \kappa^2 D_{44}\left(\phi_{yij} + \sum_{l=1}^{N} A_{jl}^y w_{il}\right) = 0 \tag{10.63}$$
$$(i = 2, 3, ..., N-1, j = 2, 3, ..., N-1)$$

where A_{ij}^x, B_{ij}^x are the weighting coefficients of the first- and the second-order derivatives with respect to x, A_{ij}^y, B_{ij}^y are the weighting coefficients of the first- and the second-order derivatives with respect to y, $\sigma_{xx} = \xi_1\sigma$, $\sigma_{yy} = \xi_2\sigma$, $\tau_{xy} = \xi_3\sigma$, where $\sigma(> 0)$ is the buckling coefficient.

The generalized forces, i.e., Eqs. (10.50)–(10.54), can be expressed in terms of the differential quadrature as

$$(Q_x)_{ij} = h\kappa^2 D_{44}\left(\phi_{xij} + \sum_{k=1}^{N} A_{ik}^x w_{kj}\right) \tag{10.64}$$

$$(Q_y)_{ij} = h\kappa^2 D_{44}\left(\phi_{yij} + \sum_{l=1}^{N} A_{jl}^y w_{il}\right) \tag{10.65}$$

$$(M_{xx})_{ij} = \frac{h^3}{12}\left(D_{11}\sum_{k=1}^{N} A_{ik}^x \phi_{xkj} + D_{12}\sum_{l=1}^{N} A_{jl}^y \phi_{yil} + D_{16}\sum_{l=1}^{N} A_{jl}^y \phi_{xil} + D_{16}\sum_{k=1}^{N} A_{ik}^x \phi_{ykj}\right) \tag{10.66}$$

$$(M_{yy})_{ij} = \frac{h^3}{12}\left(D_{12}\sum_{k=1}^{N} A_{ik}^x \phi_{xkj} + D_{22}\sum_{l=1}^{N} A_{jl}^y \phi_{yil} + D_{26}\sum_{l=1}^{N} A_{jl}^y \phi_{xil} + D_{26}\sum_{k=1}^{N} A_{ik}^x \phi_{ykj}\right) \tag{10.67}$$

$$(M_{xy})_{ij} = (M_{yx})_{ij} = \frac{h^3}{12}\left(D_{16}\sum_{k=1}^{N} A_{ik}^x \phi_{xkj} + D_{26}\sum_{l=1}^{N} A_{jl}^y \phi_{yil} + D_{66}\sum_{l=1}^{N} A_{jl}^y \phi_{xil} + D_{66}\sum_{k=1}^{N} A_{ik}^x \phi_{ykj}\right) \tag{10.68}$$

At position corresponding to the DOFs along edges $x = 0$ and $x = a$, namely, $(w)_{x=0}$, $(\phi_x)_{x=0}$, $(\phi_y)_{x=0}$, and $(w)_{x=a}$, $(\phi_x)_{x=a}$, $(\phi_y)_{x=a}$, Eqs. (10.64), (10.66), and (10.68) are placed. In other words, the subscripts i and j in $(Q_x)_{ij}, (M_{xx})_{ij}, (M_{xy})_{ij}$ are $i = 2,3,\dots,N-1$, and $j = 1, N$.

Similarly, at position corresponding to the DOFs along edges $y = 0$ and $y = b$, namely, $(w)_{y=0}$, $(\phi_x)_{y=0}$, $(\phi_y)_{y=0}$, and $(w)_{y=b}$, $(\phi_x)_{y=b}$, $(\phi_y)_{y=b}$, Eqs. (10.65), (10.68), and (10.67) are placed. In other words, the subscripts i and j in $(Q_y)_{ij}, (M_{xy})_{ij}, (M_{yy})_{ij}$ are $i = 1, N$, and $j = 2,3,\dots,N-1$.

For the four corner points, $(2M_{xy})_{ij}, (M_{xx})_{ij}, (M_{yy})_{ij}$ are placed at $w_{ij}, (\phi_x)_{ij}, (\phi_y)_{ij}$ ($i = 1,N; j = 1,N$), respectively.

For the elastoplastic buckling analysis of a rectangular plate by using the DQM, appropriate boundary equations should be applied. For clamped boundaries, $w_{ij} = 0$, $\phi_{xij} = 0$, and $\phi_{yij} = 0$ should be applied, where ij represents all grid points located on the clamped boundaries. For simply supported boundaries, besides applying $w_{ij} = 0$ and $\phi_{yij} = 0$ or $w_{ij} = 0$ and $\phi_{xij} = 0$, either $(M_{xx})_{ij} = 0$ or $(M_{yy})_{ij} = 0$ should be applied, where ij represents all grid points located on the simply supported boundaries. For free boundaries, either $(M_{xx})_{ij} = 0$, $(M_{yy})_{ij} = 0$, $(Q_x)_{ij} = 0$ or $(M_{xy})_{ij} = 0$, $(M_{yy})_{ij} = 0$, $(Q_y)_{ij} = 0$ should be applied, where $(Q_x)_{ij}$, $(Q_y)_{ij}$, $(M_{xx})_{ij}$, $(M_{yy})_{ij}$, $(M_{xy})_{ij}$ are formulated by using Eqs. (10.64)–(10.68) and ij represents all grid points located on the free boundaries. For free corners, $(2M_{xy})_{ij} = (M_{xx})_{ij} = (M_{yy})_{ij} = 0$ should be enforced, where ij represents all free corner points.

In terms of the differential quadrature, the governing equations at all inner grid points and generalized force equations at all boundary points can be written in the following partitioned matrix form,

$$\begin{bmatrix} F_{11} & F_{12} \\ F_{21} & F_{22} \end{bmatrix} \{\Delta\} = \sigma \begin{bmatrix} M_{11} & 0 \\ 0 & 0 \end{bmatrix} \{\Delta\} \tag{10.69}$$

where $\{\Delta\}^T = \lfloor w_{11}, w_{12},\dots, w_{NN}, \phi_{x11}, \phi_{x12},\dots\phi_{xNN}, \phi_{y11}, \phi_{y12},\dots\phi_{yNN} \rfloor$.

After eliminating the zero w, ϕ_x, or ϕ_y on the boundary points, Eq. (10.69) can be rewritten by

$$\begin{bmatrix} F'_{11} & F'_{12} \\ F'_{21} & F'_{22} \end{bmatrix}\{\bar{\Delta}\} = \sigma \begin{bmatrix} M'_{11} & 0 \\ 0 & 0 \end{bmatrix}\{\bar{\Delta}\} \text{ or } [F']\{\bar{\Delta}\} = \sigma[M']\{\bar{\Delta}\} \tag{10.70}$$

Note that the dimension of $[F']$ and $[M']$ may be smaller than $3N^2 \times 3N^2$ depending on the boundary conditions. After some simple manipulations, Eq. (10.70) becomes

$$([F'_{11}] - [F'_{12}][F'_{22}]^{-1}[F'_{21}])\{w\}^T = \sigma[M'_{11}]\{w\}^T \tag{10.71}$$

or

$$[\bar{F}]\{\bar{w}\} = \sigma[M'_{11}]\{\bar{w}\} \tag{10.72}$$

Equation (10.72) is a nonlinear generalized eigenvalue equation, since D_{ij} or D'_{ij} depends on the unknown load parameter σ. The solution procedures are similar to the ones of the thin plate. Direct iteration method can be used for the solution to Eq. (10.72).

To start with, set $\sigma = P_0$; P_0 should be small enough not to cause any plastic deformation. Next, compute D_{ij} or D'_{ij} according to the two plasticity theories. Once D_{ij} or D'_{ij} are obtained, Eq. (10.72) becomes a generalized eigenvalue equation that can be solved by existing generalized eigensolvers. The lowest eigenvalue corresponding to a positive σ is denoted by P_1. Check if $|(P_1 - P_0)/P_1| <$ err, where err is the prescribed error bound. If $|(P_1 - P_0)/P_1| >$ err, replace P_0 by $[(1-\eta)P_0 + \eta P_1](\eta \le 1)$. Repeat the iteration processes until $|(P_i - P_{i-1})P_i| <$ err; then P_i is the solution to be found.

10.3.3 **EXAMPLES AND DISCUSSION**

EXAMPLE 10.5

For verifications, consider first the buckling of square thick plate ($a/b = 1$) with all edges simply supported (SSSS). Material parameters in the Ramberg–Osgood relation are $E/\sigma_0 = 750$, $k_1 = 0.25$, $\mu = 0.3$, and c takes the values of 2, 3, and 20, respectively.

Figure 10.16 shows the buckling stress factor K versus the thickness to side ratio h/a for the plate under equibiaxial load ($\xi_1 = \xi_2 = 1$, $\xi_3 = 0$). K is defined by Eq. (10.36). Symbols are DQ data, and lines are theoretical predictions cited from Ref. [5]. In the DQ analysis, Grid III is used and the grid spacing is 21×21 for accuracy considerations.

As is expected, the DQ results are as accurate as the theoretical solutions. Thus, the formulations, solution procedures, and developed computer programs are verified. Similar to the thin plates cases, difference between the incremental theory and deformation theory is observed. The difference becomes larger with the increase in c and thickness to side ratio h/a.

To investigate the effect of transverse shear on the buckling load, the same problem investigated by using the thin plate theory, i.e., Example 10.1, is reanalyzed by using the thick plate theory. For comparisons, the thickness parameter α takes the values of 0.0001, 0.001, and 0.002, which correspond to h/a of 0.0110, 0.0349, and 0.0493, respectively. The plate is under constant biaxial edge loadings, i.e., $\sigma_{yy}(=\sigma)$, $\sigma_{xx}(=-\xi\sigma)$, and $\tau_{xy}(=0)$. In other words, $\xi_1 = -\xi$, $\xi_2 = 1$, and $\xi_3 = 0$. The material constants E, μ, k_2, and c are 72.4 Gpa, 0.32, 3.94×10^{21}, and 10.9, respectively.

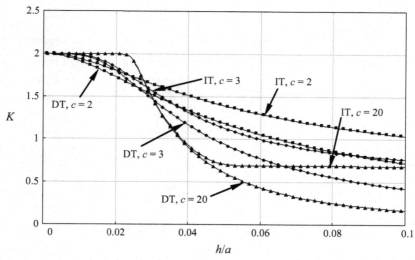

FIGURE 10.16

Buckling Stress Factors for SSSS Square Plates Under Equibiaxial Load [4]

Table 10.2 lists the DQ results by using both thin and thick plate theories. It is seen that the results obtained by the thick plate theory are slightly smaller than the ones by thin plate theory. However, the difference between the thin and thick theories is small, since the three thickness are all within the range of classical thin plate theory ($h/a < 0.05$).

EXAMPLE 10.6

Consider the buckling of square thick plate ($a/b = 1$) under general edge loadings. Material parameters in the Ramberg–Osgood relation are $k_1 = 0.25$, $\mu = 0.3$, $c = 10$, and $E/\sigma_0 = 750$. $h = 0.075a$ is considered. Three combinations of boundary conditions are investigated, namely, SSSS, CCCC, and FCSC.

In the DQ analysis, Grid III is used and the grid spacing is 21×21 for accuracy considerations. For presentation, three nondimensional parameters R_1, R_2, and R_3, defined by Eq. (10.73), are introduced.

$$
\begin{aligned}
R_1 &= \frac{\sigma_{xxcr}}{\sigma_{xx0}} \\
R_2 &= \frac{\sigma_{yycr}}{\sigma_{yy0}} \\
R_3 &= \frac{\tau_{xycr}}{\tau_{xy0}}
\end{aligned}
\qquad (10.73)
$$

where σ_{xxcr}, σ_{yycr}, and τ_{xycr} are buckling stress components under general edge loadings, σ_{xx0}, σ_{yy0}, and τ_{xy0} are buckling load under uniaxial loading in the x and y direction and pure shear loading, respectively.

Table 10.2 Comparison of DQ Results by Using Thin and Thick Plate Theories

Thickness	ξ_1	Plasticity Theory	Thin Plate Theory	Thick Plate Theory	Maximum Difference (%)
$h = 0.011a$	−1.5	IT	9.999	9.982	0.1755
	−1.5	DT	9.999	9.981	0.1754
$h = 0.0349a$	−1.5	IT	9.253	9.101	1.666
	1.0	DT	1.999	1.985	0.7062
$h = 0.0493a$	−1.5	IT	9.253	8.955	3.325
	1.0	DT	1.881	1.865	0.8653

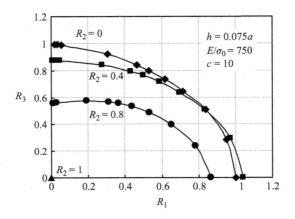

FIGURE 10.17

Buckling Stress Parameters of SSSS Square Plates ($R_2 = 0, 0.4, 0.8, 1$)

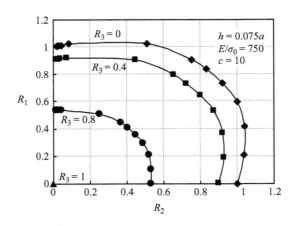

FIGURE 10.18

Buckling Stress Parameters of SSSS Square Plates ($R_3 = 0, 0.4, 0.8, 1$)

Since the results obtained by the deformation theory of plasticity are more reliable than the ones by the incremental plasticity theory; thus, only results employing deformation theory of plasticity are presented in Figs. 10.17–10.22.

Figures 10.17 and 10.18 show the results for SSSS square plates. In Fig. 10.17, each curve is for a fixed R_2. In Fig. 10.18, each curve is for a fixed R_3.

Figures 10.19 and 10.20 show the results for CCCC square plates. In Fig. 10.19, each curve is for a fixed R_2. In Fig. 10.20, each curve is for a fixed R_3. Similar variations as for the SSSS square plates are observed.

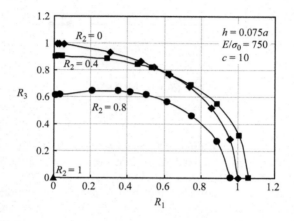

FIGURE 10.19

Buckling Stress Parameters of CCCC Square Plates ($R_2 = 0, 0.4, 0.8, 1$)

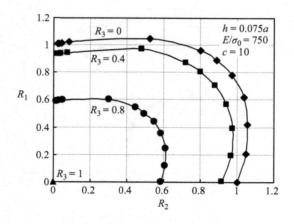

FIGURE 10.20

Buckling Stress Parameters of CCCC Square Plates ($R_3 = 0, 0.4, 0.8, 1$)

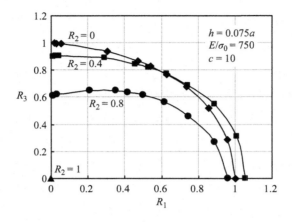

FIGURE 10.21

Buckling Stress Parameters of FCFC Square Plates ($R_2 = 0$, 0.4, 0.8, 1)

Figures 10.21 and 10.22 show the results for FCFC square plates. In Fig. 10.21, each curve is for a fixed R_2. In Fig. 10.22, each curve is for a fixed R_3. Similar variations as for the SSSS and CCCC square plates are observed.

Since analytical solutions are not available and the DQ results are accurate enough, the DQ data shown in Figs. 10.17–10.22 may be useful for reference purposes. More results can be found in Refs. [3–4].

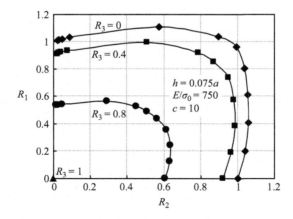

FIGURE 10.22

Buckling Stress Parameters of FCFC Square Plates ($R_3 = 0$, 0.4, 0.8, 1)

10.4 SUMMARY

In this chapter, the DQM is used to successfully perform the elastoplastic buckling analysis for both thin and thick plates under various combined edge loadings. Ramberg–Osgood material is considered. Two plasticity theories are used: one is the incremental plasticity (IT) with Prandtl–Reuss constitutive equation and the other is the deformation plasticity (DT) with Hencky constitutive equation. Several examples are investigated.

Numerical results show that the DQM can yield accurate results with relatively small number of grid points. Thus, the attractive features of rapid convergence, high accuracy, and computational efficiency of the DQM are illustrated. It is pointed out for the first time that the assumption of small deformation is violated by using the IT if the difference of the results obtained by the IT and the DT is large. Thus, in such cases the results obtained by using the IT should not be trusted. Some new results are also presented for rectangular plates under various combined loadings and can be used for references.

REFERENCES

[1] D. Durban, Z. Zuckerman, Elastoplastic buckling of rectangular plates in biaxial compression/tension, Int. J. Mech. Sci. 41 (1999) 751–765.
[2] X. Wang, J. Huang, Elastoplastic buckling analyses of rectangular plates under biaxial loadings by the differential quadrature method, Thin Wall. Struct. 47 (2009) 12–20.
[3] W. Zhang, Elastoplastic buckling analysis of rectangular plates by using the differential quadrature method. Master Thesis. Nanjing University of Aeronautics and Astronautics, China, 2010 (in Chinese).
[4] W. Zhang, X. Wang, Elastoplastic buckling analysis of thick rectangular plates by using the differential quadrature method, Comput. Math. Appl. 61 (2011) 44–61.
[5] C.M. Wang, Y. Xiang, J. Charkrabarty, Elastic/plastic buckling of thick plates, Int. J. Solids Struct. 38 (2001) 8617–8640.

11

STRUCTURAL ANALYSIS BY THE QEM

11.1 INTRODUCTION

This chapter demonstrates the dynamic analysis, free vibration, and buckling analysis by using the QEM. Two kinds of quadrature elements, i.e., quadrature bar element and quadrature thin plate element, are used. Formulating an N-node quadrature element is simple and flexible and details on its formulation are worked out. A few applications are given for illustration.

11.2 DYNAMIC ANALYSIS OF A FLEXIBLE ROD HIT BY RIGID BALL

A cantilever flexible rod hit by a rigid ball with velocity V at its free end is schematically shown in Fig. 11.1. The change in cross-sectional area due to Poisson's effect is neglected; thus, the equation of motion that governs the propagation of axial elastic waves in a bar is given by

$$EA\frac{\partial^2 u(x,t)}{\partial x^2} + p(x,t) = \rho A \frac{\partial^2 u(x,t)}{\partial t^2} \tag{11.1}$$

where E and ρ are Young's modulus and mass density of the bar material, L and A are the length and cross-sectional area of the bar, respectively, $u(x,t)$ is the displacement, $p(x,t)$ is the distributed axial force per unit length, x is the Cartesian coordinate located in the center axis of the bar, and t is the time.

Boundary conditions are $\partial u(0,t)/\partial x = f(t)/EA$ and $u(L,t) = 0$, where $f(t) = m\ddot{u}(0,t)$ during impact, $f(t) = 0$ during the separation of the rigid ball from the flexible bar, and m is the mass of the rigid ball. The initial conditions are $u(x,0) = \dot{u}(x,0) = 0$, where the over dot denotes the first-order derivative with respect to time t.

An eight-node quadrature bar element, shown in Fig. 11.2, is used in the analysis. GLL points are used as the element nodes, and GLL quadrature is used to obtain the stiffness matrix $[k]$ and the mass matrix $[m]$, which can be obtained by using Eq. (4.6) and Eq. (4.7), namely,

$$\begin{cases} k_{ij} = \dfrac{2EA}{l}\displaystyle\sum_{k=1}^{8} H_k A_{ki} A_{kj} \\ m_{ij} = \dfrac{\rho Al}{2}\displaystyle\sum_{k=1}^{8} H_k \delta_{ik}\delta_{jk} = \dfrac{\rho Al}{2}H_j\delta_{ij} \end{cases} \quad (i,j=1,2,...,8) \tag{11.2}$$

where ξ_i and H_i are abscissas and weights of the 8-point GLL quadrature, which can be found in Appendix I, A_{ik} or A_{jk} are the weighting coefficients of the first-order derivative with respect to ξ in the ordinary DQM, computed by Eq. (1.29), and l is the element length. Due to the usage of the GLL nodes and GLL quadrature, the element mass matrix is in diagonal form.

Differential Quadrature and Differential Quadrature Based Element Methods. 978-0-12-803081-3

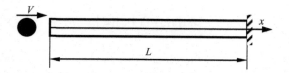

FIGURE 11.1

Axial Impact of a Cantilever Bar by Rigid Ball

If $p(x,t)$ is not zero, then the work equivalent force vector is given by

$$F(x_i,t) = F_i(t) = \int_{-1}^{1} \frac{l}{2} p(\xi,t) l_i(\xi) \, d\xi = \frac{l}{2} \sum_{k=1}^{8} H_k p(\xi_k,t) l_i(\xi_k)$$
$$= \frac{l}{2} \sum_{k=1}^{N} H_k p(\xi_k,t) \delta_{ki} = \frac{l}{2} H_i p(\xi_i,t) \qquad (i=1,2,...,8) \tag{11.3}$$

The assemblage procedures are exactly the same as the ones in conventional finite element method or the time domain spectral element method. After assemblage, the following matrix equation is obtained,

$$[M]\{\ddot{u}(t)\} + [K]\{u(t)\} = \{F(t)\} \tag{11.4}$$

where $[M]$ and $[K]$ are the structural mass matrix and stiffness matrix, $\{F\}$ is the applied force vector, and the double over dots denote the second-order derivative with respect to time t.

Due to the diagonal form of the structural mass matrix $[M]$, the set of the second-order ordinary differential equations in time can be conveniently integrated by using the central finite difference method. Detailed procedures are given below for readers' reference.

Let $v(t) = \dot{u}(x,t)$ be the velocity and $a(t) = \ddot{u}(x,t)$ the acceleration. To start with, compute $\{u\}_{-1}$ first by using the given initial conditions, namely,

$$\{u\}_{-1} = \{u\}_0 - \Delta t \{v\}_0 + \Delta t^2 / 2 \{a\}_0 \tag{11.5}$$

where Δt is the time increment, and $\{u\}_0$ and $\{v\}_0$ are the known initial displacement and velocity; the initial acceleration $\{a\}_0$ can be determined by the equation of motion.

1. Compute the "effective load" at time t_n, namely,

$$\{\bar{F}\}_n = \{F\}_n - [K]\{u\}_n \tag{11.6}$$

where $[K]$ is the banded structural stiffness matrix. In programming, this banded property should be used to save the CPU time. Alternatively, $[K]\{u\}_n$ can be formulated by summing the nodal forces element by element, namely,

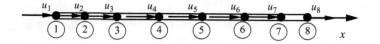

FIGURE 11.2

An Eight-Node Bar Quadrature Element

$$[K]\{u\}_n = \sum_{i=1}^{N}[k_i]\{u_i\}_n \qquad (11.7)$$

where $[k_i]$ is the stiffness matrix of the ith element and $\{u_i\}_n$ is the nodal displacement vector of the ith element at time t_n. Therefore, the formulation of the structural stiffness matrix $[K]$ is not needed and the programming is much simpler.

2. Determine the displacement at time t_{n+1}, namely,

$$\{u\}_{n+1} = \Delta t^2 [M]^{-1}\{\overline{F}\}_n + 2\{u\}_n - \{u\}_{n-1} \qquad (11.8)$$

where $[M]^{-1}$ can be easily obtained, since $[M]$ is a diagonal matrix.

3. Compute the acceleration and velocity at time t_n, namely,

$$\{a\}_n = \Delta t^{-2}\left(\{u\}_{n-1} - 2\{u\}_n + \{u\}_{n+1}\right) \qquad (11.9)$$

$$\{v\}_n = \left(-\{u\}_{n-1} + \{u\}_{n+1}\right)/(2\Delta t) \qquad (11.10)$$

Repeat steps 1–3 until the time reaches the specified value.

For illustration and comparison, the material and geometrical parameters are chosen as Ref. [1]: $E = 2.1 \times 10^{11}$ N/ m^2, $\rho = 7.9 \times 10^3$ kg/ m^3, $L = 0.3$ m, $A = 9 \times 10^{-4}$ m^2, $m = 2.133$ kg. The initial velocity of the rigid ball is $V = 1.0$ m/ s. One hundred eight-node quadrature bar elements are used to model the entire bar. To ensure the stable time integration by using the central finite difference method, the time increment Δt is taken as 10^{-8} s in the numerical simulations.

Figure 11.3 shows the comparison of contact stress time-history at the free end obtained by the QEM to the analytical solutions. From Fig. 11.3, it is seen that although the average results are closed

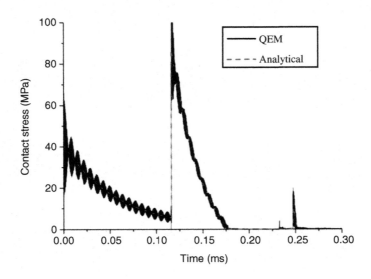

FIGURE 11.3

The Contact Stress Time-History by the QEM

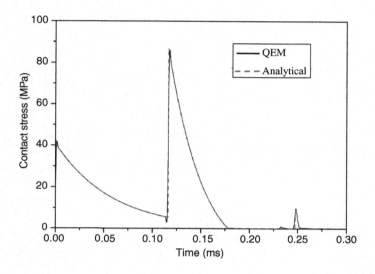

FIGURE 11.4

The Filtered Contact Stress Time-History by the QEM

to the analytical solution and the second impact instant is correctly captured, the simulations are polluted by the computational noise.

The simulated data is then filtered by using a Butterworth filter. The results are shown in Fig. 11.4. It is seen that the agreement with the analytical solution is very good. According to the analytical solution, the first impact ends at time of 0.1785 ms and the second impact starts at time of 0.2476 ms. The numerical simulations, obtained by the QEM, are that the first impact ends at time of 0.1767 ms and the second impact starts at time of 0.2487 ms. The relative percentage error is about −1.0 and 0.4%, respectively [1].

The time history of the displacement at the contact point ($x = 0$) is shown in Fig. 11.5. It is seen that numerical results obtained by the QEM agree very well with the analytical solution. The displacement is not as sensitive as the contact stress or acceleration, since it is obtained by double integrations of the acceleration.

11.3 FREE VIBRATION OF THIN PLATES

This section demonstrates the free vibration analysis by using the QEM. Figure 11.6 shows a 25-node quadrature rectangular plate element. The total number of degrees of freedom (DOFs) for the 25-node element is either 45 or 49 depending on the shape functions. For example, if Lagrange interpolation functions are used as the shape functions, the total number of DOF is 45. Details may be found in Chapter 4.

Since the QEM is similar to the higher-order finite element method, the assemblage and solution procedures are exactly the same as the conventional finite element method. For simplicity in presentation, only one quadrature rectangular plate element is used in the analysis; thus, the assemblage is not needed.

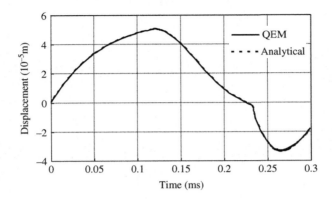

FIGURE 11.5

Displacement History at the Free End of the Cantilever Bar

If only one quadrature rectangular plate element is used, the matrix equation for free vibration of plates by the QEM is given by

$$
\begin{bmatrix} K_{bb} & K_{bi} \\ K_{ib} & K_{ii} \end{bmatrix} \begin{Bmatrix} \{\Delta_b\} \\ \{W_i\} \end{Bmatrix} = \begin{Bmatrix} \{F_b\} \\ \lambda[I]\{W_i\} \end{Bmatrix}
\tag{11.11}
$$

where subscripts b and i denote the quantities related to the derivative DOF at boundary points and displacement at all grid points, $\{F_b\}$ is generalized bending moment vector and $\lambda = \rho h \omega^2$.

Although the form of Eq. (11.11) is the same as Eq. (8.19) and also contains $(N+2)\times(N+2)-4$ equations, some differences exist. The element stiffness matrix is obtained either by Eq. (4.109) or by Eq. (4.129) and the element mass matrix is obtained either by Eq. (4.111) or by Eq. (4.131). Since the diagonal terms in the mass matrix are not the same, the corresponding row of the stiffness matrix in

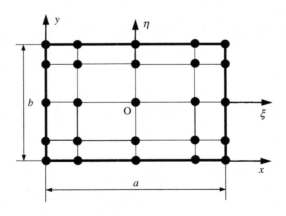

FIGURE 11.6

Sketch of a 25-Node Quadrature Rectangular Plate Element

Eq. (11.11) has been divided by the diagonal term to make the equation a standard eigenvalue matrix equation. Besides, $\{\Delta_b\}$ contains only the derivative DOF. In other words, the dimension of $\{\Delta_b\}$ is $4N \times 1$ and the dimension of $\{W_i\}$ is $N^2 \times 1$. Note that the modified stiffness matrix in Eq. (11.11) is no longer symmetric.

Similar to the finite element method, any combinations of boundary conditions can be easily applied. Only the essential boundary conditions need to be specified. In other words, only the zero deflection or the zero slope at the boundary points needs to be specified. If the slope is zero, the corresponding element in $\{F_b\}$ is unknown, otherwise the element in $\{F_b\}$ is zero for free vibration analysis if the corresponding derivative DOF in $\{\Delta_b\}$ is unknown. Applying the zero deflection and the zero slope boundary conditions is also easy and done by simply eliminating the corresponding rows and columns. After imposing the essential boundary conditions, the equation of rectangular plate for free vibration analysis becomes

$$\begin{bmatrix} \bar{K}_{bb} & \bar{K}_{bi} \\ \bar{K}_{ib} & \bar{K}_{ii} \end{bmatrix} \begin{Bmatrix} \{\bar{\Delta}_b\} \\ \{\bar{W}_i\} \end{Bmatrix} = \begin{Bmatrix} \{0\} \\ \lambda[I]\{\bar{W}_i\} \end{Bmatrix} \tag{11.12}$$

where the over bar means that the quantities have been modified, since the zero generalized displacements have been removed.

After eliminating the nonzero $\{\bar{\Delta}_b\}$, Eq. (11.12) can be rewritten as

$$\left[\bar{K}_{ii} - \bar{K}_{ib}\bar{K}_{bb}^{-1}\bar{K}_{bi} \right]\{\bar{W}_i\} = \lambda[I]\{\bar{W}_i\} \tag{11.13}$$

Equation (11.13) can be further simplified to

$$\left[\bar{K} \right]\{\bar{W}_i\} = \lambda[I]\{\bar{W}_i\} \tag{11.14}$$

Note that Eq. (11.12) is slightly different from Eq. (8.20) for rectangular plate with free edges, i.e., $\{\bar{W}_i\}$ contains not only the deflection at all inner nodal points but also the deflection at all boundary nodal points on the free edges. A standard eigensolver can be used for obtaining the frequencies and mode shapes.

A FORTRAN program and its converted MATLAB file are written for analyzing free vibration of isotropic and anisotropic rectangular plates. The program is attached in Appendix VI. Any combinations of boundary conditions can be applied by the written program with only a few modifications. Some examples have been already presented in Chapter 4 and more applications will be given herein.

It is worthy noting that the zero deflection or zero slope at boundary points can also be applied by simply replacing the corresponding diagonal term in the stiffness matrix with a big number, say 10^{30}.

11.3.1 FREE VIBRATION OF ISOTROPIC RECTANGULAR PLATE

EXAMPLE 11.1

Consider free vibration of isotropic rectangular plate with length a, width b, and uniform thickness h. Three combinations of boundary conditions, i.e., SCSC, SCFC, and CFFF, and five aspect ratios are investigated by the QEM. Poisson's ratio μ is 0.3. The results obtained by the QEM with nodal number of 21×21 are listed in Tables 11.1–11.3. For comparisons, the frequency parameter $\bar{\lambda}$ defined by Eq. (11.150 is introduced).

$$\bar{\lambda} = \omega a^2 \sqrt{\rho h / D} \qquad (11.15)$$

where $D = Eh^3 / 12(1 - \mu^2)$ is the flexural rigidity of the plate and E is Young's modulus.

It has been demonstrated in Chapter 4 that the quadrature plate element with mixed interpolations is the best one, especially when clamped edges are involved, since Hermite interpolation is slightly better than the Lagrange interpolation for clamped edges. Therefore, one quadrature plate element with Lagrange interpolation in the x direction and Hermite interpolation in the y direction is used in the analysis. The exact or accurate upper bound solutions given by Leissa [2] are cited for comparisons. For simplicity, if the results obtained by the QEM are exactly the same as the ones given by Leissa [2], then only one of them is listed in the table. Otherwise upper bound solutions, data in parenthesis, are included. The edge numbering sequence is the same as the one used by Leissa [2], i.e., starting from edge at $x = 0$, then edge at $y = 0$, edge at $x = a$, and finally edge at $y = b$. For example, the symbol SCFC denotes that edge at $x = 0$ is simply supported, edges at $y = 0$ and $y = b$ are clamped, and edge at $x = a$ is free.

It is seen that the frequency parameters obtained by the QEM with mixed interpolations are exactly the same as the ones given by Leissa [2] for most cases. For one case, the ninth-mode frequency parameter given by Leissa [2] for aspect ratio of 2/3 is slightly different from the result obtained by the QEM. The datum in bold should be caused by a misprint obviously, since the modified DQM yields exactly the same mode frequency parameters as the ones listed in Table 11.1.

For free vibration of SCFC rectangular plates, the frequency parameters obtained by the QEM, listed in Table 11.2, are slightly different from the ones given by Leissa [2]; thus, the upper bound solutions, data in parenthesis, are also included for comparisons. It is seen that the data obtained by the QEM are all a little smaller than the upper bound solutions. It should be pointed out again that the modified DQM can also yield similar accurate mode frequency parameters as the ones listed in Table 11.2.

For free vibration of CFFF rectangular plates, the first six frequency parameters obtained by the QEM are listed in Table 11.3. Since they are slightly different from the ones given by Leissa [2], the upper bound solutions (in parenthesis) are also included in Table 11.3 for comparisons. It is seen that the data obtained by the QEM are all a little smaller than the upper bound solutions. Again the modified DQM can yield similar accurate mode frequency parameters as the ones listed in Table 11.3.

From Tables 11.1–11.3 , it is seen that the QEM with nodal points of 21×21 can yield very accurate lower-order mode frequencies. Due to space limitations, the results obtained by the QEM for other combinations of boundary conditions are omitted. With a few modifications on the provided FORTRAN program or MATLAB file, solutions for the rectangular plate with any combinations of boundary conditions and with any different materials can be obtained.

Table 11.1 Frequency Parameter $\bar{\lambda}$ for SCSC Plates

Mode Sequence	a/b				
	0.4	**2/3**	**1.0**	**1.5**	**2.5**
1	12.1347	17.3730	28.9509	56.3481	145.4839
2	18.3647	35.3445	54.7431	78.9836	164.7387
3	27.9657	45.4294	69.3270	123.1719	202.2271
4	40.7500	62.0544	94.5853	146.2677	261.1053
5	41.3782	62.3131	102.2162	170.1112	342.1442
6	47.0009	88.8047	129.0955	189.1219	392.8746
7	56.1782	94.2131	140.2045	212.8169	415.6906
8	56.6756	97.4254	154.7757	276.0012	444.9682
9	68.7486 (68.7486)	110.0788 (**101.0788**)	170.3465 (170.3465)	276.0125 (276.0125)	455.3054 (455.3054)

Table 11.2 Frequency Parameter $\bar{\lambda}$ for SCFC Plates ($\mu = 0.3$)

Mode Sequence	a/b				
	0.4	**2/3**	**1.0**	**1.5**	**2.5**
1	22.480 (22.544)	22.776 (22.855)	23.3690 (23.460)	24.676 (24.775)	28.463 (28.564)
2	24.258 (24.296)	27.928 (27.971)	35.5689 (35.612)	53.680 (53.731)	70.262 (70.561)
3	28.310 (28.341)	40.648 (40.683)	62.8709 (63.126)	64.676 (64.959)	113.89 (114.00)
4	35.297 (35.345)	62.088 (62.310)	66.7599 (66.808)	97.116 (97.257)	130.24 (130.84)
5	45.629 (45.710)	62.640 (62.695)	77.3700 (77.502)	123.93 (124.48)	159.29 (159.54)
6	59.426 (59.562)	68.557 (68.683)	108.86 (108.99)	127.83 (127.92)	209.32 (210.32)

Table 11.3 Frequency Parameter $\bar{\lambda}$ for CFFF Plates ($\mu = 0.3$)

Mode Sequence	a/b				
	0.4	**2/3**	**1.0**	**1.5**	**2.5**
1	3.4975 (3.5107)	3.4851 (3.5024)	3.4710 (3.4917)	3.4534 (3.4772)	3.4280 (3.4562)
2	4.7674 (4.7861)	6.3880 (6.4062)	8.5063 (8.5246)	11.656 (11.676)	17.962 (17.988)
3	8.0670(8.1146)	14.466 (14.538)	21.284 (21.429)	21.465 (21.618)	21.396 (21.563)
4	13.803 (13.882)	21.915 (22.038)	27.199 (27.331)	39.325 (39.492)	57.217 (57.458)
5	21.520 (21.638)	25.911 (26.073)	30.955 (31.111)	53.541 (53.876)	60.122 (60.581)
6	23.046 (23.731)	31.448 (31.618)	54.184 (54.443)	61.613 (61.994)	105.93 (106.34)

11.3.2 FREE VIBRATION OF ISOTROPIC SKEW PLATE

EXAMPLE 11.2

Consider the free vibration of isotropic rhombic plate shown in Fig. 11.7. Three combinations of boundary conditions, i.e., CSCS, CSCF, and CFFF, are considered. Poisson's ratio μ is 0.3. The edge numbering sequence shown in Fig. 11.7 is adopted for comparison purpose. CSCF stands that edges at $\eta = 0$ and $\eta = a$ are clamped, edge at $\xi = a$ is simply supported, and edge at $\xi = 0$ is free. The side length is a. Thickness is h and only uniform thickness is considered.

Define

$$
\begin{aligned}
D_{11} &= D_{22} = D = \frac{Eh^3}{12(1-\mu^2)} \\
D_{16} &= D_{26} = -D\sin\theta \\
D_{12} &= D(\mu\cos^2\theta + \sin^2\theta) \\
D_{66} &= \frac{D(1+\sin^2\theta - \mu\cos^2\theta)}{2} \\
\lambda &= \cos^4\theta\rho h\omega^2
\end{aligned}
\tag{11.16}
$$

where E is the modulus of elasticity and ρ is the mass density of the material.

According to Fig. 11.7, skew angle $\theta = 0°$ corresponds to the isotropic square plate. With Eq. (11.16), the FORTRAN program or MATLAB file for analyzing free vibration of anisotropic rectangular plate by using one quadrature rectangular plate element can be used to obtain the frequencies of isotropic skew plate.

Accurate results are obtained by the QEM with nodal points of 21×21. The results are listed in Tables 11.4–11.6 for rhombic plates with CSCS, CSCF, and CFFF boundary conditions. Skew angle θ varies from $15°$ to $75°$. One quadrature plate element with mixed interpolations is used in the analysis, namely, Lagrange interpolation is used in the ξ direction and Hermite interpolation is used in the η direction. GLL points are used as the element nodes, and GLL quadrature is used to obtain the stiffness matrix and mass matrix. In other words, the stiffness matrix of the element is computed by Eq. (4.129) and the mass matrix is computed by Eq. (4.131).

In Table 11.4, the results obtained by the QEM for the CSCS or SCSC rhombic plate are summarized. The accurate upper bound solutions cited from Ref. [3] are also included for comparison.

From Table 11.4, it is seen that the data obtained by the QEM ($\theta \leq 30°$) are exactly the same as the accurate upper bound solutions reported in Ref. [3]. However, slightly larger relative differences occur when skew angle is large ($\theta = 75°$). The 3rd to 5th frequency parameters (bold numbers) cited from Ref. [3] seem not the accurate upper bound solutions pointed by the author and his research associates in Ref. [4]. Similar to the QEM, the modified DQM can also yield the same accurate mode frequency parameters as the ones listed in Table 11.4.

In Table 11.5, the results obtained by the QEM for the CSCF rhombic plate are summarized. The accurate results obtained by the finite element method (FEM) with fine meshes (100×100) are included for comparisons. It is seen that the results obtained by the QEM agree well with the finite element data. The results presented in Table 11.5 are new and waiting for comparison with future data obtained by other investigators. Note that finite element data are not the upper bound solutions.

In Table 11.6, the results obtained by the QEM for the CFFF rhombic plate are summarized. The accurate upper bound solutions cited from Ref. [5] are also included for comparison. It is seen that the results obtained by the QEM agree very well with the upper bound solutions. It should be pointed out that the modified DQM can yield similar accurate mode frequency parameters as the ones listed in Tables 11.5–11.6. However, the mixed method should be used in the DQM for applying the boundary conditions, namely, MMWC-3 is used in the ξ direction and the DQEM is used in the η direction. Besides, the DQM with only Grid V can yield reliable solutions. Generally speaking, the modified DQM yields slightly less accurate mode frequency parameters as compared to the QEM for skew plates with free edges, especially when the skew angle is large.

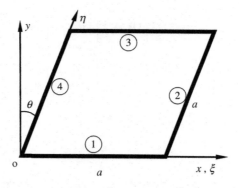

FIGURE 11.7

Sketch of a Rhombic Plate

Table 11.4 Frequency Parameter $\bar{\lambda}$ for CSCS (SCSC) Rhombic Plates

$\theta°$	15		30		45		60		75	
Mode	QEM	[3]	QEM	[3]	QEM	[3]	QEM	[3]	QEM	[3]
1	30.697	30.697	36.954	36.954	52.379	52.375	96.264	96.209	318.07	316.40
2	56.703	56.703	64.264	64.264	83.542	83.541	137.24	137.23	390.19	390.69
3	74.621	74.621	92.972	92.972	123.25	123.25	188.12	188.11	485.26	**478.82**
4	94.046	94.046	100.78	100.78	136.97	136.96	237.78	237.76	569.25	**582.83**
5	111.99	111.99	137.68	137.68	167.96	167.97	267.95	267.82	667.22	**704.75**

Table 11.5 Frequency Parameter $\bar{\lambda}$ for CSCF Rhombic Plates

$\theta°$	15		30		45		60		75	
Mode	QEM	FEM	QEM	FEM	QEM	FEM	QEM	FEM	QEM	FEM
1	24.533	24.524	28.478	28.462	37.055	37.016	57.279	57.100	138.75	137.07
2	37.068	37.639	42.575	42.532	56.658	56.588	96.461	96.271	230.92	228.56
3	66.019	65.981	71.906	71.840	85.763	85.665	122.46	122.12	318.64	316.30
4	68.004	67.939	79.458	79.390	106.38	106.21	165.84	165.50	359.46	356.08
5	83.218	83.122	101.65	101.51	130.69	130.54	196.49	195.68	443.10	441.35

Table 11.6 Frequency Parameter $\bar{\lambda}$ for CFFF Rhombic Plates

$\theta°$	15		30		45		60		75	
Mode	QEM	[5]	QEM	[5]	QEM	[5]	QEM	[5]	QEM	[5]
1	3.5832	3.5831	3.9281	3.9279	4.5054	4.5052	5.2416	5.2431	6.0153	6.0235
2	8.6964	8.6971	9.4096	9.4100	11.248	11.247	16.037	16.023	24.808	24.823
3	22.229	22.230	25.286	25.287	26.964	26.968	30.360	30.362	49.058	48.754
4	26.332	26.334	25.930	25.931	31.499	31.505	45.287	45.300	72.819	72.641
5	33.860	33.864	41.331	41.338	50.709	50.739	59.025	59.123	95.341	95.416

11.3.3 FREE VIBRATION OF ANISOTROPIC RECTANGULAR PLATE

EXAMPLE 11.3

Consider the free vibration of laminated composite plate or anisotropic rectangular plate. For comparisons, the nondimensional frequency parameter is introduced for the thin laminated rectangular plates, namely,

$$\bar{\lambda} = \omega a^2 \sqrt{\rho h / D_0} \qquad (11.17)$$

where $D_0 = E_{11} h^3 / 12(1 - \mu_{12}\mu_{21})$, h is the total thickness of the laminated plate, E_{11} is the modulus of elasticity in the principal direction 1, μ_{12}, μ_{21} are the major and minor Poisson's ratios, a is the plate length, and ρ is the mass density of the material.

The material properties of a lamina are $E_{11}/E_{22} = 2.45$, $G_{12} = 0.48\,E_{22}$, $\mu_{12} = 0.23$. The geometric parameters are, $h/a = 0.006$ and $a/b = 1.0$. For the three-layer angle-ply $(\theta, -\theta, \theta)$ symmetric laminated square plates, Eq. (7.35) yields

$$D_{ij} = \frac{h^3}{12}\bar{Q}_{ij} \quad (i,j=1,2)$$

$$D_{66} = \frac{h^3}{12}\bar{Q}_{ij}$$

$$D_{16} = \frac{h^3}{12.96}\bar{Q}_{16}$$

$$D_{26} = \frac{h^3}{12.96}\bar{Q}_{26}$$

(11.18)

where $\bar{Q}_{ij}$ $(i,j=1,2,6)$ are computed by Eq. (7.36).

Table 11.7 lists the first eight-mode frequency parameters for the three-layer angle-ply $(\theta,-\theta,\theta)$ symmetric laminated cantilever square plate (CFFF). The thickness of each layer is the same, and the total thickness of the laminated plate is h. The results listed in Table 11.7 are obtained by the QEM with one element of 21×21 GLL nodal points. Mixed interpolations are used. For comparisons, the results obtained by NASTRAN with fine mesh of 100×100 are included. It is seen that the results obtained by the QEM agree quite well with the finite element data.

Table 11.7 Frequency Parameters $\bar{\lambda}$ for Three-Layer Angle-ply $(\theta / -\theta / \theta)$ Symmetric Laminated Square Plate (CFFF)

Ply-angle	Method	Mode Number							
		1	2	3	4	5	6	7	8
0°	QEM	3.507	6.879	18.377	22.067	27.337	42.048	42.690	61.558
	FEM	3.507	6.878	18.377	22.069	27.336	42.043	42.702	61.582
15°	QEM	3.336	7.009	18.610	21.189	27.072	41.872	44.060	58.919
	FEM	3.329	7.016	18.601	21.179	27.086	41.771	44.055	59.066
30°	QEM	2.965	7.208	18.151	20.136	26.521	42.125	46.802	52.982
	FEM	2.964	7.207	18.152	20.136	26.518	42.124	46.807	53.001
45°	QEM	2.603	7.171	16.122	20.649	25.984	42.282	46.575	50.048
	FEM	2.603	7.170	16.123	20.648	25.980	42.278	46.592	50.055

11.4 BUCKLING OF THIN RECTANGULAR PLATE

Buckling of thin rectangular plate is considered. For simplicity in presentation, only one quadrature plate element is used in the analysis. The matrix equation for buckling analysis of rectangular plate by the QEM is given by

$$\begin{bmatrix} K_{bb} & K_{bi} \\ K_{ib} & K_{ii} \end{bmatrix} \begin{Bmatrix} \{\Delta_b\} \\ \{W_i\} \end{Bmatrix} = \lambda \begin{bmatrix} G_{bb} & G_{bi} \\ G_{ib} & G_{ii} \end{bmatrix} \begin{Bmatrix} \{\Delta_b\} \\ \{W_i\} \end{Bmatrix}$$

(11.19)

where subscripts b and i denote the quantities related to the derivative DOFs at boundary points and displacement at all grid points. Note that the stiffness matrix and the geometrical stiffness matrix in Eq. (11.19) are symmetric.

It is seen that the form of Eq. (11.19) is only slightly different from Eq. (7.19); the equation obtained by the modified DQM. Eq. (11.19) also contains $(N+2)\times(N+2)-4$ equations if Lagrange interpolation or mixed method is used in the formulations of the stiffness matrix and geometrical stiffness matrix. However, the stiffness matrix and the geometric matrix are obtained differently. The stiffness matrix is obtained either by Eq. (4.109) or by Eq. (4.129), and the geometric stiffness matrix is obtained either by Eq. (4.113) or by Eq. (4.133). Besides, $\{\Delta_b\}$ contains only the derivative DOFs. In other words, the dimension of $\{\Delta_b\}$ is $4N\times1$ and the dimension of $\{W_i\}$ is $N^2\times1$.

If MMWC-3 is used to introduce the derivative DOFs at boundary points, then only $[G_{ii}]$ in Eq. (11.19) is nonzero, namely,

$$\begin{bmatrix} K_{bb} & K_{bi} \\ K_{ib} & K_{ii} \end{bmatrix} \begin{Bmatrix} \{\Delta_b\} \\ \{W_i\} \end{Bmatrix} = \lambda \begin{bmatrix} 0 & 0 \\ 0 & G_{ii} \end{bmatrix} \begin{Bmatrix} \{\Delta_b\} \\ \{W_i\} \end{Bmatrix} \tag{11.20}$$

Take an SSSS rectangular plate as an example. After eliminating the nonzero $\{\Delta_b\}$ and zero displacements at all boundary nodes, Eq. (11.20) can be rewritten as

$$\left[\bar{K}_{ii} - \bar{K}_{ib} K_{bb}^{-1} \bar{K}_{bi} \right] \{\bar{W}_i\} = \lambda \left[\bar{G}_{ii} \right] \{\bar{W}_i\} \tag{11.21}$$

where the over bar denotes that the zero displacements at all boundary nodes have been removed and the corresponding matrix has been modified.

Equation (11.21) can be simplified further, namely,

$$p\{\bar{W}_i\} = \left[\bar{K}_{ii} - \bar{K}_{ib} K_{bb}^{-1} \bar{K}_{bi} \right]^{-1} \left[\bar{G}_{ii} \right] \{\bar{W}_i\} \tag{11.22}$$

or

$$\left[\bar{K} \right] \{\bar{W}_i\} = p[I]\{\bar{W}_i\} \tag{11.23}$$

Eigenvalues can be obtained by a standard eigensolver and the inverse of the largest eigenvalue corresponds to the buckling load.

The geometric matrix can also be formulated by using the Hermite interpolation. Then all submatrices of $[G]$, i.e., $[G_{bb}],[G_{bi}],[G_{ib}],[G_{ii}]$ in Eq. (11.19), are nonzero. Take an SSSS rectangular plate as example. After eliminating the zero displacement at all boundary nodes, Eq. (11.19) is modified as

$$\begin{bmatrix} K_{bb} & \bar{K}_{bi} \\ \bar{K}_{ib} & \bar{K}_{ii} \end{bmatrix} \begin{Bmatrix} \{\Delta_b\} \\ \{\bar{W}_i\} \end{Bmatrix} = \lambda \begin{bmatrix} G_{bb} & \bar{G}_{bi} \\ \bar{G}_{ib} & \bar{G}_{ii} \end{bmatrix} \begin{Bmatrix} \{\Delta_b\} \\ \{\bar{W}_i\} \end{Bmatrix} \tag{11.24}$$

where the over bar means that the zero displacement at all boundary nodes has been removed; thus, the corresponding matrix has been modified.

Equation (11.24) can be solved by a general eigensolver. Alternatively, it can be transformed into a standard eigenvalue equation as follows:

$$p \begin{bmatrix} I & 0 \\ 0 & I \end{bmatrix} \begin{Bmatrix} \{\Delta_b\} \\ \{\bar{W}_i\} \end{Bmatrix} = \begin{bmatrix} K_{bb} & \bar{K}_{bi} \\ \bar{K}_{ib} & \bar{K}_{ii} \end{bmatrix}^{-1} \begin{bmatrix} G_{bb} & \bar{G}_{bi} \\ \bar{G}_{ib} & \bar{G}_{ii} \end{bmatrix} \begin{Bmatrix} \{\Delta_b\} \\ \{\bar{W}_i\} \end{Bmatrix} \tag{11.25}$$

or in short,

$$[\bar{K}]\{\bar{W}\} = p[I]\{\bar{W}\} \tag{11.26}$$

Eigenvalues can be obtained by a standard eigensolver and the inverse of the largest eigenvalue corresponds to the buckling load.

EXAMPLE 11.4

Consider an SSSS square graphite/epoxy plate under biaxial uniformly distributed edge compression, i.e., $\sigma_{x0} = -\sigma_0$, $\sigma_{y0} = -\sigma_0$, $\tau_{xy0} = 0$. The plate length and width are a and b ($a/b = 1$), the thickness is h, and uniform thickness is considered for simplicity. For this loading case, $P_x = P_y = -\sigma_0 h$, and $P_{xy} = 0$ within the entire plate. The material properties are $Q_{11}/Q_{22} = 25, Q_{12}/Q_{22} = 0.25$, and $Q_{66}/Q_{22} = 0.5$ with principal material direction at 45° to the plate side (x axis). For comparison purposes, the buckling coefficient $\bar{P}_x$ is introduced and defined by

$$\bar{P}_x = \frac{\sigma_0 b^2}{Q_{22} h^2} \tag{11.27}$$

It is known that for isotropic rectangular plate, the rate of convergence for both the QEM and the DQM is very high; both methods with grid spacing of 11 × 11 can yield numerically exact buckling load for an SSSS square plate. Due to strong anisotropy, however, the rate of convergence for various approximate and numerical methods is relatively low; thus, larger number of grid points should be used for the QEM and the DQM to ensure the solution accuracy.

Table 11.8 shows the buckling coefficient obtained by the QEM. One element is used in the analysis. This example is also investigated by the DQM at the same time. The results obtained by the DQM with Grid III and MMWC-3 are also listed in Table 11.8. Some existing results as well as finite element results obtained by NASTRAN with fine meshes are also included in Table 11.8 for comparison and verifications. Figure 11.8 is the buckling mode obtained by NASTRAN. The deflection is symmetric about the principal material directions at ±45° from the plate edges.

Three kinds of QEM have been used. QEM (1) uses MMWC-3 to introduce the additional DOF in both x and y directions. QEM (2) used the Hermite interpolation for formulating both stiffness and geometrical stiffness matrices. To avoid introduction of the mixed second-order derivative at four corners, the Lagrange interpolation is used to formulate the weighting coefficient of w_{xy} in QEM (2). QEM (3) uses MMWC-3 to introduce the additional DOF in the x direction and the Hermite interpolation in the y direction.

It is seen from Tables 11.8 and 7.10 that the FEM converges from the above, and that the DQM as well as all three QEMs converges from below. The difference of the results obtained by the three QEMs is negligible. Besides, the data obtained by both the QEM and the DQM are close to each other.

As mentioned in Chapter 7, the results obtained by Whitney are not quite accurate, since the number of terms in series is not large enough. The results obtained by Ashton converge to a wrong solution due to the introduction of extra constraints, since double sine series was used as the trial displacement functions in the Ritz method.

In using the DQM, the method of applying the multiple boundary conditions also affects the solution accuracy. Although the accuracy of the DQ results cited from Ref. [6] is close to the result given by Whitney, the convergence trend seems wrong. If a larger grid number of points were used in the analysis, the DQM with MMWC-1 to apply the multiple boundary conditions would have yielded much smaller buckling coefficients. This demonstrates the importance of the way to apply the multiple boundary conditions in using the DQM.

Table 11.8 Buckling Coefficient $\bar{P}_x$ of Square Plate Under Biaxial Uniformly Distributed Compressive Loads

QEM (1)	8.75279 (15 × 15)	8.78848 (21 × 21)
QEM (2)	8.75420 (15 × 15)	8.78892 (21 × 21)
QEM (3)	8.78101 (15 × 15)	8.80720 (21 × 21)
DQM	8.72514 (15 × 15)	8.79054 (25 × 25)
DQM (MMWC-1) [6]	8.740 (7 × 7)	8.574 (9 × 9)
Ashton [6]	11.565 ($m = n = 5$)	11.060 ($m = n = 7$)
Whitney [6]	8.418 ($m = n = 7$)	8.556 ($m = n = 9$)
NASTRAN	9.101 (100 × 100)	9.092 (150 × 150)

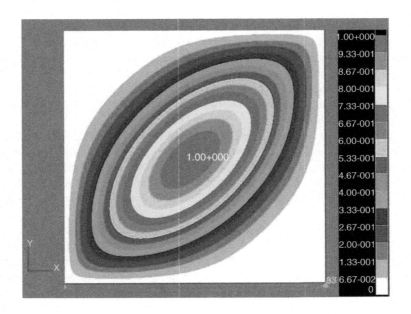

FIGURE 11.8

Sketch of the Buckling Mode

11.5 SUMMARY

In this chapter, the QEM is used to successfully perform the free vibration analysis of thin plates and the dynamic analysis of a cantilever flexible bar hit by a rigid ball. Two kinds of quadrature elements, i.e., quadrature bar element and quadrature thin plate element, are used. It is seen that the formulation of an N-node quadrature element is simple and flexible. Several examples are investigated.

Numerical results show that similar to the modified DQM, the QEM can also yield accurate frequencies with relatively small number of grid points. Besides, the QEM can yield accurate contact stress-time history. Thus, the attractive features of rapid convergence, high accuracy, and computational efficiency of the QEM are illustrated.

The QEM is essentially the same as the time domain spectral element method (SPE) and thus sometimes is also called the SPE or spectral finite element method in literature, especially when GLL points are used as the element nodal points and GLL quadrature is used to formulate the stiffness and mass matrices. Thus, more applications by using the QEM can be found in Refs. [7–13].

REFERENCES

[1] C. Xu, X. Wang, Efficient numerical method for dynamic analysis of flexible rod hit by rigid ball, T. NUAA 29 (4) (2012) 338–344.

[2] A.W. Leissa, The free vibration of rectangular plates, J. Sound Vib. 31 (1973) 257–293.

[3] C.S. Huang, O.G. McGee, J.W. Kim, et al., Corner stress singularity effects on the vibration of rhombic plates with combinations of clamped and simply supported edges, J. Sound Vib. 193 (1996) 555–580.

[4] X. Wang, Y. Wang, Z. Yuan, Accurate vibration analysis of skew plates by the new version of the differential quadrature method, Appl. Math. Model. 38 (2014) 926–937.

[5] O.G. McGee, A.W. Leissa, C.S. Huang, Vibrations of cantilevered skewed plates with corner stress singularities, Int. J. Numer. Meth. Eng. 35 (1992) 409–424.

[6] C.W. Bert, X. Wang, A.G. Striz, Differential quadrature for static and free vibration analyses of anisotropic plates, Int. J. Solids Struct. 30(13) (1993) 1737–1744.

[7] X. Wang, F. Wang, C. Xu, L. Ge, New spectral plate element for simulating Lamb wave propagation in plate structures, J. NUAA 44 (5) (2012) 645–651 (in Chinese).

[8] F. Wang, X. Wang, Z. Feng, Simulation of wave propagation in plate structures by using new spectral element with piezoelectric coupling, J. Vibroeng. 15 (1) (2013) 268–276.

[9] Y. Wang, X. Wang, Static analysis of higher order sandwich beams by weak form quadrature element method, Compos. Struct. 116 (2014) 841–848.

[10] L. Ge, X. Wang, F. Wang, Accurate modeling of PZT-induced Lamb wave propagation in structures by using a novel spectral finite element method, Smart Mater. Struct. 23 (2014) 095018.

[11] L. Ge, X. Wang, C. Jin, Numerical modeling of PZT-induced Lamb wave-based crack detection in plate-like structures, Wave Motion 51 (2014) 867–885.

[12] C. Jin, X. Wang, L. Ge, Novel weak form quadrature element method with expanded Chebyshev nodes, Appl. Math. Lett. 34 (2014) 51–59.

[13] C. Jin, X. Wang, Dynamic analysis of functionally graded material bars by using novel weak form quadrature element method, J. Vibroeng. 16 (6) (2014) 2790–2799.

Appendix I

I GLL QUADRATURE

The abscissas and corresponding weights, used in Gauss–Lobatto–Legendre quadrature or simply called GLL quadrature, are given in Table I.1 for readers' reference. The Maple program to compute the abscissas and weights are also given if one needs the data for N other than the ones given in Table I.1. The abscissas can be used as grid points in the DQM, i.e., the Grid VII given by Eq. (1.40).

I.1 ABSCISSAS AND WEIGHTS IN GLL QUADRATURE

Table I.1 contains the abscissas and corresponding weights used in the GLL quadrature, the DQM, and the QEM for N varying from 3 to 21.

Table I.1 Abscissas and Weights in GLL Quadrature

N	Abscissa, x_i	Weight, H_i
3	± 1	1/3
	0	4/3
4	± 1	1/6
	$\pm\sqrt{5}/5$	5/6
5	± 1	1/10
	± 0.654653670707977	0.544444444444444
	0	0.711111111111111
6	± 1	1/15
	± 0.765055323929465	0.378474956297844
	± 0.285231516480645	0.554858377035486
7	± 1	1/21
	± 0.830223896278567	0.276826047361565
	± 0.468848793470714	0.431745381209868
	0	0.487619047619048

(Continued)

Differential Quadrature and Differential Quadrature Based Element Methods. 978-0-12-803081-3

Table I.1 Abscissas and Weights in GLL Quadrature *(cont.)*

N	Abscissa, x_i	Weight, H_i
8	±1	1/28
	±0.871740148509606	0.210704227143594
	±0.591700181433144	0.341122692483510
	±0.209299217902474	0.412458794658706
9	±1	1/36
	±0.899757995411459	0.165495361560769
	±0.677186279510741	0.274538712500240
	±0.363117463826176	0.346428510973052
	0	0.371519274376418
10	±1	1/45
	±0.919533908166459	0.133305990851307
	±0.738773865105505	0.224889342062968
	±0.477924949810445	0.292042683679636
	±0.165278957666387	0.327539761183898
11	±1	1/55
	±0.934001430408059	0.109612273266783
	±0.784483473663144	0.187169881780646
	±0.565235326996205	0.248048104263872
	±0.295758135586936	0.286879124779000
	0	0.300217595455690
12	±1	1/66
	±0.944899272222882	0.091684517413605
	±0.819279321644007	0.157974705565153
	±0.632876153031861	0.212508417760924
	±0.399530940965349	0.251275603199198
	±0.136552962854928	0.271405240910698
13	±1	1/78
	±0.953309846642164	0.077801686748071
	±0.846347564651872	0.134981926690461
	±0.686188469081757	0.183646865204071
	±0.482909821091336	0.220767793566050
	±0.249286930106240	0.244015790306674
	0	0.251930849333446

(Continued)

Table I.1 Abscissas and Weights in GLL Quadrature *(cont.)*

N	Abscissa, x_i	Weight, H_i
14	±1	1/91
	±0.959935045267261	0.066837284497167
	±0.867801053830347	0.116586655893506
	±0.728868599091326	0.160021851764257
	±0.550639402928647	0.194826149373626
	±0.342724013342713	0.219126253009796
	±0.116331868883704	0.231612794468458
15	±1	1/105
	±0.965245926503839	0.058029893034830
	±0.885082044222977	0.101660070340631
	±0.763519689951815	0.140511699803666
	±0.606253205469846	0.172789647252341
	±0.420638054713673	0.196987235964736
	±0.215353955363794	0.211973585926826
	0	0.217048116348816
16	±1	1/120
	±0.969568046270218	0.050850360991482
	±0.899200533093472	0.089393697309843
	±0.792008291861815	0.124255382131622
	±0.652388702882493	0.154026980806829
	±0.486059421887137	0.177491913391692
	±0.299830468900763	0.193690023825203
	±0.101326273521949	0.201958308178232
17	±1	1/136
	±0.973132176631418	0.044921940528120
	±0.910879995915573	0.079198270540947
	±0.815696251221770	0.110592909015537
	±0.691028980627685	0.137987746197338
	±0.541385399330102	0.160394661997501
	±0.372174433565477	0.177004253515666
	±0.189511973518317	0.187216339677624
	0	0.190661874753469

(Continued)

Table I.1 Abscissas and Weights in GLL Quadrature *(cont.)*

N	Abscissa, x_i	Weight, H_i
18	±1	1/153
	±0.976105557412199	0.039970628823495
	±0.920649185347534	0.070637166823643
	±0.835593535218090	0.099016271784270
	±0.723679329283243	0.124210533134307
	±0.588504834318662	0.145411961572047
	±0.434415036912124	0.161939517237548
	±0.266362652878281	0.173262109489448
	±0.089749093484652	0.179015863439703
19	±1	1/171
	±0.978611766222080	0.035793365308575
	±0.928901528152586	0.063381891862863
	±0.852460577796646	0.089131757062884
	±0.751494202552613	0.112315341489282
	±0.628908137265220	0.132267280443705
	±0.488229285680714	0.148413942597320
	±0.333504847824499	0.160290924044052
	±0.169186023409282	0.167556584527144
	0	0.170001919284826
20	±1	1/190
	±0.980743704893914	0.032237122922469
	±0.935934498812665	0.057181801842336
	±0.866877978089950	0.080631763964959
	±0.775368260952056	0.101991499892940
	±0.663776402290311	0.120709227638607
	±0.534992864031886	0.136300482360499
	±0.392353183713910	0.148361554071064
	±0.239551705922986	0.156580102647493
	±0.080545937238822	0.160743286387847

(Continued)

Table I.1 Abscissas and Weights in GLL Quadrature *(cont.)*		
N	Abscissa, x_i	Weight, H_i
21	±1	1/210
	±0.982572296604548	0.029184840634708
	±0.941976296959746	0.051843169097629
	±0.879294755323591	0.073273919239746
	±0.796001926077712	0.092985467916404
	±0.694051026062223	0.110517083199618
	±0.575831960261831	0.125458121178042
	±0.444115783279003	0.137458462859923
	±0.301989856508765	0.146236862448100
	±0.152785515802186	0.151587575111684
	0.0	0.153385190332175

I.2 EXAMPLE OF MAPLE PROGRAM

1. Find the abscissa, x_i for $N = 5$.
```
>with(orthopoly):
>N: = 5: # N is the degree of Legendre polynomial
>N1: = N−1:
>solve( diff(orthopoly[P](N1, x),x) = 0,x):
>evalf({%},15); # roots
```

$$\{0., 0.6546536707007978, \ 0.654653670707978\}$$

2. Find the weights H_i for $N = 5$.
```
>with(orthopoly):
>N: = 5: # N is the degree of Legendre polynomial
>N1: = N−1:
>y: = 0.0: evalf(2/N1/N/(orthopoly[P](N1, y)^2),15); # weight
```

$$0.711111111111112$$

```
>y: = 0.654653670707978:evalf(2/N1/N/(orthopoly[P](N1, y)^2),15); # weight
```

$$0.544444444444444$$

```
>y: = 1.0:evalf(2/N1/N,15); # weight =2/N/(N−1) for y = 1 and y = −1
```

$$0.100000000000000$$

Appendix II

II SUBROUTINES AND FUNCTIONS

II.1 FOUR COMMONLY USED SUBROUTINES

Four commonly used subroutines, SUBROUTINE INVMAT (N,A,D,EP,ISW,B,C,L), SUBROUTINE TZZ (NM,N,A,SCALE,WR,WI,INT), SUBROUTINE MTM(A,B,C,N,M,L), and SUBROUTINE GRULE(N,X,W,Y), are called by the FORTRAN programs presented in this book.

Subroutine INVMAT is called to find the inverse of matrix [A] by Gauss–Jordan method. The FORTRAN source codes can be found in many published books, e.g., subroutine BSSGJ (.) in Ref. [1] and subroutine BRINV (.) in Ref. [2]. IMSL [3] contains a similar subroutine named DLINRG which can be used to compute the inverse of a real general matrix. If one uses the Digital Visual Fortran to run the program, simply replace subroutine INVMAT by DLINRG. Therefore, the FORTRAN program for this subroutine is not provided to save the space. MATLAB has this function named "inv(A)". Note that if BSSGJ (.) is to be used, small modifications are needed since the variables in the subroutine are in single precision. The variables in the included programs are in double precisions.

Subroutine "TZZ" is a standard eigenvalue solver and called to find all eigenvalues of a real general matrix. Its FORTRAN codes are listed below.

```
C-------------------------------------------------------------------------------------------------------
C       CALLED BY PROGRAMS 1-4
C-------------------------------------------------------------------------------------------------------
C
        SUBROUTINE TZZ(NM,N,A,SCALE,WR,WI,INT)
        IMPLICIT REAL*8 (A-H,O-Z)
        DIMENSION A(NM,N),SCALE(N),WR(N),WI(N),INT(N)
        CALL BALANC(NM,N,A,LOW,IGH,SCALE)
        CALL ELMHES(NM,N,LOW,IGH,A,INT)
        CALL HQR(NM,N,LOW,IGH,A,WR,WI,IERR)
        RETURN
        END
```

The FORTRAN source codes of the three subroutines called by subroutine TZZ, i.e., subroutines named BALANC, ELMHES, and HQR, can be found in many published books, e.g., in Refs. [1,2]. IMSL [3] contains a similar subroutine named DEVLRG which can be used to compute all eigenvalues of a real general matrix. If one uses the Digital Visual Fortran to run the program, simply replace subroutine TZZ by DEVLRG. Therefore, the FORTRAN source codes of these three subroutines are not provided to save the space. MATLAB has this function named "eig(A)". Again, small modifications are needed if the codes from Ref. [1] are to be used, since variables in the included programs are in double precisions.

Subroutine "MTM" is called to find the product of two matrices, i.e., C[N,L] = A[N,M] × B[M,L]. The subroutine is short, thus given below. Since MATLAB has this function, the FORTRAN program is thus not converted to MATLAB file.

```
C------------------------------------------------------------------------------------------
C      CALLED BY PROGRAMS 3 & 4
C------------------------------------------------------------------------------------------
C
       SUBROUTINE MTM(A,B,C,N,M,L)
       IMPLICIT REAL*8(A-H,O-Z)
       DIMENSION A(N,M),B(M,L),C(N,L)       ![C]=[A][B]
       DO 1 I=1,N
       DO 1 J=1,L
       C(I,J)=0.0D0
       DO 2 K=1,M
2      C(I,J)=C(I,J)+A(I,K)*B(K,J)
1      CONTINUE
       RETURN
       END
```

Subroutine "GRULE" is called to compute the abscissas and weights used in Gaussian quadrature as well as to obtain Grid V for the DQM. The subroutine, partly reproduced from Ref. [4], is given below and also converted to the MATLAB function.

```
C------------------------------------------------------------------------------------------
C      Below is the FORTRAN subroutine
C------------------------------------------------------------------------------------------
C
       SUBROUTINE GRULE(N,X,W,Y)
       IMPLICIT REAL*8 (A-H,O-Z)
       DIMENSION X(N),W(N),Y(N)
C
C   SEE: PHILIP J DAVIS, PHILIP ROBINOWITZ, METHODS OF NUMERICAL INTEGRATION,
C   ACADEMIC PRESS, INC, (LONDON) LTD, 1975[4].
C
C   COMPUTE THE (N+1)/2 NONNEGATIVE ABSCISSAS X(I) AND CORRESPONDING WEIGHTS
C   W(I) OF THE N-POINT GAUSS-LEGENDRE INTEGRATION RULE.
C
       M=(N+1)/2
       E1=N*(N+1)
       PI=DATAN(1.0D0)*4.0D0
       DO 1 I=1,M
       T=(4*I-1)*PI/(4*N+2)
       X0=(1.D0-(1.D0-1.D0/N)/(8.D0*N*N))*DCOS(T)
       PKM1=1.D0
       PK=X0
       DO 3 K=2,N
       T1=X0*PK
       PKP1=T1-PKM1-(T1-PKM1)/K+T1
       PKM1=PK
```

```
3     PK=PKP1
      DEN=1.D0-X0*X0
      D1=N*(PKM1-X0*PK)
      DPN=D1/DEN
      D2PN=(2.D0*X0*DPN-E1*PK)/DEN
      D3PN=(4.D0*X0*D2PN+(2.D0-E1)*DPN)/DEN
      D4PN=(6.D0*X0*D3PN+(6.D0-E1)*D2PN)/DEN
      U=PK/DPN
      V=D2PN/DPN
      H=-U*(1.D0+0.5D0*U*(V+U*(V*V-U*D3PN/(3.D0*DPN))))
      P=PK+H*(DPN+0.5D0*H*(D2PN+H/3.0D0*(D3PN+0.25D0*H*D4PN)))
      DP=DPN+H*(D2PN+0.5D0*H*(D3PN+H*D4PN/3.0D0))
      H=H-P/DP
      X(I)=X0+H
      FX=D1-H*E1*(PK+0.5D0*H*(DPN+H/3.D0*(D2PN+0.25D0*H*
     1    (D3PN+0.2D0*H*D4PN))))
1     W(I)=2.D0*(1.D0-X(I)*X(I))/(FX*FX)
      IF (M+M .GT. N) X(M)=0.0D0
C
C     ADD: COMPUTE ALL (N) ABSCISSAS X(I) AND THEIR CORRESPONDING WEIGHTS W(I) OF
C          THE N-POINT GAUSS-LEGENDRE INTEGRATION RULE.
C
      IF (M+M .EQ. N) THEN
      DO I=1,M
      Y(I)=-X(I)
      Y(M+I)=X(M-I+1)
      W(M+I)=W(M-I+1)
      ENDDO
      ELSE
      DO I=1,M-1
      Y(I)=-X(I)
      Y(M)=X(M)
      Y(M+I)=X(M-I)
      W(M+I)=W(M-I)
      ENDDO
      ENDIF
      DO I=1,N
      X(I)=Y(I)
      ENDDO
      RETURN
      END
C------------------------------------------------------------------------
C                    The end of the FORTRAN subroutine GRULE
C------------------------------------------------------------------------


%------------------------------------------------------------------------
%     Below is the converted MATLAB function named GRULE
%------------------------------------------------------------------------
```

function [X,W]=GRULE(N)
```
%
%  SEE: PHILIP J DAVIS, PHILIP ROBINOWITZ, METHODS OF NUMERICAL INTEGRATION,
%  ACADEMIC PRESS, INC, (LONDON) LTD, 1975[4].
%
```

```
%    COMPUTE THE (N+1)/2 NONNEGATIVE ABSCISSAS X(I) AND CORRESPONDING WEIGHTS
%    W(I) OF THE N-POINT GAUSS-LEGENDRE INTEGRATION RULE.
%
M=fix((N+1)/2);
E1=N*(N+1);

for I=1:M
T=(4*I-1)*pi/(4*N+2);
X0=(1.0-(1.0-1.0/N)/(8.0*N*N))*cos(T);
PKM1=1.0;
PK=X0;

for K=2:N
      T1=X0*PK;
      PKP1=T1-PKM1-(T1-PKM1)/K+T1;
      PKM1=PK;
      PK=PKP1;
end

DEN=1.0-X0*X0;
D1=N*(PKM1-X0*PK);
DPN=D1/DEN;
D2PN=(2.0*X0*DPN-E1*PK)/DEN;
D3PN=(4.0*X0*D2PN+(2.0-E1)*DPN)/DEN;
D4PN=(6.0*X0*D3PN+(6.0-E1)*D2PN)/DEN;
U=PK/DPN;
V=D2PN/DPN;
H=-U*(1.0+0.5*U*(V+U*(V*V-U*D3PN/(3.0*DPN))));
P=PK+H*(DPN+0.5*H*(D2PN+H/3.0*(D3PN+0.25*H*D4PN)));
DP=DPN+H*(D2PN+0.5*H*(D3PN+H*D4PN/3.0));
H=H-P/DP;
X(I)=X0+H;
FX=D1-H*E1*(PK+0.5*H*(DPN+H/3.0*(D2PN+0.25*H*(D3PN+0.2*H*D4PN))));
W(I)=2.0*(1.0-X(I)*X(I))/(FX*FX);
end

if (M+M)>N
X(M)=0.0;
end
%
%    ADD: COMPUTE ALL (N) ABSCISSAS X(I) AND THEIR CORRESPONDING WEIGHTS W(I) OF
%           THE N-POINT GAUSS-LEGENDRE INTEGRATION RULE.
%
if (M+M)==N
I=1:M;
Y(I)=-X(I);
Y(M+I)=X(M-I+1);
W(M+I)=W(M-I+1);
else
I=1:M-1;
Y(I)=-X(I);
Y(M)=X(M);
Y(M+I)=X(M-I);
W(M+I)=W(M-I);
end

X=Y;
%-------------------------------------------------------------------------------------------------
%    The end of the MATLAB function GRULE
%-------------------------------------------------------------------------------------------------
```

II.2 SUBROUTINES AND CONVERTED FUNCTIONS TO COMPUTE THE WEIGHTING COEFFICIENTS

In this section, 11 FORTRAN subroutines and their converted MATLAB functions are included. These subroutines or functions are used to compute the weighting coefficients of the first- to fourth-order derivatives with respect to x. The weighting coefficients of the ordinary DQM, the modified DQM, and the DQEM are obtained by calling these subroutines or functions. The input data of x in the first eight subroutines or functions can be in the range of either $[0,L]$ or $[-L/2, L/2]$, where L is the length. The input data of x in the last three subroutines or functions are in the range of $[-1,1]$. The 11 FORTRAN subroutines are presented first, and then the corresponding converted MATLAB functions. The weighting coefficients of the first- to fourth-order derivatives with respect to x are stored in matrices A, B, C, and D, respectively.

The first four subroutines or converted functions, i.e., DQSS, DQCC, DQSC, and DQCS, are used in Program 1. MMWC-1 is used to build in the condition of $w_{xx} = 0$ and MMWC-3 or MMWC-4 is used to build in the condition of $w_x = 0$. Letters DQ denote the DQM, letters S and C stand for simply support end and clamped end. Currently A_{ij} is explicitly computed by using Eq. (1.30) thus the method is the DQM. One can also use Eq. (1.33) to compute A_{ij}, and then the method is called the HDQM or more precisely the HDQMNEW.

Subroutine or converted function named DQM is used in Program 6. A_{ij} is explicitly computed by using Eq. (1.30) and B_{ij} is computed by using Eq. (1.21). The method is the ordinary DQM. One can also use Eq. (1.33) to compute A_{ij}, and then the method is called the HDQM. The weighting coefficients of the first- and the second-order derivatives with respect to x are stored in matrices A and B, respectively.

Subroutine or the converted function named DQEM uses the Hermite interpolation to determine the weighting coefficients. Equations (2.13)–(2.18) are used to compute the weighting coefficients of the first- to the fourth-order derivatives with respect to x explicitly. They are stored in matrices AA, BB, CC, and DD, respectively.

Subroutine or the converted function DQMNEW is the differential quadrature method (DQM) with MMWC-3 to introduce the additional degrees of freedom (DOF) at the end points. A_{ij} is explicitly computed by using Eq. (1.30). The weighting coefficients of the first- to the fourth-order derivatives with respect to x are stored in matrices A, B, C, and D, respectively. BB is the weighting coefficient of the second-order derivative with respect to x in the conventional DQM and used to compute the in-plane stress components in Program 5 only.

Subroutine or the converted function HDQMNEW is the harmonic differential quadrature method (HDQM) with MMWC-3 to introduce the additional DOFs at the end points. The other new feature is that Eq. (1.33) is used to compute the weighting coefficients of the first-order derivative with respect to x explicitly. The weighting coefficients of the first- to the fourth-order derivatives with respect to x are stored in matrices A, B, C, and D, respectively. BB is the weighting coefficient of the second-order derivative with respect to x in the conventional HDQM and used to compute the in-plane stress components in Program 5 only.

The last three subroutines or converted functions, i.e., DQELE2, DQMNEW2, and HDQMNEW2, are used in the QEM. The three subroutines or converted functions are based on the DQEM, the modified DQM, and the HDQM to compute the weighting coefficients, respectively. Two differences from subroutines or converted functions named DQEM, DQMNEW, and HDQMNEW exist. One difference is that the input data of x are in the range of $[-1,1]$, the other is that only the weighting coefficients of the first- and second-order derivatives with respect to x^* ($x^* = xL/2$) are computed, where L is the length. The variable COE equals to $L/2$.

```
C-------------------------------------------------------------------------------------------
C     Eleven FORTRAN subroutines to compute weighting coefficients by various ways are listed below.
C-------------------------------------------------------------------------------------------
C        THE SUBROUTINE BELOW IS CALLED BY PROGRAM 1
C-------------------------------------------------------------------------------------------
C
       SUBROUTINE DQSS(N,X,A,B,C,D,Y)
       IMPLICIT REAL*8 (A-H,O-Z)
       DIMENSION A(N,N),B(N,N),C(N,N),D(N,N),X(N),Y(N)
C
C     X[0,L] WHERE L IS THE LENGTH;   MMWC-1
C
       DO 4 I=1,N                     ! EQ. (1.30)
       Y(I)=1.0D0
       DO 3 J=1,N
       IF (I .EQ.J) GOTO 3
       A(I,J)=1.0D0
       Y(I)=Y(I)*(X(I)-X(J))
3      CONTINUE
4      CONTINUE
       DO 6 I=1,N
       DO 5 J=1,N
       IF (I .NE.J) A(I,J)=Y(I)/(X(I)-X(J))/Y(J)
5      CONTINUE
6      CONTINUE
       DO 8 I=1,N
       A(I,I)=0.0D0
       DO 7 J=1,N
       IF (I .NE.J) A(I,I)=A(I,I)+1.0D0/(X(I)-X(J))
7      CONTINUE
8      CONTINUE

       DO 12 I=1,N                    ! EQ (1.21)
       DO 12 J=1,N
       B(I,J)=0.0D0
       DO 11 K=1,N
       B(I,J)=B(I,J)+A(I,K)*A(K,J)
11     CONTINUE
12     CONTINUE
       DO   I=1,N                     ! EQS (1.22) & (1.23)
       DO   J=1,N
       C(I,J)=0.0D0
       D(I,J)=0.0D0
       DO   K=2,N-1                       ! W"=0   AT 1 (K=1) AND N (K=N) HAVE BEEN BUILT IN
       C(I,J)=C(I,J)+A(I,K)*B(K,J)
       D(I,J)=D(I,J)+B(I,K)*B(K,J)
       ENDDO
       ENDDO
       ENDDO
       RETURN
       END
```

```
C-------------------------------------------------------------------------------------------------------------------
C       THE SUBROUTINE BELOW IS CALLED BY PROGRAM 1
C-------------------------------------------------------------------------------------------------------------------
C
        SUBROUTINE DQCC(N,X,A,B,C,D,Y)
        IMPLICIT REAL*8 (A-H,O-Z)
        DIMENSION A(N,N),B(N,N),C(N,N),D(N,N),X(N),Y(N),BB(N,N)
C
C    X[0,L] WHERE L IS THE LENGTH;   MMWC-3
C
        DO 4 I=1,N                      ! EQ. (1.30)
        Y(I)=1.0D0
        DO 3 J=1,N
        IF (I .EQ.J) GOTO 3
        A(I,J)=1.0D0
        Y(I)=Y(I)*(X(I)-X(J))
3       CONTINUE
4       CONTINUE
        DO 6 I=1,N
        DO 5 J=1,N
        IF (I .NE.J) A(I,J)=Y(I)/(X(I)-X(J))/Y(J)
5       CONTINUE
6       CONTINUE
        DO 8 I=1,N
        A(I,I)=0.0D0
        DO 7 J=1,N
        IF (I .NE.J) A(I,I)=A(I,I)+1.0D0/(X(I)-X(J))
7       CONTINUE
8       CONTINUE

        DO 12 I=1,N                     ! EQ (1.21)
        DO 12 J=1,N
        B(I,J)=0.0D0
        DO 11 K=1,N
        B(I,J)=B(I,J)+A(I,K)*A(K,J)
11      CONTINUE
12      CONTINUE
C
        DO    I=1,N,N-1
        DO    J=1,N
        BB(I,J)=0.0D0
        DO    K=2,N-1                   ! W'=0 AT 1 (K=1) AND N (K=N) HAVE BEEN BUILT IN
        BB(I,J)=BB(I,J)+A(I,K)*A(K,J)
        ENDDO
        ENDDO
        ENDDO

        DO    I=2,N-1
        DO    J=1,N
        BB(I,J)=B(I,J)
        ENDDO
        ENDDO
C
        DO    I=1,N                     ! EQS (1.22) & (1.23)
        DO    J=1,N
```

```
          C(I,J)=0.0D0
          D(I,J)=0.0D0
          DO   K=1,N
          C(I,J)=C(I,J)+A(I,K)*BB(K,J)
          D(I,J)=D(I,J)+B(I,K)*BB(K,J)
          ENDDO
          ENDDO
          ENDDO
          RETURN
          END



C---------------------------------------------------------------------------------------------------------------------
C       THE SUBROUTINE BELOW IS CALLED BY PROGRAM 1
C---------------------------------------------------------------------------------------------------------------------
C
          SUBROUTINE DQSC(N,X,A,B,C,D,Y)
          IMPLICIT REAL*8 (A-H,O-Z)
          DIMENSION A(N,N),B(N,N),C(N,N),D(N,N),X(N),Y(N),BB(N,N)
C
C       X[0,L] WHERE L IS THE LENGTH;    MIXED MMWC-1 & MMWC-4
C
          DO 4 I=1,N                        ! EQ. (1.30)
          Y(I)=1.0D0
          DO 3 J=1,N
          IF (I .EQ.J) GOTO 3
          A(I,J)=1.0D0
          Y(I)=Y(I)*(X(I)-X(J))
3         CONTINUE
4         CONTINUE
          DO 6 I=1,N
          DO 5 J=1,N
          IF (I .NE.J) A(I,J)=Y(I)/(X(I)-X(J))/Y(J)
5         CONTINUE
6         CONTINUE
          DO 8 I=1,N
          A(I,I)=0.0D0
          DO 7 J=1,N
          IF (I .NE.J) A(I,I)=A(I,I)+1.0D0/(X(I)-X(J))
7         CONTINUE
8         CONTINUE

          DO 12 I=1,N                       ! EQ (1.21)
          DO 12 J=1,N
          B(I,J)=0.0D0
          DO 11 K=1,N
          B(I,J)=B(I,J)+A(I,K)*A(K,J)
11        CONTINUE
12        CONTINUE
C
          DO   J=1,N                        ! EQS (1.22) & (1.23)
          BB(N,J)=0.0D0
          DO   K=1,N-1                      ! W'=0 AT N (K=N) HAS BEEN BUILT IN (MMWC-4)
          BB(N,J)=BB(N,J)+A(N,K)*A(K,J)
```

```
         ENDDO
         ENDDO

         DO   I=1,N-1
         DO   J=1,N
         BB(I,J)=B(I,J)
         ENDDO
         ENDDO
C
         DO   I=1,N
         DO   J=1,N
         C(I,J)=0.0D0
         D(I,J)=0.0D0
         DO   K=2,N                        ! W"=0 AT 1 HAS BEEN BUILT IN (MMWC-1)
         C(I,J)=C(I,J)+A(I,K)*BB(K,J)
         D(I,J)=D(I,J)+B(I,K)*BB(K,J)
         ENDDO
         ENDDO
         ENDDO
         RETURN
         END

C--------------------------------------------------------------------------------------------------------
C       THE SUBROUTINE BELOW IS CALLED BY PROGRAM 1
C--------------------------------------------------------------------------------------------------------
C
         SUBROUTINE DQCS(N,X,A,B,C,D,Y)
         IMPLICIT REAL*8 (A-H,O-Z)
         DIMENSION A(N,N),B(N,N),C(N,N),D(N,N),X(N),Y(N),BB(N,N)
C
C    X[0,L] WHERE L IS THE LENGTH;   MIXED MMWC-4 & MMWC-1
C
         DO 4 I=1,N                       ! EQ. (1.30)
         Y(I)=1.0D0
         DO 3 J=1,N
         IF (I .EQ.J) GOTO 3
         A(I,J)=1.0D0
         Y(I)=Y(I)*(X(I)-X(J))
3        CONTINUE
4        CONTINUE
         DO 6 I=1,N
         DO 5 J=1,N
         IF (I .NE.J) A(I,J)=Y(I)/(X(I)-X(J))/Y(J)
5        CONTINUE
6        CONTINUE
         DO 8 I=1,N
         A(I,I)=0.0D0
         DO 7 J=1,N
         IF (I .NE.J) A(I,I)=A(I,I)+1.0D0/(X(I)-X(J))
7        CONTINUE
8        CONTINUE

         DO 12 I=1,N                      ! EQ (1.21)
         DO 12 J=1,N
         B(I,J)=0.0D0
         DO 11 K=1,N
         B(I,J)=B(I,J)+A(I,K)*A(K,J)
```

```
11       CONTINUE
12       CONTINUE
C
         DO   J=1,N
         BB(1,J)=0.0D0
         DO   K=2,N                          ! W'=0 AT 1 (K=1) HAS BEEN BUILT IN (MMWC-4)
         BB(1,J)=BB(1,J)+A(1,K)*A(K,J)
         ENDDO
         ENDDO
         DO   I=2,N
         DO   J=1,N
         BB(I,J)=B(I,J)
         ENDDO
         ENDDO
C
         DO   I=1,N                          ! EQS (1.22) & (1.23)
         DO   J=1,N
         C(I,J)=0.0D0
         D(I,J)=0.0D0
         DO   K=1,N-1                         ! W"=0 AT N HAS BEEN BUILT IN (MMWC-1)
         C(I,J)=C(I,J)+A(I,K)*BB(K,J)
         D(I,J)=D(I,J)+B(I,K)*BB(K,J)
         ENDDO
         ENDDO
         ENDDO
         RETURN
         END


C----------------------------------------------------------------------------------------------------------------
C       THE SUBROUTINE BELOW IS CALLED BY PROGRAM 6
C----------------------------------------------------------------------------------------------------------------
C
         SUBROUTINE DQM(N,X,A,B)
         IMPLICIT REAL*8 (A-H,O-Z)
         DIMENSION A(N,N),B(N,N),X(N),Y(N)
C
C    X[-L/2,L/2] OR [0,L], L IS THE LENGTH
C
         DO 4 I=1,N                     ! EQ. (1.30)
         Y(I)=1.0D0
         DO 3 J=1,N
         IF (I .EQ.J) GOTO 3
         A(I,J)=1.0D0
         Y(I)=Y(I)*(X(I)-X(J))
3        CONTINUE
4        CONTINUE
         DO 6 I=1,N
         DO 5 J=1,N
         IF (I .NE.J) A(I,J)=Y(I)/(X(I)-X(J))/Y(J)
5        CONTINUE
6        CONTINUE
         DO 8 I=1,N
         A(I,I)=0.0D0
         DO 7 J=1,N
         IF (I .NE.J) A(I,I)=A(I,I)+1.0D0/(X(I)-X(J))
7        CONTINUE
8        CONTINUE
```

```
      DO 12 I=1,N              ! EQ (1.21)
      DO 12 J=1,N
      B(I,J)=0.0D0
      DO 11 K=1,N
      B(I,J)=B(I,J)+A(I,K)*A(K,J)
11    CONTINUE
12    CONTINUE

      RETURN
      END


C---------------------------------------------------------------------------
C              Subroutine DQEM below is included for reference.
C---------------------------------------------------------------------------
C
      SUBROUTINE DQEM(N,X,AA,BB,CC,DD)
      IMPLICIT REAL*8 (A-H,O-Z)
      DIMENSION X(N), A(N,N),B(N,N),C(N,N),D(N,N),Y(N)
      DIMENSION AA(N,N+2),CC(N,N+2),BB(N,N+2),DD(N,N+2)
C
C   X[0, L] , L IS THE LENGTH; WX0 AND WXN ARE PUT AFTER WN
C
      DO 4 I=1,N                ! EQ (1.21)
      Y(I)=1.0D0
      DO 3 J=1,N
      IF (I .EQ.J) GOTO 3
      A(I,J)=1.0D0
      Y(I)=Y(I)*(X(I)-X(J))
3     CONTINUE
4     CONTINUE
      DO 6 I=1,N
      DO 5 J=1,N
      IF (I .NE.J) A(I,J)=Y(I)/(X(I)-X(J))/Y(J)
5     CONTINUE
6     CONTINUE
      DO 8 I=1,N
      A(I,I)=0.0D0
      DO 7 J=1,N
      IF (I .NE.J) A(I,I)=A(I,I)+1.0D0/(X(I)-X(J))
7     CONTINUE
8     CONTINUE

      DO 12 I=1,N               ! EQ (1.21)
      DO 12 J=1,N
      B(I,J)=0.0D0
      DO 11 K=1,N
      B(I,J)=B(I,J)+A(I,K)*A(K,J)
11    CONTINUE
12    CONTINUE
C
```

```
      DO   I=1,N                          ! EQS (1.22) & (1.23)
      DO   J=1,N
      C(I,J)=0.0D0
      D(I,J)=0.0D0
      DO   K=1,N
      C(I,J)=C(I,J)+A(I,K)*B(K,J)
      D(I,J)=D(I,J)+B(I,K)*B(K,J)
      ENDDO
      ENDDO
      ENDDO
C
C     DQEM
C
      DO I=1,N                                              ! EQ. (2.14)
      AA(I,N+1)=A(I,1)*(X(I)-X(1))*(X(I)-X(N))/(X(1)-X(N))
      AA(I,N+2)=A(I,N)*(X(I)-X(1))*(X(I)-X(N))/(X(N)-X(1))
      ENDDO
      AA(1,N+1)=AA(1,N+1)+1.0D0
      AA(N,N+2)=AA(N,N+2)+1.0D0
      DO I=1,N
      BB(I,N+1)=B(I,1)*(X(I)-X(1))*(X(I)-X(N))/(X(1)-X(N))
     1          +2.0D0*A(I,1)*(X(I)-X(1)+X(I)-X(N))/(X(1)-X(N))
      BB(I,N+2)=B(I,N)*(X(I)-X(1))*(X(I)-X(N))/(X(N)-X(1))
     1          +2.0D0*A(I,N)*(X(I)-X(1)+X(I)-X(N))/(X(N)-X(1))
      ENDDO
      BB(1,N+1)=BB(1,N+1)+2.0D0/(X(1)-X(N))
      BB(N,N+2)=BB(N,N+2)+2.0D0/(X(N)-X(1))
C
      DO I=1,N
      CC(I,N+1)=C(I,1)*(X(I)-X(1))*(X(I)-X(N))/(X(1)-X(N))
     1          +3.0D0*B(I,1)*(X(I)-X(1)+X(I)-X(N))/(X(1)-X(N))
     2          +6.0D0*A(I,1)/(X(1)-X(N))
      CC(I,N+2)=C(I,N)*(X(I)-X(1))*(X(I)-X(N))/(X(N)-X(1))
     1          +3.0D0*B(I,N)*(X(I)-X(1)+X(I)-X(N))/(X(N)-X(1))
     2          +6.0D0*A(I,N)/(X(N)-X(1))
      ENDDO
      DO I=1,N
      DD(I,N+1)=D(I,1)*(X(I)-X(1))*(X(I)-X(N))/(X(1)-X(N))
     1          +4.0D0*C(I,1)*(X(I)-X(1)+X(I)-X(N))/(X(1)-X(N))
     2          +12.0D0*B(I,1)/(X(1)-X(N))
      DD(I,N+2)=D(I,N)*(X(I)-X(1))*(X(I)-X(N))/(X(N)-X(1))
     1          +4.0D0*C(I,N)*(X(I)-X(1)+X(I)-X(N))/(X(N)-X(1))
     2          +12.0D0*B(I,N)/(X(N)-X(1))
      ENDDO
C
      DO I=1,N                                              ! EQS (2.15) & (2.16)
      AA(I,1)=A(I,1)*(X(I)-X(N))/(X(1)-X(N))-(A(1,1)+1.0D0/(X(1)-X(N)))
     1       *AA(I,N+1)
      AA(I,N)=A(I,N)*(X(I)-X(1))/(X(N)-X(1))-(A(N,N)+1.0D0/(X(N)-X(1)))
     1       *AA(I,N+2)
```

```
      ENDDO
      AA(1,1)=AA(1,1)+1.0D0/(X(1)-X(N))
      AA(N,N)=AA(N,N)+1.0D0/(X(N)-X(1))
      DO I=1,N
      BB(I,1)=(B(I,1)*(X(I)-X(N))+2.0D0*A(I,1))/(X(1)-X(N))-
     1       (A(1,1)+1.0D0/(X(1)-X(N)))*BB(I,N+1)
      BB(I,N)=(B(I,N)*(X(I)-X(1))+2.0D0*A(I,N))/(X(N)-X(1))-
     1       (A(N,N)+1.0D0/(X(N)-X(1)))*BB(I,N+2)
      ENDDO
      DO I=1,N
      CC(I,1)=(C(I,1)*(X(I)-X(N))+3.0D0*B(I,1))/(X(1)-X(N))-
     1       (A(1,1)+1.0D0/(X(1)-X(N)))*CC(I,N+1)
      CC(I,N)=(C(I,N)*(X(I)-X(1))+3.0D0*B(I,N))/(X(N)-X(1))-
     1       (A(N,N)+1.0D0/(X(N)-X(1)))*CC(I,N+2)
      ENDDO
      DO I=1,N
      DD(I,1)=(D(I,1)*(X(I)-X(N))+4.0D0*C(I,1))/(X(1)-X(N))-
     1       (A(1,1)+1.0D0/(X(1)-X(N)))*DD(I,N+1)
      DD(I,N)=(D(I,N)*(X(I)-X(1))+4.0D0*C(I,N))/(X(N)-X(1))-
     1       (A(N,N)+1.0D0/(X(N)-X(1)))*DD(I,N+2)
      ENDDO
C
      DO J=2,N-1
      DO I=1,N
      AA(I,J)=A(I,J)*(X(I)-X(1))*(X(I)-X(N))/(X(J)-X(1))/(X(J)-X(N))
      ENDDO
      AA(J,J)=AA(J,J)+(X(J)-X(1)+X(J)-X(N))/(X(J)-X(1))/(X(J)-X(N))
      ENDDO
      DO J=2,N-1
      DO I=1,N
      BB(I,J)=B(I,J)*(X(I)-X(1))*(X(I)-X(N))/(X(J)-X(1))/(X(J)-X(N))
     1      +2.0D0*A(I,J)*(2.0D0*X(I)-X(1)-X(N))/(X(J)-X(1))/(X(J)-X(N))
      ENDDO
      BB(J,J)=BB(J,J)+2.0D0/(X(J)-X(1))/(X(J)-X(N))
      ENDDO
C
      DO J=2,N-1
      DO I=1,N
      CC(I,J)=C(I,J)*(X(I)-X(1))*(X(I)-X(N))/(X(J)-X(1))/(X(J)-X(N))
     1       +3.0D0*B(I,J)*(2.0D0*X(I)-X(1)-X(N))/(X(J)-X(1))/(X(J)-X(N))
     2       +6.0D0*A(I,J)/(X(J)-X(1))/(X(J)-X(N))
      ENDDO
      ENDDO

      DO J=2,N-1
      DO I=1,N
      DD(I,J)=D(I,J)*(X(I)-X(1))*(X(I)-X(N))/(X(J)-X(1))/(X(J)-X(N))
     1       +4.0D0*C(I,J)*(2.0D0*X(I)-X(1)-X(N))/(X(J)-X(1))/(X(J)-X(N))
     2       +12.0D0*B(I,J)/(X(J)-X(1))/(X(J)-X(N))
```

```
      ENDDO
      ENDDO
      RETURN
      END
C-------------------------------------------------------------------------------------------------------
C      THE SUBROUTINE BELOW IS CALLED BY PROGRAMS 2 & 5
C-------------------------------------------------------------------------------------------------------
C
      SUBROUTINE DQMNEW(M,X,A,B,C,D,BB)
      IMPLICIT REAL*8 (A-H,O-Z)
      DIMENSION A(M,M+2),B(M,M+2),C(M,M+2),D(M,M+2),X(M)
      DIMENSION Y(M),BB(M,M)
C
C   X[0,L] OR [-L/2,L/2], WHERE L IS THE LENGTH; MMWC-3 IS USED FOR B,C,D
C
      DO 4 I=1,M                    ! EQ. (1.30)
      Y(I)=1.0D0
      DO 3 J=1,M
      IF (I .EQ.J) GOTO 3
      Y(I)=Y(I)*(X(I)-X(J))
3     CONTINUE
4     CONTINUE
      DO 6 I=1,M
      DO 5 J=1,M
      IF (I .NE.J) A(I,J)=Y(I)/(X(I)-X(J))/Y(J)
5     CONTINUE
6     CONTINUE
      DO 8 I=1,M
      A(I,I)=0.0D0
      DO 7 J=1,M
      IF (I .NE.J) A(I,I)=A(I,I)+1.0D0/(X(I)-X(J))
7     CONTINUE
8     CONTINUE

      DO 10 I=1,M                   ! EQ (1.21)
      DO 10 J=1,M
      BB(I,J)=0.0D0
      DO 9 K=1,M
      BB(I,J)=BB(I,J)+A(I,K)*A(K,J)
9     CONTINUE
      B(I,J)=BB(I,J)
10    CONTINUE
C
      DO   I=1,M,M-1                ! MMWC-3 FOR TWO END POINTS, EQ (3.17)
      DO   J=1,M
      B(I,J)=0.0D0
      DO   K=2,M   -1
      B(I,J)=B(I,J)+A(I,K)*A(K,J)
      ENDDO
```

```
      ENDDO
      ENDDO
C
      DO I=1,M
      DO J=M+1,M+2                    ! EXPAND
      A (I,J)=0.0D0
      B (I,J)=0.0D0
      ENDDO
      ENDDO
C
      B(1,M+1)=A(1,1)                 ! MMWC-3
      B(1,M+2)=A(1,M)
      B(M,M+1)=A(M,1)
      B(M,M+2)=A(M,M)

      DO 12 I=1,M                     ! MMWC-3, EQS(3.18)-(3.19)
      DO 12 J=1,M+2
      C(I,J)=0.0D0
      D(I,J)=0.0D0
      DO 11 K=1,M
      C(I,J)=C(I,J)+ A(I,K)*B(K,J)
      D(I,J)=D(I,J)+BB(I,K)*B(K,J)
11    CONTINUE
12    CONTINUE
      RETURN
      END



C-------------------------------------------------------------------------
C           THE SUBROUTINE BELOWIS CAN BE  CALLED BY PROGRAMS 2 & 5
C-------------------------------------------------------------------------
C
      SUBROUTINE HDQMNEW(M,X,A,B,C,D,BB)
      IMPLICIT REAL*8 (A-H,O-Z)
      DIMENSION A(M,M+2),B(M,M+2),C(M,M+2),D(M,M+2),X(M)
      DIMENSION Y(M),BB(M,M)
C
C  X[0, L] , L IS THE LENGTH, NOTE: EQ(1.32) FOR X[-1,1], THUS PI/8.
C
      PI4=DATAN(1.0D0)  !PI/4
      DO 4 I=1,M                      ! SIMILAR TO EQ. (1.32)
      Y(I)=1.0D0
      DO 3 J=1,M
      IF (I .EQ.J) GOTO 3
      Y(I)=Y(I)*DSIN((X(I)-X(J))*PI4)
3     CONTINUE
4     CONTINUE
      DO 6 I=1,M
      DO 5 J=1,M
      IF (I .NE.J) A(I,J)=PI4*Y(I)/DSIN(PI4*(X(I)-X(J)))/Y(J)
```

```
5       CONTINUE
6       CONTINUE
        DO 8 I=1,M
        A(I,I)=0.0D0
        DO 7 J=1,M
        IF (I .NE.J) A(I,I)=A(I,I)+PI4*DCOS((X(I)-X(J))*PI4)/
    1                        DSIN(PI4*(X(I)-X(J)))
7       CONTINUE
8       CONTINUE
        DO 10 I=1,M                      ! EQ (1.21)
        DO 10 J=1,M
        BB(I,J)=0.0D0
        DO 9 K=1,M
        BB(I,J)=BB(I,J)+A(I,K)*A(K,J)
9       CONTINUE
        B(I,J)=BB(I,J)
10      CONTINUE
        DO    I=1,M,M-1                  ! MMWC-3, EQ (3.17)
        DO    J=1,M
        B(I,J)=0.0D0
        DO    K=2,M-1
        B(I,J)=B(I,J)+A(I,K)*A(K,J)
        ENDDO
        ENDDO
        ENDDO
        DO I=1,M
        DO J=M+1,M+2
        A (I,J)=0.0D0
        B (I,J)=0.0D0
        ENDDO
        ENDDO
        B(1,M+1)=A(1,1)
        B(1,M+2)=A(1,M)
        B(M,M+1)=A(M,1)
        B(M,M+2)=A(M,M)
C
        DO 12 I=1,M                      ! EQS (3.18)-(3.19)
        DO 12 J=1,M+2
        C(I,J)=0.0D0
        D(I,J)=0.0D0
        DO 11 K=1,M
        C(I,J)=C(I,J)+ A(I,K)*B(K,J)
        D(I,J)=D(I,J)+BB(I,K)*B(K,J)
11      CONTINUE
12      CONTINUE
        RETURN
        END
C-------------------------------------------------------------------------
C       THE SUBROUTINE BELOW IS CALLED BY PROGRAMS 3 & 4
C-------------------------------------------------------------------------
C
```

```
         SUBROUTINE DQELE2(N,X,AA,BB,COE)
         IMPLICIT REAL*8 (A-H,O-Z)
         DIMENSION X(N)
         DIMENSION A(N,N),B(N,N),Y(N)
         DIMENSION AA(N,N+2),BB(N,N+2)
C
C     X[-1,1], COE-- LENGTH/2; DQEM
C
         DO 4 I=1,N                    ! EQ. (1.30)
         Y(I)=1.0D0
         DO 3 J=1,N
         IF (I .EQ.J) GOTO 3
         A(I,J)=1.0D0
         Y(I)=Y(I)*(X(I)-X(J))
3        CONTINUE
4        CONTINUE
         DO 6 I=1,N
         DO 5 J=1,N
         IF (I .NE.J) A(I,J)=Y(I)/(X(I)-X(J))/Y(J)
5        CONTINUE
6        CONTINUE
         DO 8 I=1,N
         A(I,I)=0.0D0
         DO 7 J=1,N
         IF (I .NE.J) A(I,I)=A(I,I)+1.0D0/(X(I)-X(J))
7        CONTINUE
8        CONTINUE
         DO I=1,N
         DO J=1,N
         A(I,J)=A(I,J)/1.0D0
         ENDDO
         ENDDO

         DO 12 I=1,N                   ! EQ. (1.21)
         DO 12 J=1,N
         B(I,J)=0.0D0
         DO 11 K=1,N
         B(I,J)=B(I,J)+A(I,K)*A(K,J)
11       CONTINUE
12       CONTINUE
C
C          DQEM
C
         DO I=1,N                                        ! EQ. (2.14)
         AA(I,N+1)=A(I,1)*(X(I)-X(1))*(X(I)-X(N))/(X(1)-X(N))
         AA(I,N+2)=A(I,N)*(X(I)-X(1))*(X(I)-X(N))/(X(N)-X(1))
         ENDDO
         AA(1,N+1)=AA(1,N+1)+1.0D0
         AA(N,N+2)=AA(N,N+2)+1.0D0
         DO I=1,N
```

```
      BB(I,N+1)=B(I,1)*(X(I)-X(1))*(X(I)-X(N))/(X(1)-X(N))
  1            +2.0D0*A(I,1)*(X(I)-X(1)+X(I)-X(N))/(X(1)-X(N))
      BB(I,N+2)=B(I,N)*(X(I)-X(1))*(X(I)-X(N))/(X(N)-X(1))
  1            +2.0D0*A(I,N)*(X(I)-X(1)+X(I)-X(N))/(X(N)-X(1))
      ENDDO
       BB(1,N+1)=BB(1,N+1)+2.0D0/(X(1)-X(N))
       BB(N,N+2)=BB(N,N+2)+2.0D0/(X(N)-X(1))
C
      DO I=1,N                                                    ! EQS (2.15) & (2.16)
      AA(I,1)=A(I,1)*(X(I)-X(N))/(X(1)-X(N))-(A(1,1)+1.0D0/(X(1)-X(N)))
  1         *AA(I,N+1)
      AA(I,N)=A(I,N)*(X(I)-X(1))/(X(N)-X(1))-(A(N,N)+1.0D0/(X(N)-X(1)))
  1         *AA(I,N+2)
      ENDDO
      AA(1,1)=AA(1,1)+1.0D0/(X(1)-X(N))
      AA(N,N)=AA(N,N)+1.0D0/(X(N)-X(1))
      DO I=1,N
      BB(I,1)=(B(I,1)*(X(I)-X(N))+2.0D0*A(I,1))/(X(1)-X(N))-
  1          (A(1,1)+1.0D0/(X(1)-X(N)))*BB(I,N+1)
      BB(I,N)=(B(I,N)*(X(I)-X(1))+2.0D0*A(I,N))/(X(N)-X(1))-
  1          (A(N,N)+1.0D0/(X(N)-X(1)))*BB(I,N+2)
      ENDDO
C
      DO J=2,N-1
      DO I=1,N
      AA(I,J)=A(I,J)*(X(I)-X(1))*(X(I)-X(N))/(X(J)-X(1))/(X(J)-X(N))
      ENDDO
      AA(J,J)=AA(J,J)+(X(J)-X(1)+X(J)-X(N))/(X(J)-X(1))/(X(J)-X(N))
      ENDDO
      DO J=2,N-1
      DO I=1,N
      BB(I,J)=B(I,J)*(X(I)-X(1))*(X(I)-X(N))/(X(J)-X(1))/(X(J)-X(N))
  1         +2.0D0*A(I,J)*(2.0D0*X(I)-X(1)-X(N))/(X(J)-X(1))/(X(J)-X(N))
      ENDDO
      BB(J,J)=BB(J,J)+2.0D0/(X(J)-X(1))/(X(J)-X(N))
      ENDDO
C
      DO I=1,N
      DO J=1,N+2
      AA(I,J)=AA(I,J)/COE
      BB(I,J)=BB(I,J)/COE/COE
      ENDDO
      ENDDO
      RETURN
      END


C-------------------------------------------------------------------------------------------------------------
C      THE SUBROUTINE BELOW IS CALLED BY PROGRAMS 3 & 4
C-------------------------------------------------------------------------------------------------------------
C
```

```
         SUBROUTINE DQMNEW2(M,X,A,B,COE)
         IMPLICIT REAL*8 (A-H,O-Z)
         DIMENSION A(M,M+2),B(M,M+2),X(M)
         DIMENSION Y(M)
C
C   COE -- LENGTH/2; X[-1,1]; MMWC-3
C
         DO 4 I=1,M                        ! EQ. (1.30)
         Y(I)=1.0D0
         DO 3 J=1,M
         IF (I .EQ.J) GOTO 3
         Y(I)=Y(I)*(X(I)-X(J))
3        CONTINUE
4        CONTINUE
         DO 6 I=1,M
         DO 5 J=1,M
         IF (I .NE.J) A(I,J)=Y(I)/(X(I)-X(J))/Y(J)
5        CONTINUE
6        CONTINUE
         DO 8 I=1,M
         A(I,I)=0.0D0
         DO 7 J=1,M
         IF (I .NE.J) A(I,I)=A(I,I)+1.0D0/(X(I)-X(J))
7        CONTINUE
8        CONTINUE
         DO I=1,M
         DO J=1,M
         A(I,J)=A(I,J)/COE
         ENDDO
         ENDDO
         DO 10 I=2,M-1                            ! EQ. (3.17)
         DO 10 J=1,M
         B(I,J)=0.0D0
         DO 9 K=1,M
         B(I,J)=B(I,J)+A(I,K)*A(K,J)
9        CONTINUE
10       CONTINUE
         DO   I=1,M,M-1
         DO   J=1,M
         B(I,J)=0.0D0
         DO   K=2,M-1                    ! MMWC-3,  EQ. (3.17)
         B(I,J)=B(I,J)+A(I,K)*A(K,J)
         ENDDO
         ENDDO
         ENDDO
         DO I=1,M
         DO J=M+1,M+2
         A (I,J)=0.0D0
         B (I,J)=0.0D0
         ENDDO
```

```
        ENDDO
        B(1,M+1)=A(1,1)
        B(1,M+2)=A(1,M)
        B(M,M+1)=A(M,1)
        B(M,M+2)=A(M,M)
        RETURN
        END


C------------------------------------------------------------------------------------------------------------------------
C       THE SUBROUTINE BELOW IS CALLED BY PROGRAMS 3 & 4
C------------------------------------------------------------------------------------------------------------------------
C
        SUBROUTINE HDQMNEW2(M,X,A,B,COE)
        IMPLICIT REAL*8 (A-H,O-Z)
        DIMENSION A(M,M+2),B(M,M+2),X(M)
        DIMENSION Y(M)
C
C    X [-1,1]; COE-- LENGTH/2;   HDQMNEW
C
        PI8=DATAN(1.0D0)/2.0D0    ! PI/8
        DO 4 I=1,M                             ! EQ. (1.32)
        Y(I)=1.0D0
        DO 3 J=1,M
        IF (I .EQ.J) GOTO 3
        Y(I)=Y(I)*DSIN((X(I)-X(J))*PI8)
3       CONTINUE
4       CONTINUE
        DO 6 I=1,M
        DO 5 J=1,M
        IF (I .NE.J) A(I,J)=PI8*Y(I)/DSIN(PI8*(X(I)-X(J)))/Y(J)
5       CONTINUE
6       CONTINUE
        DO 8 I=1,M
        A(I,I)=0.0D0
        DO 7 J=1,M
        IF (I .NE.J) A(I,I)=A(I,I)+PI8*DCOS((X(I)-X(J))*PI8)/
     1                    DSIN(PI8*(X(I)-X(J)))
7       CONTINUE
8       CONTINUE

        DO I=1,M
        DO J=1,M
        A(I,J)=A(I,J)/COE
        ENDDO
        ENDDO

        DO 10 I=2,M-1                            ! EQ. (3.17)
        DO 10 J=1,M
        B(I,J)=0.0D0
        DO 9 K=1,M
        B(I,J)=B(I,J)+A(I,K)*A(K,J)
9       CONTINUE
10      CONTINUE
```

```
        DO     I=1,M,M-1
        DO     J=1,M
        B(I,J)=0.0D0
        DO     K=2,M-1
        B(I,J)=B(I,J)+A(I,K)*A(K,J)
        ENDDO
        ENDDO
        ENDDO
        DO I=1,M
        DO J=M+1,M+2
        A (I,J)=0.0D0
        B (I,J)=0.0D0
        ENDDO
        ENDDO
        B(1,M+1)=A(1,1)
        B(1,M+2)=A(1,M)
        B(M,M+1)=A(M,1)
        B(M,M+2)=A(M,M)

        RETURN
        END


C-------------------------------------------------------------------------------------------------------------
C                        The end of the eleven FORTRAN subroutines
C-------------------------------------------------------------------------------------------------------------



%-------------------------------------------------------------------------------------------------------------
%     The converted MATLAB functions from the FORTRAN subroutines are listed below.
%     The names are the same as the corresponding FORTRAN subroutines.
%-------------------------------------------------------------------------------------------------------------


%-------------------------------------------------------------------------------------------------------------
%     The function below is used in Program1_Main
%-------------------------------------------------------------------------------------------------------------
%
function [A, B, C, D]=DQSS(N,X)
%
%     X[0,L], L is the length; MMWC-1
%
Y=ones(N);
for I=1:N                          % EQ. (1.30)
        for J=1:N
                if I~=J
                        Y(I)=Y(I)*(X(I)-X(J));
                end
        end
end

A=zeros(N,N);
for I=1:N
        for J=1:N
                if I~=J
                        A(I,J)=Y(I)/(X(I)-X(J))/Y(J);
                        A(I,I)=A(I,I)+1.0/(X(I)-X(J));
```

```
            end
        end
end

B=A*A;                        % EQ. (1.21)

K=2:N-1;           % W''=0    AT 1 (K=1) AND N (K=N) HAVE BEEN BUILT IN
C=A(:,K)*B(K,:);                        % EQ (1.22)
D=B(:,K)*B(K,:);                        % EQ (1.23)
```

```
%--------------------------------------------------------------------------------------------------------------------------
%    The function below is used in Program1_Main
%--------------------------------------------------------------------------------------------------------------------------
%
function [A, B, C, D]=DQCC(N,X)
%
%    X[0,L], L is the length; MMWC-3
%
Y=ones(N);
for I=1:N                          % EQ. (1.30)
        for J=1:N
            if I~=J
                    Y(I)=Y(I)*(X(I)-X(J));
            end
        end
end

A=zeros(N,N);
for I=1:N
        for J=1:N
            if I~=J
                    A(I,J)=Y(I)/(X(I)-X(J))/Y(J);
                    A(I,I)=A(I,I)+1.0/(X(I)-X(J));
            end
        end
end

B=A*A;                        % EQ. (1.21)

BB=zeros(N,N);
I=[1,N]; K=2:N-1;
BB(I,:)=A(I,K)*A(K,:);           % W'=0 AT 1 (K=1) AND N (K=N) HAVE BEEN BUILT IN

I=2:N-1;
BB(I,:)=B(I,:);

C=A*BB;                        % EQ (1.22)
D=B*BB;                        % EQ. (1.23)
```

```
%--------------------------------------------------------------------------------------------------------------------------
%    The function below is used in Program1_Main
%--------------------------------------------------------------------------------------------------------------------------
%
function [A, B, C, D]=DQSC(N,X)
%
%    X[0,L], L is the length;    MIXED MMWC-1 & MMWC-4
```

```
%
Y=ones(N);
for I=1:N                        % EQ. (1.30)
        for J=1:N
                if I~=J
                        Y(I)=Y(I)*(X(I)-X(J));
                end
        end
end

A=zeros(N,N);
for I=1:N
        for J=1:N
                if I~=J
                        A(I,J)=Y(I)/(X(I)-X(J))/Y(J);
                        A(I,I)=A(I,I)+1.0/(X(I)-X(J));
                end
        end
end

B=A*A;                           % EQ. (1.21)

BB=zeros(N,N);
K=1:N-1;
BB(N,:)=A(N,K)*A(K,:);           % W'=0 AT N (K=N) HAS BEEN BUILT IN (MMWC-4)

I=1:N-1;
BB(I,:)=B(I,:);

K=2:N;                           % W''=0 AT 1 HAS BEEN BUILT IN (MMWC-1)
C=A(:,K)*BB(K,:);                % EQ. (1.22)
D=B(:,K)*BB(K,:);                % EQ. (1.23)

%-----------------------------------------------------------------------------------------------------------------------
%    The function below is used in Program1_Main
%-----------------------------------------------------------------------------------------------------------------------
%
function [A, B, C, D]=DQCS(N,X)
%
%    X[0,L], L is the length;    MIXED MMWC-4 & MMWC-1
%
Y=ones(N);
for I=1:N                        % EQ. (1.30)
        for J=1:N
                if I~=J
                        Y(I)=Y(I)*(X(I)-X(J));
                end
        end
end
A=zeros(N,N);
for I=1:N
        for J=1:N
                if I~=J
                        A(I,J)=Y(I)/(X(I)-X(J))/Y(J);
                        A(I,I)=A(I,I)+1.0/(X(I)-X(J));
                end
```

```
      end
end
B=A*A;                      % EQ. (1.21)

BB=zeros(N,N);
K=2:N;
BB(1,:)=A(1,K)*A(K,:);      % W'=0 AT 1 (K=1) HAS BEEN BUILT IN (MMWC-4)
I=2:N;
BB(I,:)=B(I,:);

K=1:N-1;                    % W"=0 AT N (K=N) HAS BEEN BUILT IN (MMWC-1)
C=A(:,K)*BB(K,:);           % EQ. (1.22)
D=B(:,K)*BB(K,:);           % EQ. (1.23)
```

```
%-----------------------------------------------------------------------------------
%     The function below is used in Program6_Main
%-----------------------------------------------------------------------------------
%
function [ A,B ] = DQM( N,X )
%
%     LENGTH IS INCLUDED IN THE X(I)
%
      A=zeros(N,N);
      B=zeros(N,N);

      Y=ones(1,N);
       for I=1:N              % EQ. (1.30)
          for J=1:N
              if I ~=J
                   A(I,J)=1.0D0;
                   Y(I)=Y(I)*(X(I)-X(J));
              end
          end
       end
       for I=1:N
          for J=1:N
              if I ~=J
                   A(I,J)=Y(I)/(X(I)-X(J))/Y(J);
              end
          end
       end

       for I=1:N
          A(I,I)=0.0D0;
          for J=1:N
              if I ~= J
                   A(I,I)=A(I,I)+1.0D0/(X(I)-X(J));
              end
          end
       end

       B=A*A;                 % EQ. (1.21)
end
```

```
%-----------------------------------------------------------------------------------
%     The function DQEM is included for reference.
%-----------------------------------------------------------------------------------
%
```

```
function [AA, BB, CC, DD]=DQEM(N, X)
%
%     X[0,L], L is the length; WX0 AND WXN ARE PUT AFTER WN
%
Y=ones(N);
for I=1:N                    % EQ. (1.30) OR EQ (2.17)
        for J=1:N
            if I~=J
                    Y(I)=Y(I)*(X(I)-X(J));
            end
        end
end

A=zeros(N,N);
for I=1:N
        for J=1:N
            if I~=J
                    A(I,J)=Y(I)/(X(I)-X(J))/Y(J);
                    A(I,I)=A(I,I)+1.0/(X(I)-X(J));
            end
        end

end

B=A*A;              % EQ. (1.21)
C=A*B;              % EQ. (1.22)
D=B*B;              % EQ. (1.23)

%
%       DQEM
%
for I=1:N                                                % EQ. (2.14)
        AA(I,N+1)=A(I,1)*(X(I)-X(1))*(X(I)-X(N))/(X(1)-X(N));
        AA(I,N+2)=A(I,N)*(X(I)-X(1))*(X(I)-X(N))/(X(N)-X(1));
end

AA(1,N+1)=AA(1,N+1)+1.0;
AA(N,N+2)=AA(N,N+2)+1.0;

for I=1:N
        BB(I,N+1)=B(I,1)*(X(I)-X(1))*(X(I)-X(N))/(X(1)-X(N))+2.0*A(I,1)*(X(I)-X(1)+X(I)-X(N))/(X(1)-X(N));
        BB(I,N+2)=B(I,N)*(X(I)-X(1))*(X(I)-X(N))/(X(N)-X(1))+2.0*A(I,N)*(X(I)-X(1)+X(I)-X(N))/(X(N)-X(1));
end

BB(1,N+1)=BB(1,N+1)+2.0/(X(1)-X(N));
BB(N,N+2)=BB(N,N+2)+2.0/(X(N)-X(1));

for I=1:N
        CC(I,N+1)=C(I,1)*(X(I)-X(1))*(X(I)-X(N))/(X(1)-X(N))+3.0*B(I,1)*(X(I)-X(1)+X(I)-X(N))/(X(1)-X(N))+...
        6.0*A(I,1)/(X(1)-X(N));
        CC(I,N+2)=C(I,N)*(X(I)-X(1))*(X(I)-X(N))/(X(N)-X(1))+3.0*B(I,N)*(X(I)-X(1)+X(I)-X(N))/(X(N)-X(1))+...
        6.0*A(I,N)/(X(N)-X(1));
end

for I=1:N
        DD(I,N+1)=D(I,1)*(X(I)-X(1))*(X(I)-X(N))/(X(1)-X(N))+4.0*C(I,1)*(X(I)-X(1)+X(I)-X(N))/(X(1)-X(N))+...
        12.0*B(I,1)/(X(1)-X(N));
        DD(I,N+2)=D(I,N)*(X(I)-X(1))*(X(I)-X(N))/(X(N)-X(1))+4.0*C(I,N)*(X(I)-X(1)+X(I)-X(N))/(X(N)-X(1))+...
        12.0*B(I,N)/(X(N)-X(1));
end
```

```
for I=1:N                                                               % EQS. (2.15)-(2.16)
        AA(I,1)=A(I,1)*(X(I)-X(N))/(X(1)-X(N))-(A(1,1)+1.0/(X(1)-X(N)))*AA(I,N+1);
        AA(I,N)=A(I,N)*(X(I)-X(1))/(X(N)-X(1))-(A(N,N)+1.0/(X(N)-X(1)))*AA(I,N+2);
end

AA(1,1)=AA(1,1)+1.0/(X(1)-X(N));
AA(N,N)=AA(N,N)+1.0/(X(N)-X(1));

for I=1:N
        BB(I,1)=(B(I,1)*(X(I)-X(N))+2.0*A(I,1))/(X(1)-X(N))-(A(1,1)+1.0/(X(1)-X(N)))*BB(I,N+1);
        BB(I,N)=(B(I,N)*(X(I)-X(1))+2.0*A(I,N))/(X(N)-X(1))-(A(N,N)+1.0/(X(N)-X(1)))*BB(I,N+2);
end

for I=1:N
        CC(I,1)=(C(I,1)*(X(I)-X(N))+3.0*B(I,1))/(X(1)-X(N))-(A(1,1)+1.0/(X(1)-X(N)))*CC(I,N+1);
        CC(I,N)=(C(I,N)*(X(I)-X(1))+3.0*B(I,N))/(X(N)-X(1))-(A(N,N)+1.0/(X(N)-X(1)))*CC(I,N+2);
end

for I=1:N
        DD(I,1)=(D(I,1)*(X(I)-X(N))+4.0*C(I,1))/(X(1)-X(N))-(A(1,1)+1.0/(X(1)-X(N)))*DD(I,N+1);
        DD(I,N)=(D(I,N)*(X(I)-X(1))+4.0*C(I,N))/(X(N)-X(1))-(A(N,N)+1.0/(X(N)-X(1)))*DD(I,N+2);
end

for J=2:N-1
        for I=1:N
        AA(I,J)=A(I,J)*(X(I)-X(1))*(X(I)-X(N))/(X(J)-X(1))/(X(J)-X(N));
        end
        AA(J,J)=AA(J,J)+(X(J)-X(1)+X(J)-X(N))/(X(J)-X(1))/(X(J)-X(N));
end

for J=2:N-1
        for I=1:N

        BB(I,J)=B(I,J)*(X(I)-X(1))*(X(I)-X(N))/(X(J)-X(1))/(X(J)-X(N))+2.0*A(I,J)*(2.0*X(I)-X(1)-X(N))/(X(J)-X(1))/...
        (X(J)-X(N));
        end

        BB(J,J)=BB(J,J)+2.0/(X(J)-X(1))/(X(J)-X(N));
end

for J=2:N-1
        for I=1:N

        CC(I,J)=C(I,J)*(X(I)-X(1))*(X(I)-X(N))/(X(J)-X(1))/(X(J)-X(N))+3.0*B(I,J)*(2.0*X(I)-X(1)-X(N))/(X(J)-X(1))/...
        (X(J)-X(N))+6.0*A(I,J)/(X(J)-X(1))/(X(J)-X(N));
        end
end

for J=2:N-1
        for I=1:N

        DD(I,J)=D(I,J)*(X(I)-X(1))*(X(I)-X(N))/(X(J)-X(1))/(X(J)-X(N))+4.0*C(I,J)*(2.0*X(I)-X(1)-X(N))/(X(J)-X(1))/...
        (X(J)-X(N))+12.0*B(I,J)/(X(J)-X(1))/(X(J)-X(N));
        end
end
```

```
%-------------------------------------------------------------------------------------------------------------------
%      The function below is used in Program2_Main and Program5_Main
%-------------------------------------------------------------------------------------------------------------------
%
function [A, B, C, D, BB]=DQMNEW(M, X)

% The length is in the X(I), namely, X[0,L]; L is the length
% MMWC-3 is used for B,C,D.

Y=ones(M);
for I=1:M                      % EQ. (1.30)
      for J=1:M
            if I~=J
                  Y(I)=Y(I)*(X(I)-X(J));
            end
      end
end

A=zeros(M,M);
for I=1:M
      for J=1:M
            if I~=J
                  A(I,J)=Y(I)/(X(I)-X(J))/Y(J);
                  A(I,I)=A(I,I)+1.0/(X(I)-X(J));
            end
      end
end

BB=A*A;                        % EQ. (1.21)
B=BB;

I=[1,M]; J=1:M; K=2:M-1;       % MMWC-3 for two end points, EQ (3.17)
B(I,J)=A(I,K)*A(K,J);

A(1:M,M+1:M+2)=0.0;
B(1:M,M+1:M+2)=0.0;

B(1,M+1)=A(1,1);               % MMWC-3
B(1,M+2)=A(1,M);
B(M,M+1)=A(M,1);
B(M,M+2)=A(M,M);

C=A(:,1:M)*B;                  % MMWC-3, EQ (3.18)
D=BB*B;                        % EQ. (3.19)


%-------------------------------------------------------------------------------------------------------------------
%      The function below can be used by Program2_Main and Program5_Main
%-------------------------------------------------------------------------------------------------------------------
%
function [A, B, C, D, BB]=HDQMNEW(M, X)
%-------------------------------------------------------------------------------------------------------------------
%            HDQMNEW with MMWC-3
%            Definition: HDQMNEW is the harmonic differential quadrature method using Eq.(1.33)
%                              to compute the weighting coefficient explicitly.
%-------------------------------------------------------------------------------------------------------------------
%
%      X[0, L], L is the length;
%
```

```
        PI4=pi/4;                      % The slightly difference is caused by the range of x.

        Y=ones(M);
        for I=1:M                      % Similar to EQ. (1.33)
            for J=1:M
                if I~=J
                    Y(I)=Y(I)*sin((X(I)-X(J))*PI4);
                end
            end
        end
A=zeros(M,M);
for I=1:M
    for J=1:M
        if I~=J
            A(I,J)=PI4*Y(I)/sin(PI4*(X(I)-X(J)))/Y(J);
            A(I,I)=A(I,I)+PI4*cos((X(I)-X(J))*PI4)/sin(PI4*(X(I)-X(J)));
        end
    end
end

BB=A*A;                  % EQ. (1.21)
B=BB;

I=[1,M]; J=1:M; K=2:M-1;    % EQ. (3.17)
B(I,J)=A(I,K)*A(K,J);

A(1:M,M+1:M+2)=0.0;
B(1:M,M+1:M+2)=0.0;

B(1,M+1)=A(1,1);
B(1,M+2)=A(1,M);
B(M,M+1)=A(M,1);
B(M,M+2)=A(M,M);

C=A(:,1:M)*B;            % EQ. (3.18)
D=BB*B;                 % EQ. (3.19)


%------------------------------------------------------------------------------------------------------------------
%    The function below is used in Program3_Main and Program4_Main
%------------------------------------------------------------------------------------------------------------------
%
function [ AA,BB ] = DQELE2( N,X,COE )
%
%    COE= LENGTH/2;   X[-1,1]
%
    AA=zeros(N,N+2);
    BB=zeros(N,N+2);

    A=zeros(N,N);
    B=zeros(N,N);

    Y=ones(1,N);            % EQ. (1.30)
    for I=1:N
        for J=1:N
            if I ~= J
                A(I,J)=1.0D0;
                Y(I)=Y(I)*(X(I)-X(J));
```

```
                    end
                end
            end
        for   I=1:N
            for J=1:N
                if I ~= J
                    A(I,J)=Y(I)/(X(I)-X(J))/Y(J);
                end
            end
        end

        for   I=1:N
            A(I,I)=0.0D0;
            for   J=1:N
                if I ~= J
                    A(I,I)=A(I,I)+1.0D0/(X(I)-X(J));
                end
            end
        end


        A=A/1.0D0;
        B=A*A;                    % EQ. (1.21)

%       DQEM
%
        for I=1:N                                                            % EQ. (2.14)
            AA(I,N+1)=A(I,1)*(X(I)-X(1))*(X(I)-X(N))/(X(1)-X(N));
            AA(I,N+2)=A(I,N)*(X(I)-X(1))*(X(I)-X(N))/(X(N)-X(1));
        end
        AA(1,N+1)=AA(1,N+1)+1.0D0;
        AA(N,N+2)=AA(N,N+2)+1.0D0;
        for I=1:N
            BB(I,N+1)=B(I,1)*(X(I)-X(1))*(X(I)-X(N))/(X(1)-X(N))...
                    +2.0D0*A(I,1)*(X(I)-X(1)+X(I)-X(N))/(X(1)-X(N));
            BB(I,N+2)=B(I,N)*(X(I)-X(1))*(X(I)-X(N))/(X(N)-X(1))...
                    +2.0D0*A(I,N)*(X(I)-X(1)+X(I)-X(N))/(X(N)-X(1));
        end
        BB(1,N+1)=BB(1,N+1)+2.0D0/(X(1)-X(N));
        BB(N,N+2)=BB(N,N+2)+2.0D0/(X(N)-X(1));

%
%
        for I=1:N                                                            % EQS. (2.15)-(2.16)
            AA(I,1)=A(I,1)*(X(I)-X(N))/(X(1)-X(N))-(A(1,1)+1.0D0/(X(1)-X(N)))...
                *AA(I,N+1);
            AA(I,N)=A(I,N)*(X(I)-X(1))/(X(N)-X(1))-(A(N,N)+1.0D0/(X(N)-X(1)))...
                *AA(I,N+2);
        end
        AA(1,1)=AA(1,1)+1.0D0/(X(1)-X(N));
        AA(N,N)=AA(N,N)+1.0D0/(X(N)-X(1));
        for I=1:N
            BB(I,1)=(B(I,1)*(X(I)-X(N))+2.0D0*A(I,1))/(X(1)-X(N))-...
                (A(1,1)+1.0D0/(X(1)-X(N)))*BB(I,N+1);
            BB(I,N)=(B(I,N)*(X(I)-X(1))+2.0D0*A(I,N))/(X(N)-X(1))-...
                (A(N,N)+1.0D0/(X(N)-X(1)))*BB(I,N+2);
        end
```

```
%
     for J=2:N-1
        for I=1:N
             AA(I,J)=A(I,J)*(X(I)-X(1))*(X(I)-X(N))/(X(J)-X(1))/(X(J)-X(N));
        end
        AA(J,J)=AA(J,J)+(X(J)-X(1)+X(J)-X(N))/(X(J)-X(1))/(X(J)-X(N));
     end

     for J=2:N-1
        for I=1:N
             BB(I,J)=B(I,J)*(X(I)-X(1))*(X(I)-X(N))/(X(J)-X(1))/(X(J)-X(N))...
                 +2.0D0*A(I,J)*(2.0D0*X(I)-X(1)-X(N))/(X(J)-X(1))/(X(J)-X(N));
        end
        BB(J,J)=BB(J,J)+2.0D0/(X(J)-X(1))/(X(J)-X(N));
     end

     AA=AA/COE;
     BB=BB/COE/COE;

end

%------------------------------------------------------------------------------------------------------------------
%    The function below is used in Program3_Main and Program4_Main
%------------------------------------------------------------------------------------------------------------------
%
function [ A,B ] = DQMNEW2( M,X,COE )
%
%    COE= LENGTH/2;   X[-1,1];
%
     A=zeros(M,M+2);
     B=zeros(M,M+2);

     Y=ones(1,M);
     for I=1:M                            % EQ. (1.30)
        for J=1:M
            if I ~= J
                Y(I)=Y(I)*(X(I)-X(J));
            end
        end
     end

     for I=1:M
        for J=1:M
            if I ~=J
                A(I,J)=Y(I)/(X(I)-X(J))/Y(J);
            end
        end
     end

     for I=1:M
        A(I,I)=0.0d0;
        for J=1:M
            if I~=J
                A(I,I)=A(I,I)+1.0d0/(X(I)-X(J));
            end
        end
     end
```

```
    A(1:M,1:M)=A(1:M,1:M)/COE;

    B(2:M-1,1:M)=A(2:M-1,1:M)*A(1:M,1:M);              % EQ. (3.17)
    B(1:M-1:M,1:M)=A(1:M-1:M,2:M-1)*A(2:M-1,1:M);

    A (1:M,M+1:M+2)=0.0D0;
    B (1:M,M+1:M+2)=0.0D0;


    B(1,M+1)=A(1,1);
    B(1,M+2)=A(1,M);
    B(M,M+1)=A(M,1);
    B(M,M+2)=A(M,M);

end

%-------------------------------------------------------------------------------------------------------------------------------
%    The function below is used in Program3_Main and Program4_Main
%-------------------------------------------------------------------------------------------------------------------------------
%
function [ A,B ] = HDQMNEW2( M,X,COE )

    A=zeros(M,M+2);
    B=zeros(M,M+2);
%
%    COE= LENGTH/2;   X[-1,1];
%
    PI8=atan(1.0D0)/2.0D0;   % PI/8
    Y=ones(1,M);

    for   I=1:M                        % EQ. (1.32)
        for J=1:M
            if I ~= J
                Y(I)=Y(I)*sin((X(I)-X(J))*PI8);
            end
        end
    end
    for I=1:M
        for J=1:M
            if I ~= J
                A(I,J)=PI8*Y(I)/sin(PI8*(X(I)-X(J)))/Y(J);
            end
        end
    end

    for I=1:M
        A(I,I)=0.0D0;
        for J=1:M
            if I ~= J
                A(I,I)=A(I,I)+PI8*cos((X(I)-X(J))*PI8)/sin(PI8*(X(I)-X(J)));
            end
        end
    end
```

```
    A(1:M,1:M)=A(1:M,1:M)/COE;

    B(2:M-1,1:M)=A(2:M-1,1:M)*A(1:M,1:M);              % EQ. (3.17)
    B(1:M-1:M,1:M)=A(1:M-1:M,2:M-1)*A(2:M-1,1:M);

    A(1:M,M+1:M+2)=0.0D0;
    B(1:M,M+1:M+2)=0.0D0;

    B(1,M+1)=A(1,1);
    B(1,M+2)=A(1,M);
    B(M,M+1)=A(M,1);
    B(M,M+2)=A(M,M);

end
```

```
%--------------------------------------------------------------------------------------------------------------------
%    The end of the eleven MATLAB functions
%--------------------------------------------------------------------------------------------------------------------
%
```

REFERENCES

[1] W.H. Press, S.A. Teukolsky, W.T. Vetterling, B.P. Flannery, Numerical Recipes: The Art of Scientific Computing, second ed., Cambridge University Press, New York, 1992.

[2] S.L. Xu, FORTRAN on the Commonly Used Algorithms, second ed., Tsinghua University Press, Beijing, China, 1992 (in Chinese).

[3] IMSL MATH/LIBRARY, FORTRAN Subroutines for Mathematical Applications, IMSL, Inc., 1989.

[4] P.J. Davis, P. Robinowitz, Methods of Numerical Integration, Academic Press, Inc., London, 1975.

Appendix III

III PROGRAM 1: FREE VIBRATION OF ISOTROPIC OR ORTHOTROPIC RECTANGULAR PLATE WITHOUT FREE EDGES

Program 1 is used for free vibration analysis of isotropic or orthotropic rectangular plates without free edges by the DQM. MMWC-1 is used for applying the simply supported boundary conditions and MMWC-3 (or MMWC-4) is used for applying the clamped boundary conditions. Since the deflection at all boundary points is zero and the other boundary condition of either $w_{xx} = 0$ or $w_x = 0$ has been built-in during formulations of the weighting coefficients in the subroutines, thus only DQ equations at all inner grid points are formulated. The program is very simple.

Note that the program can be used for any combinations of simply supported and clamped boundary conditions, but cannot be used for anisotropic materials (except for the CCCC rectangular plate). Otherwise inaccurate results would be obtained, since the simply supported boundary condition is only approximately satisfied by MMWC-1 for anisotropic materials.

It should be mentioned that subroutine TZZ is the standard eigenvalue solver to find all eigenvalues of the real matrix [GS]. The solver is not included in Program 1.

To run "Program 1", the first 4 FORTRAN subroutines in Section II.2 and a standard eigenvalue solver should be included. To run the converted MATLAB file "Prgram1_Main," the first 4 functions in Section II.2 should be put in the same directory where "Prgram1_Main" is.

Description of some variables in Program 1

X(I): x coordinate of a grid point, x_i [0,A]
Y(I): y coordinate of a grid point, y_i [0,B]

Input data

N: number of grid points in both directions
A: plate length in x direction
B: plate length in y direction
VV: Poisson's ratio
D11, D12, D22, D66, D16 = D26 = 0 : the flexural stiffness (isotropic or orthotropic materials)
Note that all input data have been assigned currently.

Output data

TT: real part of the eigenvalues
CK(I): imaginary part of the eigenvalues
DSQRT(TT): frequency parameter

Differential Quadrature and Differential Quadrature Based Element Methods. 978-0-12-803081-3

```
C
C-------------------------------------------------------------------------------------
C     Program 1 (FORTRAN source codes)
C     Free vibration of isotropic or orthotropic rectangular plates without free edges by the modified DQM.
C     N----The number of grid points in x and y directions. For different combinations of boundary conditions,
C          the two CALL statements should be changed accordingly. Currently the program is for CSCS rectangular plate.
C     Simply supported boundary conditions----MMWC-1 is used.
C     Clamped boundary conditions----MMWC-3 or MMWC-4 is used.
C     Grid III is used in the program.
C-------------------------------------------------------------------------------------
C
      IMPLICIT REAL*8 (A-H,O-Z)
      PARAMETER (N=21,N2=N-2,NN=N2*N2,MM=N*N)
      CHARACTER FILE_MODE*80
      DIMENSION X(N),Y(N),AX(N,N),BX(N,N),CX(N,N),DX(N,N)
      DIMENSION AY(N,N),BY(N,N),CY(N,N),DY(N,N),Z(NN)
      DIMENSION GS(NN,NN),BK(NN),CK(NN),LM(NN)
      DIMENSION GG(MM,MM),LI(NN)
C
C-------------------------------------------------------------------------------------
C     MMWC-3 (OR MMWC-4) OR THE MIXED METHOD (MMWC-1 & MMWC-4) IS USED;
C     ONLY FOR ISOTROPIC OR ORTHOTROPIC RECTANGULAR PLATES WITHOUT FREE EDGES
C     SIX COMBINATIONS OF CLAMPED AND SIMPLY SUPPORTED BOUNDARY CONDITIONS
C
C     INPUT THE FILENAME OF OUTPUT
C-------------------------------------------------------------------------------------
C
      WRITE(*,'  (A)   ') ' ENTER OUTPUT FILE NAME'
      READ(*,'  (A)   ') FILE_MODE
      OPEN(UNIT=9,STATUS='UNKNOWN',FILE=FILE_MODE)      ! File to store output data
C
      PI=DATAN(1.0D0)*4.0D0
      NX=N                    !  TOTAL POINTS IN THE X-DIRECTION
      NY=N                    !  TOTAL POINTS IN THE Y-DIRECTION
      A=1.D0                  !  LENGTH A
      RATIO=1.0D0/1.0D0
      B=A/RATIO               !  LENGTH B
C-------------------------------------------------------------------------------------
C     MATERIAL CONSTANTS, THE SYMBOLS D11,D12, ..., ETC ARE ELEMENTS IN [D], WHICH ARE
C     COMMONLY USED FOR ANISOTROPIC PLATE.
C     FOR ISOTROPIC OR ORTHOTROPIC MATERIALS, D16=D26=0, OTHERS SHOULD BE
C     SPECIFIED FOR DIFFERENT MATERIALS
C     CURRENTLY THE ISOTROPIC MATERIAL IS SPECIFIED: VV IS POISSON'S RATIO
C     AND D11 IS THE BENDING RIGIDITY.
C-------------------------------------------------------------------------------------
```

```
      VV=0.3D0                          !   POISSON'S RATIO
      D11=1.0D0                         !    FOR NON-DIMENSIONAL FREQUENCY PARAMETER
      D16=0.0D0
      D22=1.0D0
      D26=0.0D0
      D12=VV
      D66=(1.0D0-VV)/2.0D0
C
      DO I=1,N
       X(I)=(1.0D0-DCOS((I-1)*PI/(N-1)))/2.0D0*A          ! GRID III
      ENDDO
      DO I=1,N
       Y(I)=X(I)*B/A
      ENDDO
C
C-------------------------------------------------------------------------------------------
C     SSSS RECTANGULAR PLATE
C-------------------------------------------------------------------------------------------
C
!     CALL DQSS(N,X,AX,BX,CX,DX,Z)
!     CALL DQSS(N,Y,AY,BY,CY,DY,Z)
C
C-------------------------------------------------------------------------------------------
C     CCCC RECTANGULAR PLATE
C-------------------------------------------------------------------------------------------
C
!     CALL DQCC(N,X,AX,BX,CX,DX,Z)
!     CALL DQCC(N,Y,AY,BY,CY,DY,Z)
C
C-------------------------------------------------------------------------------------------
C     SCCS RECTANGULAR PLATE
C-------------------------------------------------------------------------------------------
C
!     CALL DQSC(N,X,AX,BX,CX,DX,Z)
!     CALL DQCS (N,Y,AY,BY,CY,DY,Z)
C
C-------------------------------------------------------------------------------------------
C     SSCS RECTANGULAR PLATE
C-------------------------------------------------------------------------------------------
C
!     CALL DQSC (N,X,AX,BX,CX,DX,Z)
!     CALL DQSS (N,Y,AY,BY,CY,DY,Z)
C
C-------------------------------------------------------------------------------------------
```

```
C       SCCC RECTANGULAR PLATE
C----------------------------------------------------------------------------------------------------------------------------
C
!       CALL DQSC (N,X,AX,BX,CX,DX,Z)
!       CALL DQCC (N,Y,AY,BY,CY,DY,Z)
C
C----------------------------------------------------------------------------------------------------------------------------
C       CSCS RECTANGULAR PLATE
C----------------------------------------------------------------------------------------------------------------------------
C
        CALL DQCC(N,X,AX,BX,CX,DX,Z)
        CALL DQSS(N,Y,AY,BY,CY,DY,Z)
C
C----------------------------------------------------------------------------------------------------------------------------
C
        DO I=1,MM
        DO J=1,MM
        GG(I,J)=0.0D0
        ENDDO
        ENDDO
C
C       DQ EQUATIONS AT INNER GRID POINTS
C
        DO I=2,NY-1                                      ! MIXED DERIVATIVES
        KI=NX*(I-1)
        DO L=1,NY
        KL=NX*(L-1)
        DO J=2,NX-1
        II=KI+J
        DO K=1,NX
        JJ=KL+K
        GG(II,JJ)=2.D0*(D12+2.0D0*D66)*BY(I,L)*BX(J,K)+
     &           +4.D0*D16*AY(I,L)*CX(J,K)
     &           +4.D0*D26*CY(I,L)*AX(J,K)
        ENDDO
        ENDDO
        ENDDO
        ENDDO

        DO I=2,NY-1                                      !   D4W/DX4
        KI=NX*(I-1)
        DO J=2,NX-1
        II=KI+J
```

```
        DO K=1,NX
        JJ=KI+K
        GG(II,JJ)=GG(II,JJ)+D11*DX(J,K)
        ENDDO
        ENDDO
        ENDDO
        DO I=2,NY-1                                    !   D4W/DY4
        KI=NX*(I-1)
        DO L=1,NY
        KL=NX*(L-1)
        DO J=2,NX-1
        II=KI+J
        JJ=KL+J
        GG(II,JJ)=GG(II,JJ)+D22*DY(I,L)
        ENDDO
        ENDDO
        ENDDO
C
C----------------------------------------------------------------------------------------------------------
C    ONE OF THE TWO BOUNDARY CONDITIONS HAS BEEN BUILT IN AND W
C    AT ALL BOUNDARY POINTS IS ZERO.
C----------------------------------------------------------------------------------------------------------
C
        DO I=2,NY-1
        KI=NX*(I-1)
        IJ= (NX-2)*(I-2)
        DO J=2,NX-1
        IJK=IJ+J-1
        LI(IJK)=KI+J
        ENDDO
        ENDDO

        DO I=1,NN
        DO J=1,NN
        GS(I,J)=GG(LI(I),LI(J))                     !   RETAINED GS=[KII]
        ENDDO
        ENDDO
C
C    SUBROUTINE TZZ IS USED TO FIND EIGEN-VALUES OF [GS]
C
        CALL TZZ (NN,NN,GS,Z,BK,CK,LM)
C
C    RE-ARRANGEMENT, THE SMALLEST BK(I) IS PUT IN THE FIRST
C
```

```
        ZZZ=1.D30
        K=1
440     DO 450 I=K,NN
        IF(BK(I) .GE. ZZZ) GO TO 450
        KK=I
        ZZZ=BK(I)
450     CONTINUE
        BK(KK)=BK(K)
        BK(K)=ZZZ
        ZZZ=1.D30
C
        TEMP=CK(KK)
        CK(KK)=CK(K)
        CK(K)=TEMP
C
        K=K+1
        IF(K .LE.NN) GOTO 440
C
C   PRINT OUT
C
        K=0
        DO 599 I=1,NN
        IF (DABS(CK(I)) .GE. 1.0D-2) GO TO 599
        TT=BK(I)
        IF (TT .LT. 0.0D0) GO TO 599
        K=K+1
        WRITE (*,551) I,TT,CK(I),DSQRT(TT)        !   NON-DIMENSIONAL FREQUENCY PARAMETER
        IF (K. GE. 10) GOTO 600
599     CONTINUE
600     CONTINUE

        DO 602 I=1,NN
        TT=BK(I)
        IF (TT .LT. 0.0D0) GO TO 601
        WRITE (9,551) I,TT,CK(I),DSQRT(TT)
        GOTO 602
601     WRITE (9,551) I,TT,CK(I)
602     CONTINUE
551     FORMAT (1X,I5,3G20.10)
        STOP
        END


C
C-------------------------------------------------------------------------------------------------------------
C                            THE END OF PROGRAM 1.
C-------------------------------------------------------------------------------------------------------------
C
```

```
%
%-------------------------------------------------------------------------------------------------------
%    The converted MATLAB file named Program1_Main
%-------------------------------------------------------------------------------------------------------
%
%-------------------------------------------------------------------------------------------------------
%    Program 1: Free vibration of isotropic or orthotropic rectangular plates without free edges by the DQM
%    N----The number of grid points in both directions. For different combinations of boundary conditions,
%         the two CALL statements should be changed accordingly. Currently the program is for SCCS square plate.
%    Simply supported boundary conditions----MMWC-1 is used.
%    Clamped boundary conditions----MMWC-3 or MMWC-4 is used.
%    Grid III is used in the program.
%-------------------------------------------------------------------------------------------------------
%

clear; clc;
format long;

N=21; N2=N-2; NN=N2*N2; MM=N*N;

%
%-------------------------------------------------------------------------------------------------------
%    MMWC-3 (OR MMWC-4) OR THE MIXED METHOD (MMWC-1 & MMWC-4) IS USED;
%    FOR ISOTROPIC OR ORTHOTROPIC RECTANGULAR PLATES WITHOUT FREE EDGES;
%    SIX COMBINATIONS OF CLAMPED AND SIMPLY SUPPORTED BOUNDARY CONDITIONS.
%-------------------------------------------------------------------------------------------------------
%

NX=N;                           % TOTAL POINTS IN THE X-DIRECTION
NY=N;                           % TOTAL POINTS IN THE Y-DIRECTION
A=1.0;                          % LENGTH A
RATIO=1.0/1.0;
B=A/RATIO;                      % LENGTH B
%-------------------------------------------------------------------------------------------------------
%    MATERIAL CONSTANTS, THE SYMBOLS D11, D12, ..., etc ARE ELEMENTS IN [D], WHICH ARE
%    COMMONLY USED FOR ANISOTROPIC PLATE.
%    FOR ISOTROPIC OR ORTHOTROPIC MATERIALS, D16=D26=0, OTHERS SHOULD BE
%    SPECIFIED FOR DIFFERENT MATERIALS;
%    CURRENTLY THE ISOTROPIC MATERIAL IS SPECIFIED: VV IS POISSON'S RATIO
%    AND D11 IS THE BENDING RIGIDITY.
%-------------------------------------------------------------------------------------------------------
VV=0.3;                         % POISSON'S RATIO
D11=1.0;                        % FOR NON-DIMENSIONAL FREQUENCY PARAMETER
D16=0.0;
D22=1.0;
D26=0.0;
D12=VV;
D66=(1.0-VV)/2.0;

I=1:N;
X(I)=(1.0-cos((I-1)*pi/(N-1)))/2.0*A;        % GRID III

Y=X*B/A;
```

```
%
%-------------------------------------------------------------------------------------------------------
%       SSSS RECTANGULAR PLATE
%-------------------------------------------------------------------------------------------------------
%
%       [AX,BX,CX,DX]=DQSS(N,X);
%       [AY,BY,CY,DY]=DQSS(N,Y);
%
%-------------------------------------------------------------------------------------------------------
%       CCCC RECTANGULAR PLATE
%-------------------------------------------------------------------------------------------------------
%
%       [AX,BX,CX,DX]=DQCC(N,X);
%       [AY,BY,CY,DY]=DQCC(N,Y);
%
%-------------------------------------------------------------------------------------------------------
%       SCCS RECTANGULAR PLATE
%-------------------------------------------------------------------------------------------------------
%
        [AX,BX,CX,DX]=DQSC(N,X);
        [AY,BY,CY,DY]=DQCS(N,Y);
%
%-------------------------------------------------------------------------------------------------------
%       SSCS RECTANGULAR PLATE
%-------------------------------------------------------------------------------------------------------
%
%       [AX,BX,CX,DX]=DQSC(N,X);
%       [AY,BY,CY,DY]=DQSS(N,Y);
%
%-------------------------------------------------------------------------------------------------------
%       SCCC RECTANGULAR PLATE
%-------------------------------------------------------------------------------------------------------
%
%       [AX,BX,CX,DX]=DQSC(N,X);
%       [AY,BY,CY,DY]=DQCC(N,Y);
%
%-------------------------------------------------------------------------------------------------------
%       CSCS RECTANGULAR PLATE
%-------------------------------------------------------------------------------------------------------
%
%       [AX,BX,CX,DX]=DQCC(N,X);
%       [AY,BY,CY,DY]=DQSS(N,Y);
%
%-------------------------------------------------------------------------------------------------------
%
GG=zeros(MM,MM);

%
%       DQ EQUATIONS AT INNER GRID POINTS
%
```

```
for I=2:NY-1                                    % MIXED DERIVATIVES
     KI=NX*(I-1);
     for L=1:NY
          KL=NX*(L-1);
          for J=2:NX-1
               II=KI+J;
               for K=1:NX
               JJ=KL+K;
               GG(II,JJ)=2.0*(D12+2.0*D66)*BY(I,L)*BX(J,K)+4.0*D16*AY(I,L)*CX(J,K)+4.0*D26*CY(I,L)*AX(J,K);

               end
          end
     end
end

for I=2:NY-1                                    % D4W/DX4
     KI=NX*(I-1);
     for J=2:NX-1
          II=KI+J;
          for K=1:NX
               JJ=KI+K;
               GG(II,JJ)=GG(II,JJ)+D11*DX(J,K);
          end
     end
end

for I=2:NY-1                                    % D4W/DY4
     KI=NX*(I-1);
     for L=1:NY
          KL=NX*(L-1);
          for J=2:NX-1
               II=KI+J;
               JJ=KL+J;
               GG(II,JJ)=GG(II,JJ)+D22*DY(I,L);
          end
     end
end

%
%----------------------------------------------------------------------------------------------------------------
%    ONE OF THE TWO BOUNDARY CONDITIONS HAS BEEN BUILT IN AND W
%    AT ALL BOUNDARY POINTS IS ZERO.
%----------------------------------------------------------------------------------------------------------------
%
for I=2:NY-1
     KI=NX*(I-1);
     IJ= (NX-2)*(I-2);
     for J=2:NX-1
          IJK=IJ+J-1;
          LI(IJK)=KI+J;
     end
end
```

```
for I=1:NN
        for J=1:NN
                GS(I,J)=GG(LI(I),LI(J));                % RETAINED GS=[KII]
        end
end


EIG_GS=eig(GS);                                    % Find all real eigen-values & sort them in ascending order
EIG_GS=EIG_GS(abs(imag(EIG_GS))<=1e-2);
EIG_GS=sort(real(EIG_GS));
EIG_GS=sqrt(EIG_GS);

save EigenvalueData.txt -ascii -double EIG_GS          %Save the eigen-values in the file EigenvalueData.txt

%
%-------------------------------------------------------------------------------------------------------------------
%     The end of the converted MATLAB file Program1_Main
%-------------------------------------------------------------------------------------------------------------------
%
```

Appendix IV

IV A SIMPLE WAY FOR THE PROGRAMMING

Due to introduction of additional degree of freedom (DOF) in the DQEM and in the modified DQM, the implementation seems not as simple as the conventional DQM. To overcome this difficulty, a simple way is proposed for the convenience of programming. The programs included in this book are written based on Fig. IV.1.

In Fig. IV.1, a mesh of 5×5 is illustrated, $N = N_x = N_y = 5$. The $(2N_x + 2N_y)$ additional grid points outside of the plate are introduced to settle the first-order derivative of the deflection with respect to x or y. The additional grid points are denoted by the half-filled points shown in Fig. IV.1. In detail, the DOF for the half-filled points along line $(N_x + 1)$ is $(\partial w / \partial x)_{x=0} = w_{x0}$ and along line $(N_x + 2)$ is $(\partial w / \partial x)_{x=a} = w_{xa}$, the DOF for the half-filled points along line $(N_y + 1)$ is $(\partial w / \partial y)_{y=0} = w_{y0}$ and along line $(N_y + 2)$ is $(\partial w / \partial y)_{y=b} = w_{yb}$. In Fig. IV.1, the four unfilled points are not used if MMWC-3, MMWC-4, or the mixed method is used to introduce the additional DOF at boundary points. The four unfilled points are only needed for the DQEM or LaDQM. In this way, the programming is as simple as the conventional DQM, since each grid point has only one DOF.

In the attached FORTRAN programs, e.g., Program 2 named "Free vibration of isotropic FFFF skew plate by the DQM" and Program 3 named "Free vibration of isotropic FFFF skew plate by the QEM", the arrangement of displacement vector is $w_{1j}(j=1,2,...,N), w_{x11}, w_{x1N}, w_{2j}(j=1,2,...,N)$, $w_{x21}, w_{x2N}, ..., w_{Nj}(j=1,2,...,N), w_{xN1}, w_{xNN}, w_{yj1}(j=1,2,...,N), w_{yjN}(j=1,2,...,N)$. In detail, the displacement vector for the case of $N_x = N_y = 5$ shown in Fig. IV.1 is arranged as follows,

$$\{\bar{w}\}^T = \{w_{11}, w_{12}, w_{13}, w_{14}, w_{15}, w_{x11}, w_{x15}, w_{21}, w_{22}, w_{23}, w_{24}, w_{25},$$
$$w_{x21}, w_{x25}, w_{31}, w_{32}, w_{33}, w_{34}, w_{35}, w_{x31}, w_{x35}, w_{41}, w_{42},$$
$$w_{43}, w_{44}, w_{45}, w_{x41}, w_{x45}, w_{51}, w_{52}, w_{53}, w_{54}, w_{55}, w_{x51}, w_{x55},$$
$$w_{y11}, w_{y12}, w_{y13}, w_{y14}, w_{y15}, w_{y51}, w_{y52}, w_{y53}, w_{y54}, w_{y55}\} \tag{IV.1}$$

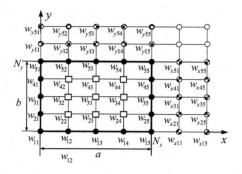

FIGURE IV.1

Sketch of a Rectangular Plate with Grid Spacing of 5×5 ($N = N_x = N_y = 5$)

Differential Quadrature and Differential Quadrature Based Element Methods. 978-0-12-803081-3

Appendix V

V PROGRAM 2: FREE VIBRATION OF ISOTROPIC FFFF SKEW PLATE BY THE DQM

V.1 FORTRAN AND MATLAB CODES

As an example, Program 2, named as "Free vibration of isotropic FFFF skew plate by the DQM," is given below for readers' reference. As before, FORTRAN codes are given first and then followed the converted MATLAB file.

It should be mentioned that subroutine INVMAT is called to find the inverse of matrix [FF] and subroutine TZZ is the standard eigenvalue solver to get all eigenvalues of a real matrix [GS]. These two FORTRAN subroutines are not included in Program 2. Besides, $NN = (N-2) \times (N-2)$ and is fixed for all combinations of boundary conditions.

To run "Program 2," besides subroutines INVMAT and TZZ, subroutines DQMNEW and GRULE in Section II.2 should also be included. To run the converted MATLAB file "Prgram2_Main," functions DQMNEW and GRULE in Section II.2 should be put in the same directory where "Prgram2_Main" is.

Description of some variables in Program 2

X(I): x coordinate of a grid point, x_i [0, A]
Y(I): y coordinate of a grid point, y_i [0, B]

Input data

N: number of grid points in both directions
A: plate length in ξ (or x) direction
B: plate length in η (or y) direction
VV: Poisson's ratio
D11, D12, D22, D66, D16, D26: the flexural stiffness (isotropic materials only)
THEATA: skew angle (degrees), the only input datum currently
Note that except THEATA, all other input data have been assigned currently.

Output data

TT: real part of the eigenvalues
CK(I): imaginary part of the eigenvalues
DSQRT(TT)/FACTOR: frequency parameter

```
C
C--------------------------------------------------------------------------------
C       Program 2 (FORTRAN)
C       Free vibration of isotropic FFFF skew plate by the DQM
C       N —The number of grid points in both directions.
C       NF—The number of non-zero DOFs associated to the boundary points, it varies for different combinations of
C              boundary conditions; e.g., NF=4*N-8 for the SSSS plate and LN (NF) should be changed accordingly.
C       THEATA—skew angle, the only input datum to run the program.
C       Only Grid V should be used, although other grids are also included in the program.
C--------------------------------------------------------------------------------
C
        IMPLICIT REAL*8 (A-H,O-Z)
        PARAMETER (N=21,N2=N-2,NN=N2*N2,MMM=(N+2)*(N+2),NF=8*N-4)
        CHARACTER FILE_MODE*80
        DIMENSION X(N),Y(N),AX(N,N+2),BX(N,N+2),CX(N,N+2),DX(N,N+2)
        DIMENSION AY(N,N+2),BY(N,N+2),CY(N,N+2),DY(N,N+2),Z(NN)
        DIMENSION GS(NN,NN),BK(NN),CK(NN),BBX(N,N),BBY(N,N) LM(NN)
        DIMENSION GF(NN,NF),FG(NF,NN),FF(NF,NF),XX(N),YY(N)
        DIMENSION GFF(NN,NF),GG(MMM,MMM),LI(NN),LN(NF)
C
C--------------------------------------------------------------------------------
C    The modified DQM, namely, MMWC-3 is used to introduce the additional DOF.
C--------------------------------------------------------------------------------
C    Input filename for storing the output data
C--------------------------------------------------------------------------------
C
        WRITE(*,'   (A)   ')' ENTER OUTPUT FILE NAME'
        READ(*,'    (A)   ') FILE_MODE
        OPEN(UNIT=9,STATUS='UNKNOWN',FILE=FILE_MODE)  ! File to store output data
C
        PI=DATAN(1.0D0)*4.0D0
        NX=N                        ! TOTAL NUMBER OF POINTS IN THE X-DIRECTION
        NY=N                        ! TOTAL NUMBER OF POINTS IN THE Y-DIRECTION
        A=1.D0                  !    LENGTH
        RATIO=1.0D0/1.0D0
        B=A/RATIO               !    WIDTH
C
C--------------------------------------------------------------------------------
C
        WRITE (*,*) 'ENTER THEATA'
        READ (*,*) THEATA              ! IN DEGREES
        VV=0.3D0
        TH=THEATA/180.D0*PI            ! SKEW ANGLES, 0 FOR RECTANGULAR PLATE

        D11=1.0D0                      ! RESULTS ARE NON-DIMENSIONAL FREQUENCY PARAMETERS
        D16=-DSIN(TH)
        D22=1.0D0
        D26=D16
        D12=VV*DCOS(TH)**2+DSIN(TH)**2
        D66=(1.0D0+DSIN(TH)**2-VV*DCOS(TH)**2)/2.0D0
```

```
C
C------------------------------------------------------------------------------------
C
      FACTOR=DCOS(TH)**2            ! RESULTS ARE NON-DIMENSIONAL FREQUENCY PARAMETERS

      X(1)=0.0D0
      X(N)=A
C------------------------------------------------------------------------------------
C          VARIOUS GRID SPACING
C------------------------------------------------------------------------------------

      NG=N-2
      CALL GRULE(NG,XX,Y,YY)     ! GAUSS INTEGRATION POINTS(XX) AND WEIGHTS(Y)
      DO I=2,N-1
      X(I)=(1.0D0+XX(I-1))/2.0D0*A    ! GRID V
      ENDDO

      DELX=A/(N-1)
      DO I=1,N
C     X(I)=DELX*(I-1)                 !   GRID I-UNIFORM
      ENDDO

      DO I=2,N-1
C       X(I)=(1.0D0-DCOS((2*I-3)*PI/(2*N-4)))/2.0D0*A        !   GRID II
C       X(I)=A*(1.0D0-DCOS((2*I-1)*PI/(2*N)))/2.0D0          !   GRID IV
      ENDDO
      DO I=1,N
C       X(I)=(1.0D0-DCOS((I-1)*PI/(N-1)))/2.0D0*A                      !   GRID III
C       X(I)=A*(1.0D0-DCOS((2*I-1)*PI/(2*N))/DCOS(PI/(2*N)))/2.0D0  !   GRID VI-EXPANDED
      ENDDO

      DO I=1,N
      Y(I)=X(I)*B/A
      ENDDO
C
C------------------------------------------------------------------------------------
C
      CALL DQMNEW(N,X,AX,BX,CX,DX,BBX)
      CALL DQMNEW(N,Y,AY,BY,CY,DY,BBY)
      NY2=NY+2
      NX2=NX+2
      DO I=1,MMM
      DO J=1,MMM
      GG(I,J)=0.0D0
      ENDDO
      ENDDO
C
C    DQ EQUATIONS AT INNER GRID POINTS
C
```

```
      DO I=2,NY-1                                    !   MIXED DERIVATIVES
      KI=NX2*(I-1)
      DO L=1,NY2
      KL=NX2*(L-1)
      DO J=2,NX-1
      II=KI+J
      DO K=1,NX2
      JJ=KL+K
      GG(II,JJ)=2.D0*(D12+2.0D0*D66)*BY(I,L)*BX(J,K)+
     &         +4.D0*D16*AY(I,L)*CX(J,K)
     &         +4.D0*D26*CY(I,L)*AX(J,K)
      ENDDO
      ENDDO
      ENDDO
      ENDDO

      DO I=2,NY-1                                    ! D4W/DX4
      KI=NX2*(I-1)
      DO J=2,NX-1
      II=KI+J
      DO K=1,NX2
      JJ=KI+K
      GG(II,JJ)=GG(II,JJ)+D11*DX(J,K)
      ENDDO
      ENDDO
      ENDDO
      DO I=2,NY-1                                    ! D4W/DY4
      KI=NX2*(I-1)
      DO L=1,NY2
      KL=NX2*(L-1)
      DO J=2,NX-1
      II=KI+J
      JJ=KL+J
      GG(II,JJ)=GG(II,JJ)+D22*DY(I,L)
      ENDDO
      ENDDO
      ENDDO
C
C-------------------------------------------------------------------------------------------------------------------
C
C     EQUATIONS OF BENDING MOMENTS AND SHEAR FORCE (Y=0, B)
C
C-------------------------------------------------------------------------------------------------------------------
C
C     SHEAR FORCE QY
C
      DO I=1,NY,NY-1              ! PUT AT W OF Y=0 (I=1) AND Y=B(I=NY)
      KI=NX2*(I-1)
      KL=NX2*(I-1)
      DO J=2,NX-1                          ! THE CORNER POINTS (J=1 AND J=NX) ARE NOT INCLUDED
```

```
      II=KI+J
      DO K=1,NX2
      JJ=KL+K
      GG(II,JJ)=2.D0*D16*CX(J,K)        ! D3W/DX3
      ENDDO
      ENDDO
      ENDDO
C
      DO I=1,NY,NY-1                    ! PUT AT W OF Y=0 (I=1) AND Y=B(I=NY)
      KI=NX2*(I-1)
      DO L=1,NY2
      KL=NX2*(L-1)
      DO J=2,NX-1                       ! THE CORNER POINTS (J=1 AND J=NX) ARE NOT INCLUDED
      II=KI+J
      DO K=1,NX2
      JJ=KL+K
      GG(II,JJ)=GG(II,JJ)+4.D0*D26*BY(I,L)*AX(J,K)        ! D3W/DX.DY2
     &      +(D12+4.D0*D66)*AY(I,L)*BX(J,K)               ! D3W/DX2.DY
      ENDDO
      ENDDO
      ENDDO
      ENDDO

      DO I=1,NY,NY-1                    ! PUT AT W OF Y=0 (I=1) AND Y=B(I=NY)
      KI=NX2*(I-1)
      DO L=1,NY2
      KL=NX2*(L-1)
      DO J=2,NX-1                       ! THE CORNER POINTS (J=1 AND J=NX) ARE NOT INCLUDED
      II=KI+J
      JJ=KL+J
      GG(II,JJ)=GG(II,JJ)+D22*CY(I,L)   ! D3W/DY3
      ENDDO
      ENDDO
      ENDDO
C
C     BENDING MOMENT MY
C
      JT=-1
      DO I=1,NY,NY-1                    ! PUT AT WY OF Y=0 (I=1) AND Y=B(I=NY)
      JT=JT+1
      KI=NX2*(NY+JT)
      KL=NX2*(I-1)
      DO J=1,NX                         ! THE CORNER POINTS (J=1 AND J=NX) ARE INCLUDED
      II=KI+J
      DO K=1,NX2
      JJ=KL+K
      GG(II,JJ)=D12*BX(J,K)             ! D2W/DX2
      ENDDO
      ENDDO
      ENDDO
```

```
C
      JT=-1
      DO I=1,NY,NY-1                        ! PUT AT WY OF Y=0 (I=1) AND Y=B(I=NY)
      JT=JT+1
      KI=NX2*(NY+JT)
      DO L=1,NY2
      KL=NX2*(L-1)
      DO J=1,NX                             ! THE CORNER POINTS (J=1 AND J=NX) ARE INCLUDED
      II=KI+J
      JJ=KL+J
      GG(II,JJ)=GG(II,JJ)+D22*BY(I,L)       ! D2W/DY2
      ENDDO
      ENDDO
      ENDDO
C
      JT=-1
      DO I=1,NY,NY-1                        ! PUT AT WY OF Y=0 (I=1) AND Y=B(I=NY)
      JT=JT+1
      KI=NX2*(NY+JT)
      DO L=1,NY2
      KL=NX2*(L-1)
      DO J=1,NX                             ! THE CORNER POINTS (J=1 AND J=NX) ARE INCLUDED
      II=KI+J
      DO K=1,NX2
      JJ=KL+K
      GG(II,JJ)=GG(II,JJ)+2.D0*D26*AY(I,L)*AX(J,K)        ! D2W/DXDY
      ENDDO
      ENDDO
      ENDDO
      ENDDO
C
C-----------------------------------------------------------------------------------------------------------
C
C     EQUATIONS OF BENDING MOMENTS AND SHEAR FORCE  (X=0, A)
C
C-----------------------------------------------------------------------------------------------------------
C
C     SHEAR FORCE QX
C
      DO I=2,NY-1                  ! THE CORNER POINTS (I=1 AND I=NY) ARE NOT INCLUDED
      KI=NX2*(I-1)
      DO L=1,NY2
      KL=NX2*(L-1)
      DO J=1,NX,NX-1               ! X=0 (J=1), X=A (J=NX)
      II=KI+J
      JJ=KL+J
      GG(II,JJ)=2.0*D26*CY(I,L)    ! D3W/DY3
      ENDDO
      ENDDO
      ENDDO
```

```
C
      DO I=2,NY-1                  ! THE CORNER POINTS (I=1 AND I=NY) ARE NOT INCLUDED
      KI=NX2*(I-1)
      DO L=1,NY2
      KL=NX2*(L-1)
      DO J=1,NX,NX-1               ! X=0 (J=1), X=A (J=NX)
      II=KI+J
      DO K=1,NX2
      JJ=KL+K
      GG(II,JJ)=GG(II,JJ)+(D12+4.D0*D66)*BY(I,L)*AX(J,K)    ! D3W/DX.DY2
     &         +4.D0*D16*AY(I,L)*BX(J,K)                    ! D3W/DX2.DY
      ENDDO
      ENDDO
      ENDDO
      ENDDO
C
      DO I=2,NY-1                  ! THE CORNER POINTS (I=1 AND I=NY) ARE NOT INCLUDED
      KI=NX2*(I-1)
      DO J=1,NX,NX-1               ! X=0 (J=1), X=A (J=NX)
      II=KI+J
      DO K=1,NX2
      JJ=KI+K
      GG(II,JJ)=GG(II,JJ)+D11*CX(J,K)          ! D3W/DX3
      ENDDO
      ENDDO
      ENDDO
C
C     BENDING MOMENT = 0
C
      DO I=1,NY                    ! THE CORNER POINTS (I=1 AND I=NY) ARE INCLUDED
      KI=NX2*(I-1)
      JT=NX
      DO J=1,NX,NX-1                            ! X=0 (J=1), X=A (J=NX)
      JT=JT+1
      II=KI+JT
      DO K=1,NX2
      JJ=KI+K
      GG(II,JJ)=D11*BX(J,K)                     ! D2W/DX2
      ENDDO
      ENDDO
      ENDDO
C
      DO I=1,NY                    ! THE CORNER POINTS (I=1 AND I=NY) ARE INCLUDED
      KI=NX2*(I-1)
      DO L=1,NY2
      KL=NX2*(L-1)
      JT=NX
      DO J=1,NX,NX-1                            ! X=0 (J=1), X=A (J=NX)
      JT=JT+1
      II=KI+JT
      JJ=KL+J
```

```
        GG(II,JJ)=GG(II,JJ)+D12*BY(I,L)                ! D2W/DY2
        ENDDO
        ENDDO
        ENDDO
C
        DO I=1,NY                        ! THE CORNER POINTS (I=1 AND I=NY) ARE INCLUDED
        KI=NX2*(I-1)
        DO L=1,NY2
        KL=NX2*(L-1)
        JT=NX
        DO J=1,NX,NX-1                                  ! X=0 (J=1), X=A (J=NX)
        JT=JT+1
        II=KI+JT
        DO K=1,NX2
        JJ=KL+K
        GG(II,JJ)=GG(II,JJ)+2.D0*D16*AY(I,L)*AX(J,K)    ! D2W/DXDY
        ENDDO
        ENDDO
        ENDDO
        ENDDO
C
C-------------------------------------------------------------------------------------------------------
C
C      CORNER CONCENTRATED FORCE R
C
C-------------------------------------------------------------------------------------------------------
C
        DO I=1,NY,NY-1                                  ! Y=0 (I=1), Y=B (I=NY)
        KI=NX2*(I-1)
        DO L=1,NY2
        KL=NX2*(L-1)
        DO J=1,NX,NX-1                                  ! X=0 (J=1), X=A (J=NX)
        II=KI+J
        DO K=1,NX2
        JJ=KL+K
        GG(II,JJ)=2.D0*D66*AY(I,L)*AX(J,K)             ! D2W/DXDY
        ENDDO
        ENDDO
        ENDDO
        ENDDO
C
        DO I=1,NY,NY-1                                  ! Y=0 (I=1), Y=B (I=NY)
        KI=NX2*(I-1)
        DO J=1,NX,NX-1                                  ! X=0 (J=1), X=A (J=NX)
        II=KI+J
        DO K=1,NX2
        JJ=KI+K
        GG(II,JJ)=GG(II,JJ)+D16*BX(J,K)                ! D2W/DX2
        ENDDO
        ENDDO
        ENDDO
```

```
C
      DO I=1,NY,NY-1                              ! Y=0 (I=1), Y=B (I=NY)
      KI=NX2*(I-1)
      DO L=1,NY2
      KL=NX2*(L-1)
      DO J=1,NX,NX-1                              ! X=0 (J=1), X=A (J=NX)
      II=KI+J
      JJ=KL+J
      GG(II,JJ)=GG(II,JJ)+D26*BY(I,L)             ! D2W/DY2
      ENDDO
      ENDDO
      ENDDO
C
C-------------------------------------------------------------------------------
C
C     APPLYING THE BOUNDARY CONDITIONS AND ELIMINATING THE BOUNDARY EQUATIONS
C     IF THE CORRESPONDING DOF IS NOT ZERO
C
C-------------------------------------------------------------------------------
C
      DO I=2,NY-1
      KI=NX2*(I-1)
      IJ= (NX-2)*(I-2)
      DO J=2,NX-1
      IJK=IJ+J-1
      LI(IJK)=KI+J
      ENDDO
      ENDDO

      DO I=1,NN
      DO J=1,NN
      GS(I,J)=GG(LI(I),LI(J))                     ! RETAINED GS=[KII]
      ENDDO
      ENDDO
C-------------------------------------------------------------------------------
C     FIND THE POSITION OF THE NON-ZERO DOFS AT BOUNDARY POINTS,
C     WHICH ARE TO BE ELIMINATED.
C     CHANGES WITH DIFFERENT COMBINATIONS OF BOUNDARY CONDITIONS.
C-------------------------------------------------------------------------------
      IJK=0
      DO J=1,NX        ! W11,...,W1N FOR QY(X,0)=0
      IJK=IJK+1        ! R(0,0)=0 AND R(A,0)=0 ARE INCLUDED
      LN(IJK)=J
      ENDDO
C
      DO I=NY,NY2      ! WN1,...,WNN (I=NY); WY11,...,WY1N(I=NY+1); WYN1,...,WYNN(I=NY2)
      KI=NX2*(I-1)     ! FOR QY(X,B)=0; MY(X,0)=0(I=NY+1); MY(X,B)=0(I=NY2)
      DO J=1,NX        ! R(0,B)=0 AND R(A,B)=0 ARE INCLUDED
      IJK=IJK+1
      LN(IJK)=KI+J
      ENDDO
      ENDDO
```

```
C
      DO I=1,NY
      KI=NX2*(I-1)
      DO J=NX+1,NX2          ! WX11,...,WXN1(J=NX+1); WX1N,...,WXNN(J=NX2)
      IJK=IJK+1              ! FOR MX(0,Y)=0(J=NX+1); MX(A,Y)=0(J=NX2)
      LN(IJK)=KI+J
      ENDDO
      ENDDO
C
      DO I=2,NY-1            ! W21,...,W(N-1)1 (J=1)   FOR QX(0,Y)=0
      KI=NX2*(I-1)          ! W2N,...,W(N-1)N (J=NX) FOR QX(A,Y)=0
      DO J=1,NX,NX-1
      IJK=IJK+1
      LN(IJK)=KI+J
      ENDDO
      ENDDO
C
C---------------------------------------------------------------------------------------------------------------------------
C      END OF FINDING POSITIONS OF THE NON-ZERO DOFS AT BOUNDARY POINTS
C---------------------------------------------------------------------------------------------------------------------------
C
      DO I=1,NF
      DO J=1,NF
      FF(I,J)=GG(LN(I),LN(J))                    ! FF=[KBB]
      ENDDO
      ENDDO
C
      DO I=1,NN
      DO J=1,NF
      GF(I,J)=GG(LI(I),LN(J))                    ! GF=[KIB]
      ENDDO
      ENDDO

      DO I=1,NF
      DO J=1,NN
      FG(I,J)=GG(LN(I),LI(J))                    ! FG=[KBI]
      ENDDO
      ENDDO
C
      CALL INVMAT(NF,FF,ZZ,1.D-30,ISW,BK,CK,LM)        ! INVERSE OF [KBB]
C
      DO I=1,NN
      DO J=1,NF
      GFF(I,J)=0.0D0
      DO K=1,NF
      GFF(I,J)=GFF(I,J)-GF(I,K)*FF(K,J)          ! -[KIB][KBB]
      ENDDO
      ENDDO
      ENDDO
```

```
C
      DO I=1,NN
      DO J=1,NN
      DO K=1,NF
      GS(I,J)=GS(I,J)+GFF(I,K)*FG(K,J) ! FINAL [K] FOR COMPUTING EIGEN-VALUES.
                                       ! PUT COMMENT ON THIS SENTENCE FOR CCCC PLATE.
      ENDDO
      ENDDO
      ENDDO
C
      CALL TZZ(NN,NN,GS,Z,BK,CK,LM)              ! FIND ALL EIGEN-VALUES
C
C      RE-ARRANGEMENT, THE SMALLEST EIGENVALUE IS THE FIRST EIGENVALUE.
C
      ZZZ=1.D30
      K=1
440   DO 450 I=K,NN
      IF(BK(I) .GE. ZZZ) GO TO 450
      KK=I
      ZZZ=BK(I)
450   CONTINUE
      BK(KK)=BK(K)
      BK(K)=ZZZ
      ZZZ=1.D30
C
      TEMP=CK(KK)
      CK(KK)=CK(K)
      CK(K)=TEMP
C
      K=K+1
      IF(K .LE.NN) GOTO 440
C
C   OUTPUT
C
      K=0
      DO 599 I=1,NN
      IF (DABS(CK(I)) .GE. 1.0D-2) GO TO 599
      TT=BK(I)
      IF (TT .LT. 0.0D0) GO TO 599
      K=K+1
      WRITE (*,551) I,TT,CK(I),DSQRT(TT)/FACTOR           ! NON-DIMENSIONAL FREQUENCIES
      IF (K. GE. 10) GOTO 600
599   CONTINUE
600   CONTINUE

      DO 602 I=1,NN
      TT=BK(I)
      IF (TT .LT. 0.0D0) GO TO 601
      WRITE (9,551) I,TT,CK(I),DSQRT(TT)/FACTOR
      GOTO 602
601   WRITE (9,551) I,TT,CK(I)
602   CONTINUE
```

```
551    FORMAT (1X,I5,3G20.10)
       STOP
       END
C
C------------------------------------------------------------------------------------------------
C                         THE END OF PROGRAM 2
C------------------------------------------------------------------------------------------------
C

%
%------------------------------------------------------------------------------------------------
%    The converted MATLAB file named Program2_Main
%------------------------------------------------------------------------------------------------
%
%    Program 2: Free vibration of isotropic FFFF skew plate by the DQM
%
%    N    — The number of grid points in both directions.
%    NF   — The number of non-zero DOFs associated to the boundary points, it varies for different combinations of
%           boundary conditions; e.g., NF=4*N-8 for the SSSS plate and LN (NF) should be changed accordingly.
%    THEATA —skew angle, the only input datum to run the program.
%
%    (a) Only Grid V should be used, although other grids are included in the program.
%    (b) The modified DQM, namely, MMWC-3 is used to introduce the additional DOF.
%------------------------------------------------------------------------------------------------

clear; clc;
format long;

N=21; N2=N-2; NN=N2*N2; MMM=(N+2)*(N+2); NF=8*N-4;

NX=N;                                % Total number of points in the x-direction
NY=N;                                % Total number of points in the y-direction
A=1.0;                               % Length
RATIO=1.0/1.0;
B=A/RATIO;                           % Width

THEATA=input('Enter THEATA [Degree]: ');
VV=0.3;
TH=THEATA/180*pi;          % Skew angles, 0 for rectangular plate
D11=1.0;                          % Results are non-dimensional frequency parameters
D16=-sin(TH);
D22=1.0;
D26=D16;
D12=VV*cos(TH)^2+sin(TH)^2;
D66=(1.0+sin(TH)^2-VV*cos(TH)^2)/2.0;

FACTOR=cos(TH)^2;          % Results are non-dimensional frequency parameters

X=zeros(N); X(1)=0.0; X(N)=A;

%------------------------------------------------------------------------------------------------
%    == VARIOUS GRID SPACING ==
%------------------------------------------------------------------------------------------------
```

```
NG=N-2;
[XX, Y]=GRULE(NG);          % GAUSS INTEGRATION POINTS(XX) AND WEIGHTS(Y)

I=2:N-1;
X(I)=(1.0+XX(I-1))/2.0*A;      % Grid V

%DELX=A/(N-1);
% I=1:N
% X(I)=DELX*(I-1);             % Grid I-uniform

% I=2:N-1;
% X(I)=(1.0-cos((2*I-3)*pi/(2*N-4)))/2.0*A;        % Grid II
% X(I)=A*(1.0-cos((2*I-1)*pi/(2*N)))/2.0;          % Grid IV

% I=1:N
% X(I)=(1.0-cos((I-1)*pi/(N-1)))/2.0*A;                    % Grid III
% X(I)=A*(1.0-cos((2*I-1)*pi/(2*N))/cos(pi/(2*N)))/2.0;    % Grid VI-expanded

Y=X*B/A;

[AX, BX, CX, DX, BBX]=DQMNEW(N, X);
[AY, BY, CY, DY, BBY]=DQMNEW(N, Y);

NY2=NY+2; NX2=NX+2;
GG=zeros(MMM,MMM);

% == DQ EQUATIONS AT INNER GRID POINTS ==
for I=2:NY-1              % Mixed derivatives
     KI=NX2*(I-1);
     for L=1:NY2
           KL=NX2*(L-1);
           for J=2:NX-1
                 II=KI+J;
                 for K=1:NX2
                       JJ=KL+K;

     GG(II,JJ)=2.0*(D12+2.0*D66)*BY(I,L)*BX(J,K)+4.0*D16*AY(I,L)*CX(J,K)+4.0*D26*CY(I,L)*AX(J,K);

                 end
           end
     end
end

for I=2:NY-1              % D4W/DX4
     KI=NX2*(I-1);
     for J=2:NX-1
           II=KI+J;
           for K=1:NX2
                 JJ=KI+K;
                 GG(II,JJ)=GG(II,JJ)+D11*DX(J,K);
           end
     end
end
```

```
for I=2:NY-1                    % D4W/DY4
      KI=NX2*(I-1);
      for L=1:NY2
            KL=NX2*(L-1);
            for J=2:NX-1
                  II=KI+J;
                  JJ=KL+J;
                  GG(II,JJ)=GG(II,JJ)+D22*DY(I,L);
            end
      end
end

%
% == EQUATIONS OF BENDING MOMENTS AND SHEAR FORCE (Y=0, B) ==
% Shear force QY
%
for I=1:NY-1:NY                 % Put at W of Y=0 (I=1) and Y=B(I=NY)
      KI=NX2*(I-1);
      KL=NX2*(I-1);
      for J=2:NX-1              % The corner points (J=1 and J=NX) are not included
            II=KI+J;
            for K=1:NX2
                  JJ=KL+K;
                  GG(II,JJ)=2.0*D16*CX(J,K);            %D3W/DX3
            end
      end
end

for I=1:NY-1:NY                 % Put at W of Y=0 (I=1) and Y=B(I=NY)
      KI=NX2*(I-1);
      for L=1:NY2
            KL=NX2*(L-1);
            for J=2:NX-1        % The corner points (J=1 and J=NX) are not included
                  II=KI+J;
                  for K=1:NX2
                        JJ=KL+K;
                        GG(II,JJ)=GG(II,JJ)+4.0*D26*BY(I,L)*AX(J,K)+(D12+4.0*D66)*AY(I,L)*BX(J,K);
                                          %D3W/DX.DY2, D3W/DX2.DY
                  end
            end
      end
end

for I=1:NY-1:NY                 % Put at W of Y=0 (I=1) and Y=B(I=NY)
      KI=NX2*(I-1);
      for L=1:NY2
            KL=NX2*(L-1);
            for J=2:NX-1        % The corner points (J=1 and J=NX) are not included
                  II=KI+J;
                  JJ=KL+J;
                  GG(II,JJ)=GG(II,JJ)+D22*CY(I,L);    % D3W/DY3
            end
      end
end

%
% Bending moment My
%
```

```
JT=-1;
for I=1:NY-1:NY                % Put at WY of Y=0 (I=1) and Y=B(I=NY)
     JT=JT+1;
     KI=NX2*(NY+JT);
     KL=NX2*(I-1);
     for J=1:NX                % The corner points (J=1 and J=NX) are included
          II=KI+J;
          for K=1:NX2
               JJ=KL+K;
               GG(II,JJ)=D12*BX(J,K);          % D2W/DX2
          end
     end
end

JT=-1;
for I=1:NY-1:NY                % Put at WY of Y=0 (I=1) and Y=B(I=NY)
     JT=JT+1;
     KI=NX2*(NY+JT);
     for L=1:NY2
          KL=NX2*(L-1);
          for J=1:NX           % The corner points (J=1 and J=NX) are included
               II=KI+J;
               JJ=KL+J;
               GG(II,JJ)=GG(II,JJ)+D22*BY(I,L);          % D2W/DY2
          end
     end
end
JT=-1;
for I=1:NY-1:NY                % Put at WY of Y=0 (I=1) and Y=B(I=NY)
     JT=JT+1;
     KI=NX2*(NY+JT);
     for L=1:NY2
          KL=NX2*(L-1);
          for J=1:NX           % The corner points (J=1 and J=NX) are included
               II=KI+J;
               for K=1:NX2
                    JJ=KL+K;
                    GG(II,JJ)=GG(II,JJ)+2.0*D26*AY(I,L)*AX(J,K);  % D2W/DXDY
               end
          end
     end
end
%
%-----------------------------------------------------------------------------------------------------------------
% == EQUATIONS OF BENDING MOMENTS AND SHEAR FORCE  (X=0, A) ==
% Shear force QX
%-----------------------------------------------------------------------------------------------------------------
%

for I=2:NY-1                   % The corner points (I=1 and I=NY) are not included
     KI=NX2*(I-1);
     for L=1:NY2
          KL=NX2*(L-1);
          for J=1:NX-1:NX      % X=0 (J=1), X=A (J=NX)
               II=KI+J;
               JJ=KL+J;
               GG(II,JJ)=2.0*D26*CY(I,L);          % D3W/DY3
```

```
                end
            end
    end
    for I=2:NY-1                        % The corner points (I=1 and I=NY) are not included
        KI=NX2*(I-1);
        for L=1:NY2
            KL=NX2*(L-1);
            for J=1:NX-1:NX             % X=0 (J=1), X=A (J=NX)
                II=KI+J;
                for K=1:NX2
                    JJ=KL+K;
                    GG(II,JJ)=GG(II,JJ)+(D12+4.0*D66)*BY(I,L)*AX(J,K)+4.0*D16*AY(I,L)*BX(J,K);
                                        % D3W/DX.DY2, D3W/DX2.DY
                end
            end
        end
    end

    for I=2:NY-1                        % The corner points (I=1 and I=NY) are not included
        KI=NX2*(I-1);
        for J=1:NX-1:NX                 % X=0 (J=1), X=A (J=NX)
            II=KI+J;
            for K=1:NX2
                JJ=KI+K;
                GG(II,JJ)=GG(II,JJ)+D11*CX(J,K);    %D3W/DX3
            end
        end
    end
    % Bending moment = 0
    for I=1:NY                          % The corner points (I=1 and I=NY) are included
        KI=NX2*(I-1);
        JT=NX;
        for J=1:NX-1:NX                 % X=0 (J=1), X=A (J=NX)
            JT=JT+1;
            II=KI+JT;
            for K=1:NX2
                JJ=KI+K;
                GG(II,JJ)=D11*BX(J,K);                  % D2W/DX2
            end
        end
    end

    for I=1:NY                          % The corner points (I=1 and I=NY) are included
        KI=NX2*(I-1);
        for L=1:NY2
            KL=NX2*(L-1);
            JT=NX;
            for J=1:NX-1:NX             % X=0 (J=1), X=A (J=NX)
                JT=JT+1;
                II=KI+JT;
                JJ=KL+J;
                GG(II,JJ)=GG(II,JJ)+D12*BY(I,L);    % D2W/DY2
            end
        end
    end
    for I=1:NY                          % The corner points (I=1 and I=NY) are included
        KI=NX2*(I-1);
```

```
        for L=1:NY2;
                KL=NX2*(L-1);
                JT=NX;
                for J=1:NX-1:NX          % X=0 (J=1), X=A (J=NX)
                        JT=JT+1;
                        II=KI+JT;
                        for K=1:NX2
                                JJ=KL+K;
                                GG(II,JJ)=GG(II,JJ)+2.0*D16*AY(I,L)*AX(J,K);          % D2W/DXDY
                        end
                end
        end
end

%-------------------------------------------------------------------------------------------------------
% == CORNER CONCENTRATED FORCE R ==
%-------------------------------------------------------------------------------------------------------

for I=1:NY-1:NY                          % Y=0 (I=1), Y=B (I=NY)
        KI=NX2*(I-1);
        for L=1:NY2
                KL=NX2*(L-1);
                for J=1:NX-1:NX          % X=0 (J=1), X=A (J=NX)
                        II=KI+J;
                        for K=1:NX2
                                JJ=KL+K;
                                GG(II,JJ)=2.0*D66*AY(I,L)*AX(J,K);                     % D2W/DXDY
                        end
                end
        end
end

for I=1:NY-1:NY                          % Y=0 (I=1), Y=B (I=NY)
        KI=NX2*(I-1);
        for J=1:NX-1:NX                  % X=0 (J=1), X=A (J=NX)
                II=KI+J;
                for K=1:NX2
                        JJ=KI+K;
                        GG(II,JJ)=GG(II,JJ)+D16*BX(J,K);          % D2W/DX2
                end
        end
end

for I=1:NY-1:NY                          % Y=0 (I=1), Y=B (I=NY)
        KI=NX2*(I-1);
        for L=1:NY2
                KL=NX2*(L-1);
                for J=1:NX-1:NX          % X=0 (J=1), X=A (J=NX)
                        II=KI+J;
                        JJ=KL+J;
                        GG(II,JJ)=GG(II,JJ)+D26*BY(I,L);                              % D2W/DY2
                end
        end
end
```

```
%--------------------------------------------------------------------------------
%    APPLYING THE BOUNDARY CONDITIONS AND ELIMINATING THE BOUNDARY EQUATIONS
%    IF THE CORRESPONDING DOF IS NOT ZERO
%--------------------------------------------------------------------------------

for I=2:NY-1
        KI=NX2*(I-1);
        IJ= (NX-2)*(I-2);
        for J=2:NX-1
                IJK=IJ+J-1;
                LI(IJK)=KI+J;
        end
end

for I=1:NN
        for J=1:NN
                GS(I,J)=GG(LI(I),LI(J));        % Retained GS=[KII]
        end
end


%--------------------------------------------------------------------------------
%    == FIND THE POSITION OF THE NON-ZERO DOFS AT BOUNDARY POINTS ==
%    Which are to be eliminated.
%    Changes with different combinations of boundary conditions.
%--------------------------------------------------------------------------------

IJK=0;
for J=1:NX              % W11,...,W1N for QY(X,0)=0
        IJK=IJK+1;      % R(0,0)=0 AND R(A,0)=0 are included
        LN(IJK)=J;
end

for I=NY:NY2            % WN1,...,WNN (I=NY); WY11,...,WY1N(I=NY+1); WYN1,...,WYNN(I=NY2)
        KI=NX2*(I-1);   % For QY(X,B)=0; MY(X,0)=0(I=NY+1); MY(X,B)=0(I=NY2)
        for J=1:NX      % R(0,B)=0 AND R(A,B)=0 are included
                IJK=IJK+1;
                LN(IJK)=KI+J;
        end
end

for I=1:NY
        KI=NX2*(I-1);
        for J=NX+1:NX2  % WX11,...,WXN1(J=NX+1); WX1N,...,WXNN(J=NX2)
                IJK=IJK+1; % For MX(0,Y)=0(J=NX+1); MX(A,Y)=0(J=NX2)
                LN(IJK)=KI+J;
        end
end

for I=2:NY-1            % W21,...,W(N-1)1 (J=1)   for QX(0,Y)=0
        KI=NX2*(I-1);   % W2N,...,W(N-1)N (J=NX) for QX(A,Y)=0
        for J=1:NX-1:NX
                IJK=IJK+1;
                LN(IJK)=KI+J;
        end
end
```

```
%-------------------------------------------------------------------------------------------------------
%    == END OF FINDING POSITIONS OF THE NON-ZERO DOFS AT BOUNDARY POINTS ==
%-------------------------------------------------------------------------------------------------------

for I=1:NF
       for J=1:NF
              FF(I,J)=GG(LN(I),LN(J));              % FF=[KBB]
       end
end

for I=1:NN
       for J=1:NF
              GF(I,J)=GG(LI(I),LN(J));        % GF=[KIB]
       end
end

for I=1:NF
       for J=1:NN
              FG(I,J)=GG(LN(I),LI(J));        % FG=[KBI]
       end
end


FF=inv(FF);              % Inverse of [KBB]
GFF=-GF*FF;              % -[KIB][KBB]
GS=GS+GFF*FG;            % Final [K] for eigen-values. For CCCC plates, this statement is not used.


EIG_GS=eig(GS);          % Find all real eigen-values & sort them in ascending order
EIG_GS=EIG_GS(abs(imag(EIG_GS))<=1e-2);
EIG_GS=sort(real(EIG_GS));
EIG_GS=sqrt(EIG_GS)/FACTOR;

save EigenvalueData.txt -ascii EIG_GS        %Save the eigen-values in the file EigenvalueData.txt


%
%-------------------------------------------------------------------------------------------------------
%    The end of the converted MATLAB file Program2_Main
%-------------------------------------------------------------------------------------------------------
%
```

V.2 ILLUSTRATION FOR APPLYING OTHER BOUNDARY CONDITIONS

For other combinations of boundary conditions, only change the NF and comment or delete some statements in the section of finding nonzero degrees of freedoms (DOFs) at boundary points. Four examples are given for illustrations, which are based on Program 2 named "Free vibration of isotropic FFFF skew plate by the DQM." Note that $NX = NY = N$.

1. CCCC skew plate
   This is the simplest case by using the modified DQM.

Since NF = 0 (modification of NF is not necessary), matrix [GS] is not modified at all. Only one modification is needed. The modification is in bold letter. Just simply put a comment (see bold letter C) at the beginning of the statement of GS(I,J) = GS(I,J) + GFF(I,K)*FG(K,J) as under, namely,

```
C------------------------------------------------------------------------------------
C       FORTRAN source codes
C------------------------------------------------------------------------------------
        DO I=1,NN
        DO J=1,NN
        DO K=1,NF                           ! Pay attention to the bold letter C, this statement is not needed.
C       GS(I,J)=GS(I,J)+GFF(I,K)*FG(K,J)    ! Final [K] for finding eigen-values
        ENDDO
        ENDDO
        ENDDO
```

```
%------------------------------------------------------------------------------------
%       Converted MATLAB codes
%------------------------------------------------------------------------------------
%       GS=GS+GFF*FG;        % Final [K] for finding eigen-values.This statement is not needed.
```

2. SSSS skew plate

Set NF = $4 \times N - 8$. Note NF is not equal to $4 \times N$, since the first-order derivatives w_ξ, w_η at corners are zero when they are expressed in terms of differential quadrature. For isotropic rectangular plate; however, w_x, w_y and M_x, M_y are all zero at corners when they are expressed in terms of differential quadrature, thus whether NF = $4 \times N - 8$ or NF = $4 \times N$ does not matter and the final result will not alter.

The statements in the section of "Find the positions of the nonzero DOFs at boundary points" are now modified as follows. The modifications are in bold letters.

```
C------------------------------------------------------------------------------------
C       FORTRAN Codes
C       Find the position of the non-zero DOFs at boundary points, which are to be eliminated.
C       Changes with different combinations of boundary conditions.
C------------------------------------------------------------------------------------
C
        IJK=0
        DO I=NY+1,NY2          ! Wy11,...,Wy1N(I=NY+1); WyN1,...,WyNN(I=NY2)
        KI=NX2*(I-1)           ! My(X,0)=0(I=NY+1); My(X,B)=0(i=NY2)
        DO J=2, NX-1           ! My(0,b)=0 and My(a,b)=0 are not included(J=1)
        IJK=IJK+1              ! My(0,0)=0 and My(a,0)=0 are not included(J=NX)
        LN(IJK)=KI+J
        ENDDO
        ENDDO
```

```
C
      DO I=2,NY-1                       ! Mx(0,0)=0 and Mx(a,0)=0 are not included(I=1)
      KI=NX2*(I-1)                      ! Mx(0,b)=0 and Mx(a,b)=0 are not included(I=NY)
      DO J=NX+1,NX2                     ! Wx11,...,WxN1(J=NX+1); Wx1N,...,WxNN(J=NX2)
      IJK=IJK+1                         ! For Mx(0,Y)=0(J=NX+1); Mx(A,Y)=0(J=NX2)
      LN(IJK)=KI+J
      ENDDO
      ENDDO
C
C------------------------------------------------------------------------
C     End of finding positions of the non-zero DOFs at boundary points
C------------------------------------------------------------------------
C
%------------------------------------------------------------------------
%     Converted MATLAB codes
%------------------------------------------------------------------------
%          Find the position of the non-zero DOFs at boundary points, which are to be eliminated.
%          Changes with different combinations of boundary conditions.
%------------------------------------------------------------------------
%
IJK=0;
for I=NY+1:NY2;            % Wy11,...,Wy1N(I=NY+1); WyN1, ..., WyNN(I=NY2)
      KI=NX2*(I-1);        % My(X,0)=0(I=NY+1); My(X,B)=0(i=NY2)
      for J=2:NX-1         % My(0,b)=0 and My(a,b)=0 are not included(J=1)
            IJK=IJK+1;     % My(0,0)=0 and My(a,0)=0 are not included(J=NX)
            LN(IJK)=KI+J;
      end
end

for I=2:NY-1              % Mx(0,0)=0 and Mx(a,0)=0 are not included(I=1)
      KI=NX2*(I-1);       % Mx(0,b)=0 and Mx(a,b)=0 are not included(I=NY)
      for J=NX+1:NX2      % Wx11,...,WxN1(J=NX+1); Wx1N,...,WxNN(J=NX2)
            IJK=IJK+1;    % For Mx(0,Y)=0(J=NX+1); Mx(A,Y)=0(J=NX2)
            LN(IJK)=KI+J;
      end
end
%
%------------------------------------------------------------------------
%     End of finding positions of the non-zero DOFs at boundary points
%------------------------------------------------------------------------
%
```

3. SFSF skew plate

In this book, two different edge-numbering sequences are used for comparisons with existing results. Refer to Fig. IV.1, symbol SFSF stands for the plate with edges at $y = 0$ and b simply supported and the other two free. In other words, the edge numbering sequence is the same as the one shown in Fig. 8.2.

For the SFSF skew plate NF = $6 \times N - 8$, since at corners only the first-order derivative w_ξ is zero and w_η is not zero. The statements in the section of "Find the positions of the nonzero DOFs at boundary points" are now modified as follows. The modifications are in bold letters.

```
C----------------------------------------------------------------------------------------
C       FORTRAN Codes
C       Find the position of the non-zero DOFs at boundary points , which are to be eliminated.
C       Changes with different combinations of boundary conditions.
C----------------------------------------------------------------------------------------
C
        IJK=0
        DO I=NY+1,NY2                           ! Wy11,...,Wy1N(I=NY+1); WyN1,...,WyNN(I=NY2)
        KI=NX2*(I-1)                            ! My(x,0)=0(I=NY+1); My(x,b)=0(i=NY2)
        DO J=1,NX
        IJK=IJK+1
        LN(IJK)=KI+J
        ENDDO
        ENDDO
C
        DO I=2,NY-1                             ! Mx(0,0)=0 and Mx(a,0)=0 are not included(I=1)
        KI=NX2*(I-1)                            ! Mx(0,b)=0 and Mx(a,b)=0 are not included(I=NY)
        DO J=NX+1,NX2                           ! Wx11,...,WxN1(J=NX+1); Wx1N,...,WxNN(J=NX2)
        IJK=IJK+1                               ! For Mx(0,y)=0(J=NX+1); Mx(a,y)=0 (J=NX2)
        LN(IJK)=KI+J
        ENDDO
        ENDDO
C
        DO I=2,NY-1                             ! W21,...,W(N-1)1 (J=1) for Qx(0,y)=0
        KI=NX2*(I-1)                            ! W2N,...,W(N-1)N (J=NX) for Qx(a,y)=0
        DO J=1,NX,NX-1                          ! The four corner force conditions are not included.
        IJK=IJK+1
        LN(IJK)=KI+J
        ENDDO
        ENDDO
C
C----------------------------------------------------------------------------------------
C       End of finding positions of the non-zero DOFs at boundary points
C----------------------------------------------------------------------------------------
C
```

```
%----------------------------------------------------------------------------------------
%       Converted MATLAB codes
%----------------------------------------------------------------------------------------
%       Find the position of the non-zero DOFs at boundary points, which are to be eliminated.
%       Changes with different combinations of boundary conditions.
%----------------------------------------------------------------------------------------
%
IJK=0;
for I=NY+1:NY2                  % Wy11,...,Wy1N(I=NY+1); WyN1,...,WyNN(I=NY2)
    KI=NX2*(I-1);               % My(x,0)=0(I=NY+1); My(x,b)=0(i=NY2)
    for J=1:NX
        IJK=IJK+1;
        LN(IJK)=KI+J;
    end
end
```

```
for I=2:NY-1                   % Mx(0,0)=0 and Mx(a,0)=0 are not included(I=1)
    KI=NX2*(I-1);              % Mx(0,b)=0 and Mx(a,b)=0 are not included(I=NY)
    for J=NX+1:NX2             % Wx11,...,WxN1(J=NX+1); Wx1N,...,WxNN(J=NX2)
        IJK=IJK+1;             % For Mx(0,y)=0(J=NX+1); Mx(a,y)=0 (J=NX2)
        LN(IJK)=KI+J;
    end
end

for I=2:NY-1                   % W21,...,W(N-1)1 (J=1) for Qx(0,y)=0
    KI=NX2*(I-1);              % W2N,...,W(N-1)N (J=NX) for Qx(a,y)=0
    for J=1:NX-1:NX            % The four corner force conditions are not included.
        IJK=IJK+1;
        LN(IJK)=KI+J;
    end
end
%
%----------------------------------------------------------------------------------------------------
%           End of finding positions of the non-zero DOFs at boundary points
%----------------------------------------------------------------------------------------------------
%
```

4. CSFF skew plate

Refer to Fig. IV.1, the symbol CSFF stands for the skew plate with edge at $y = 0$ clamped, the edge at $x = a$ simply supported and the remaining two edges free. In other words, the edge numbering sequence is the same as the one shown in Fig. 8.2.

Note that at corners $(0, 0)$ and $(a, 0)$, the first-order derivative w_ξ is zero and at corners $(a, 0)$ and (a, b) the first-order derivative w_η is zero. Thus, NF = $5 \times N - 6$. The statements in the section of "Find the positions of the nonzero DOFs at boundary points" are now modified as follows. The modifications are in bold letters.

```
C-------------------------------------------------------------------------------------------------
C       FORTRAN Codes
C       Find the position of the non-zero DOFs at boundary points, which are to be eliminated.
C       Changes with different combinations of boundary conditions.
C-------------------------------------------------------------------------------------------------
C
        IJK=0
        DO I=NY,NY2, 2      ! Qy(x,b)=0(I=NY); My(x,b)=0 (i=NY2)
        KI=NX2*(I-1)
        DO J=1,NX-1         ! Note: Nx-1, Qy(a,b) =0 & My(a,b)=0 are not included(J=Nx)
        IJK=IJK+1
        LN(IJK)=KI+J
        ENDDO
        ENDDO

        DO I=2,NY           ! Mx(0,0)=0 and Mx(a,0)=0 are not included(I=1)
        KI=NX2*(I-1)
        DO J=NX+1,NX2       ! Mx(0,y)=0(j=NX+1); Mx(a,y)=0(j=NX2)
        IJK=IJK+1
        LN(IJK)=KI+J
        ENDDO
        ENDDO
```

```
              DO I=2,NY-1
              KI=NX2*(I-1)
              DO J=1,1              ! Qx(0,y)=0
              IJK=IJK+1
              LN(IJK)=KI+J
              ENDDO
              ENDDO
C
C-------------------------------------------------------------------------------------------------------
C      End of finding positions of the non-zero DOFs at boundary points
C-------------------------------------------------------------------------------------------------------
C
```

```
%-------------------------------------------------------------------------------------------------------
%      Converted MATLAB codes
%-------------------------------------------------------------------------------------------------------
%      Find the position of the non-zero DOFs at boundary points, which are to be eliminated.
%      Changes with different combinations of boundary conditions.
%-------------------------------------------------------------------------------------------------------
%

IJK=0;
for I=NY:2:NY2                % Qy(x,b)=0(I=NY); My(x,b)=0 (i=NY2)
        KI=NX2*(I-1);
        for J=1:NX-1          % Note: Nx-1, Qy(a,b) =0 & My(a,b)=0 are not included(J=Nx)
              IJK=IJK+1;
              LN(IJK)=KI+J;
        end
end
for I=2:NY                    % Mx(0,0)=0 and Mx(a,0)=0 are not included(I=1)
        KI=NX2*(I-1);
        for J=NX+1:NX2        % Mx(0,y)=0(j=NX+1); Mx(a,y)=0 (j=NX2)
              IJK=IJK+1;
              LN(IJK)=KI+J;
        end
end

for I=2:NY-1
        KI=NX2*(I-1);
        for J=1:1             % Qx(0,y)=0
              IJK=IJK+1;
              LN(IJK)=KI+J;
        end
end
%
%-------------------------------------------------------------------------------------------------------
%      End of finding positions of the non-zero DOFs at boundary points
%-------------------------------------------------------------------------------------------------------
%
```

It is demonstrated that with only a few modifications on the program written for analyzing isotropic FFFF skew plate (Program 2 or Program2_Main), the modified program can solve the free vibration of skew plate with any combinations of boundary conditions.

Appendix VI

VI PROGRAM 3: FREE VIBRATION OF ISOTROPIC FFFF SKEW PLATE BY THE QEM

VI.1 FORTRAN AND MATLAB CODES

For readers' reference, Program 3, named as "Free vibration of isotropic FFFF skew plate by the QEM" is given below. Currently the element stiffness and mass matrices are numerically integrated by the GLL quadrature. As before, FORTRAN codes are given first and then the converted MATLAB file is followed.

Note that subroutine TZZ is the standard eigenvalue solver to find all eigenvalues of the real matrix [GS] and that subroutine INVMAT is called to find the inverse of the real matrix [FF]. The two commonly used subroutines are not included in FORTRAN Program 3.

Different from Program 2, NF is the number of nonzero first-order derivatives of w_ξ or w_η only and NN changes with different combinations of boundary conditions. The nonzero deflections at boundary points are not included in NF but included in NN. Thus, both NF and NN vary with the change of boundary conditions. For the FFFF skew plate by using QEM (Program 3), NF = $4N$ and NN = $N \times N$; however, for the FFFF skew plate by using the DQM (Program 2) NF = $8N - 4$ and NN = $(N - 2) \times (N - 2)$. NN is fixed in Program 2 for any combinations of boundary conditions.

To run "Program 3", besides subroutines INVMAT, TZZ, and MTM, subroutines DQMNEW2 and DQELE2 in Section II.2 should also be included. To run the converted MATLAB file "Prgram3_Main", functions DQMNEW2 and DQELE2 in Section II.2 should be put in the same directory where "Prgram3_Main" is. Currently the mixed method is used to introduce the additional degree of freedom at boundary points and $N = 15$. If N is to be changed, XX(I) and WI(I) should be changed accordingly.

Description of some variables in Program 3

> X(I): nondimensional x coordinate of a grid point, x_i $[-1,1]$
> Y(I): nondimensional y coordinate of a grid point, y_i $[-1,1]$

Input data

> N: number of grid points in both directions
> A: plate length in ξ (or x) direction
> B: plate length in η (or y) direction
> VV: Poisson's ratio
> D11, D12, D22, D66, D16, D26: the flexural stiffness (isotropic materials only)
> XX(I): abscissa in GLL quadrature, xx_i $[-1,1]$
> WI(I): weight in GLL quadrature
> THEATA: skew angle (degrees), the only input datum currently
> Note that except THEATA, all other input data have been assigned currently.

Output data

> TT: real part of the eigenvalues
> CK(I): imaginary part of the eigenvalues
> DSQRT(TT)/FACTOR: frequency parameter

Differential Quadrature and Differential Quadrature Based Element Methods. 978-0-12-803081-3

```
C
C-----------------------------------------------------------------------------------------------
C       Program 3    (FORTRAN Codes)
C       Free vibration of isotropic FFFF skew plate by the QEM
C       N—The number of grid points in both directions which is fixed!. NF—the number of non-zero DOFs
C           associated to the first order derivatives at the boundary points (non zero Wx and Wy only),
C           the nonzero deflections at the boundary points are not included, different from the DQM.
C           It varies for different combinations of boundary conditions and LN (NF) should be changed accordingly.
C       NW—The number of non-zero deflections, including the ones on the free edges. LI (NW) and EM
C           should be changed accordingly, different from the DQM. For the DQM, NW is fixed & equals to
C           the number of inner grid points for all combinations of boundary conditions.
C       THEATA—the only input datum to run the program, skew angle in degrees.
C       Only GLL points are used in this program currently.
C-----------------------------------------------------------------------------------------------
C
        IMPLICIT REAL*8 (A-H,O-Z)
        PARAMETER (N=15,MMM=(N+2)*(N+2),NW=N*N,NN=NW,NF=4*N)
        CHARACTER FILE_MODE*80
        DIMENSION X(N),Y(N),AX(N,N+2),BX(N,N+2),AY(N,N+2),BY(N,N+2)
        DIMENSION XX(N),WI(N),Z(NW),LI(NW),LN(NF)
        DIMENSION GS(NN,NN),BK(NW),CK(NW),LM(NW)
        DIMENSION FF(NF,NF),GF(NN,NF),FG(NF,NN),GFF(NN,NF)
        DIMENSION D(3,3),BXX(NW,MMM),BYY(NW,MMM),BXY(NW,MMM)
        DIMENSION EM(NN),EX(3,MMM-4),DEX(3,MMM-4),STIFF(MMM-4,MMM-4)
C-----------------------------------------------------------------------------------------------
C       FFFF: MMWC-3 IS USED TO INTRODUCE THE ADDITIONAL DOF
C-----------------------------------------------------------------------------------------------
C       GLL POINTS & WEIGHTS
C-----------------------------------------------------------------------------------------------
        XX(1)=-1.0D0
        XX(2)=-0.965245926503839D0
        XX(3)=-0.885082044222977D0
        XX(4)=-0.763519689951815D0
        XX(5)=-0.606253205469846D0
        XX(6)=-0.420638054713673D0
        XX(7)=-0.215353955363794D0
        XX(8)= 0.0D0
        WI(1)= 1.0D0/105.0D0
        WI(2)= 0.058029893034830D0
        WI(3)= 0.101660070340631D0
        WI(4)= 0.140511699803666D0
        WI(5)= 0.172789647252341D0
        WI(6)= 0.196987235964736D0
        WI(7)= 0.211973585926826D0
        WI(8)= 0.217048116348816D0
        DO I=9,15
        XX(I)=-XX(16-I)
        WI(I)= WI(16-I)
        ENDDO
```

```
C----------------------------------------------------------------------------------------------------
C     INPUT FILENAME OF OUTPUT
C----------------------------------------------------------------------------------------------------
      WRITE(*,'   (A)   ')' ENTER OUTPUT FILE NAME'        ! File to store output data
      READ(*,'   (A)   ') FILE_MODE
      OPEN(UNIT=9,STATUS='UNKNOWN',FILE=FILE_MODE)
C
      NX=N                           ! TOTAL POINTS IN THE X-DIRECTION
      NY=N                           ! TOTAL POINTS IN THE Y-DIRECTION
      NY2=NY+2
      NX2=NX+2
      PI=DATAN(1.0D0)*4.0D0
      A=1.D0                    !  LENGTH A
      RATIO=1.0D0/1.0D0
      B=A/RATIO                 !  LENGTH B
C----------------------------------------------------------------------------------------------------
      WRITE (*,*) 'ENTER THEATA'
      READ (*,*) THEATA                  ! SKEW ANGLE IN DEGREES
      VV=0.3D0
      TH=THEATA/180.D0*PI                ! SKEW ANGLE

      D11=1.0D0
      D16=-DSIN(TH)
      D22=1.0D0
      D26=D16
      D12=VV*DCOS(TH)**2+DSIN(TH)**2
      D66=(1.0D0+DSIN(TH)**2-VV*DCOS(TH)**2)/2.0D0
C
      D(1,1)=D11
      D(1,2)=D12
      D(1,3)=D16
      D(2,1)=D12
      D(2,2)=D22
      D(2,3)=D26
      D(3,1)=D16
      D(3,2)=D26
      D(3,3)=D66
C----------------------------------------------------------------------------------------------------
      FACTOR=DCOS(TH)**2     ! FOR NON-DIMENSIONAL FREQUENCIES
      DO I=1,N
      X(I)=XX(I)                 !XX BELONG TO [-1,1]
      ENDDO
      DO I=1,N
      Y(I)=X(I)
      ENDDO
C----------------------------------------------------------------------------------------------------
      CALL DQMNEW2(N,X,AX,BX,A/2.0D0)            ! MMWC-3
C     CALL DQMNEW2(N,Y,AY,BY,B/2.0D0)
C----------------------------------------------------------------------------------------------------
```

```
C    IF DQEM IS USED,ONLY MIXED WAY CAN BE USED SINCE THE FOUR WXY HAS
C    NOT BEEN CONSIDERED IN THIS PROGRAM.
C    AN EXAMPLE IS GIVEN. IN THE CURRENT VERSION OF PROGRAM, THE MIXED METHOD IS USED.
C    SEE THE FOUR CALL STATEMENTS FOR DETAILS.
C--------------------------------------------------------------------------------------------------------------------
C    CALL DQELE2(N,X,AX,BX,A/2.0D0)
     CALL DQELE2(N,Y,AY,BY,B/2.0D0)
C

     DO I=1,NW
     DO J=1,MMM
     BXX(I,J)=0.0D0
     BYY(I,J)=0.0D0
     BXY(I,J)=0.0D0
     ENDDO
     ENDDO
C
     DO I=1,NY                                 ! MIXED DERIVATIVES
     KI=NX*(I-1)
     DO L=1,NY2
     KL=NX2*(L-1)
     DO J=1,NX
     II=KI+J
     DO K=1,NX2
     JJ=KL+K
     BXY(II,JJ)=2.D0*AY(I,L)*AX(J,K)
     ENDDO
     ENDDO
     ENDDO
     ENDDO
C
     DO I=1,NY
     KI=NX*(I-1)
     KL=NX2*(I-1)
     DO J=1,NX
     II=KI+J
     DO K=1,NX2
     JJ=KL+K
     BXX(II,JJ)=BY(J,K)                        !  WXX
     ENDDO
     ENDDO
     ENDDO

     DO I=1,NY
     KI=NX*(I-1)
     DO L=1,NY2
     KL=NX2*(L-1)
     DO J=1,NX
     II=KI+J
     JJ=KL+J
     BYY(II,JJ)=BY(I,L)                        !  WYY
     ENDDO
     ENDDO
     ENDDO
```

```
C
C       REARRANGE NOT TO INCLUDE THE FOUR CORNER DEGREES.
C       I.E., THE DOFS AT THE FOUR OPEN CIRCLES IN FIG. IV.1 ARE NOT USED.
C
        IJK=NX2*NY+NX
        DO I=1,NW
        DO J=1,NX
        BXX(I,IJK+J)=BXX(I,IJK+J+2)              !MOVE THE COLUMN
        BYY(I,IJK+J)=BYY(I,IJK+J+2)
        BXY(I,IJK+J)=BXY(I,IJK+J+2)
        ENDDO
        ENDDO
C
        DO I=1,MMM-4
        DO J=1,MMM-4
        STIFF(I,J)=0.0D0
        ENDDO
        ENDDO
C
C       NUMERICAL INTEGRATION (GLL QUADRATURE)
C
        DO I=1,NY
        DO J=1,NX
        WIJ=WI(I)*WI(J)                 ! WEIGHT,THE DETERMINANT OF [J] IS NOT INCLUDED
        IJ=NX*(I-1)+J

        DO K=1,MMM-4
        EX(1,K)=BXX(IJ,K)
        EX(2,K)=BYY(IJ,K)
        EX(3,K)=BXY(IJ,K)
        ENDDO
        CALL MTM (D,EX,DEX,3,3,MMM-4)
        DO I1=1,MMM-4
        DO J1=1,MMM-4
        DO K=1,3
        STIFF(I1,J1)=STIFF(I1,J1)+WIJ*EX(K,I1)*DEX(K,J1)
        ENDDO
        ENDDO
        ENDDO
        ENDDO
        ENDDO
C
C    THE END OF NUMERICAL INTEGRATION
C
C    Finding the NN non-zero deflections
C
        IJK=0
        DO I=1,NY                               ! RELATED TO NON-ZERO W ONLY
        KI=NX2*(I-1)
        DO J=1,NX
        IJK=IJK+1
        LI(IJK)=KI+J
        ENDDO
        ENDDO
```

```
C
C     The end of finding the NN non-zero deflections
C
      DO I=1,NN
      DO J=1,NN
      GS(I,J)=STIFF(LI(I),LI(J))                    ! RETAINED
      ENDDO
      ENDDO
C
C     Finding the NF non-zero first order derivatives at boundary points
C
      IJK=0                                         ! TO BE ELIMINATED----RELATED TO WX OR WY
      DO I=1,NY
      KI=NX2*(I-1)
      DO J=NX+1,NX2                                 ! WX(0,Y), WX(A,Y) ARE NOT ZERO
      IJK=IJK+1
      LN(IJK)=KI+J
      ENDDO
      ENDDO

      KI=NX2*NY
      DO J=1,NX                                     ! WY(X,0) IS NOT ZERO
      IJK=IJK+1
      LN(IJK)=KI+J
      ENDDO

      KI=NX2*NY+NX
      DO J=1,NX                                     ! WY(X,B) IS NOT ZERO
      IJK=IJK+1
      LN(IJK)=KI+J
      ENDDO
C
C     The end of finding the NF non-zero first order derivatives at boundary points
C
      DO I=1,NF
      DO J=1,NF
      FF(I,J)=STIFF(LN(I),LN(J))                    ! [KBB]-TO BE ELIMINATED
      ENDDO
      ENDDO
C
      DO I=1,NN
      DO J=1,NF
      GF(I,J)=STIFF(LI(I),LN(J))                    ! [KIB]
      ENDDO
      ENDDO

      DO I=1,NF
      DO J=1,NN
      FG(I,J)=STIFF(LN(I),LI(J))                    ! [KBI]
      ENDDO
      ENDDO
```

```
C--------------------------------------------------------------------------------
C   SUBROUTINE INVMAT IS USED TO FIND INVERSE OF [FF].
C--------------------------------------------------------------------------------

        CALL INVMAT(NF,FF,ZZ,1.D-30,ISW,BK,CK,LM)      ! FIND INVERSE OF [KBB]

C
        DO I=1,NN
        DO J=1,NF
        GFF(I,J)=0.0D0
        DO K=1,NF
        GFF(I,J)=GFF(I,J)-GF(I,K)*FF(K,J)
        ENDDO
        ENDDO
        ENDDO

        DO I=1,NN
        DO J=1,NN
        DO K=1,NF
        GS(I,J)=GS(I,J)+GFF(I,K)*FG(K,J)       ! FINAL [KII]
        ENDDO
        ENDDO
        ENDDO
C--------------------------------------------------------------------------------
C
C    Diagonal terms in mass matrix
C
C--------------------------------------------------------------------------------
        IJK=0
        DO I=1,NY                              ! RELATED TO W ONLY FOR THE DIAGONAL MASS MATRIX
        KI=NX*(I-1)
        DO J=1,NX
        IJK=IJK+1
        EM(IJK)=WI(I)*WI(J)
        ENDDO
        ENDDO
C
C    The end of diagonal terms in mass matrix
C
        DO I=1,NN
        DO J=1,NN
        GS(I,J)=GS(I,J)/EM(I)                  ! IN ORDER TO USE STANDARD SOLVER FOR EIGENVALUES
        ENDDO
        ENDDO
C--------------------------------------------------------------------------------
C    SUBROUTINE TZZ IS USED TO FIND EIGENVALUES OF [GS],
C--------------------------------------------------------------------------------
        CALL TZZ(NN,NN,GS,Z,BK,CK,LM)
C
        ZZZ=1.D30                              ! REARRANGEMENT
        K=1
440     DO 450 I=K,NN
        IF(BK(I) .GE. ZZZ) GO TO 450
        KK=I
        ZZZ=BK(I)
450     CONTINUE
```

```
        BK(KK)=BK(K)
        BK(K)=ZZZ
        ZZZ=1.D30
        TEMP=CK(KK)
        CK(KK)=CK(K)
        CK(K)=TEMP
        K=K+1
        IF(K .LE.NN) GOTO 440
C
C       PRINT OUT
C
        K=0
        DO 599 I=1,NN
        IF (DABS(CK(I)) .GE.1.0D-2) GO TO 599
        TT=BK(I)
        IF (TT .LT. 0.0D0) GO TO 599
        K=K+1
        WRITE (*,551) I,TT,CK(I),DSQRT(TT)/FACTOR  ! NON-DIMENSIONAL FREQUENCY PARAMETERS
        IF (K. GE. 10) GOTO 600
599     CONTINUE
600     CONTINUE

        DO 602 I=1,NN
        TT=BK(I)
        IF (TT .LT. 0.0D0) GO TO 601
        WRITE (9,551) I,TT,CK(I),DSQRT(TT)/FACTOR
        GOTO 602
601     WRITE (9,551) I,TT,CK(I)
602     CONTINUE
551     FORMAT (1X,I5,3G20.10)
        STOP
        END
C
C-------------------------------------------------------------------------
C                       THE END OF PROGRAM 3
C-------------------------------------------------------------------------
C


%
%-------------------------------------------------------------------------
%    The converted MATLAB file named Program3_Main
%-------------------------------------------------------------------------
%       Program 3 (MATLAB Codes)
%       Free vibration of isotropic FFFF skew plate by the QEM
%       N—The number of grid points in both directions which is fixed. NF—the number of non-zero DOFs
%            associated to the first order derivatives at the boundary points (non zero Wx and Wy only),
%            the nonzero deflections at the boundary points are not included, different from the DQM.
%            It varies for different combinations of boundary conditions and LN (NF) should be changed accordingly.
%       NW—The number of non-zero deflections, including the ones on the free edges. LI (NW) and EM
%            should be changed accordingly, different from the DQM. For DQM, NW is fixed & equals to
%            the number of inner grid points for all combinations of boundary conditions.
%       THEATA—the only input datum to run the program, skew angle in degrees.
%       Only GLL points are used in this program currently.
%-------------------------------------------------------------------------
%
```

```
clear; clc;
format long;

N=15; MMM=(N+2)*(N+2); NW=N*N; NN=NW; NF=4*N;


%-------------------------------------------------------------------------------------
%        FFFF: MMWC-3 IS USED TO INTRODUCE THE ADDITIONAL DOF
%-------------------------------------------------------------------------------------
%      GLL POINTS & WEIGHTS
%-------------------------------------------------------------------------------------

XX(1)=-1.0;
XX(2)=-0.965245926503839;
XX(3)=-0.885082044222977;
XX(4)=-0.763519689951815;
XX(5)=-0.606253205469846;
XX(6)=-0.420638054713673;
XX(7)=-0.215353955363794;
XX(8)= 0.0;

WI(1)= 1.0/105.0;
WI(2)= 0.058029893034830;
WI(3)= 0.101660070340631;
WI(4)= 0.140511699803666;
WI(5)= 0.172789647252341;
WI(6)= 0.196987235964736;
WI(7)= 0.211973585926826;
WI(8)= 0.217048116348816;

I=9:15;
XX(I)=-XX(16-I);
WI(I)= WI(16-I);


%-------------------------------------------------------------------------------------
NX=N;                        % TOTAL POINTS IN THE X-DIRECTION
NY=N;                        % TOTAL POINTS IN THE Y-DIRECTION
NY2=NY+2; NX2=NX+2;


A=1.0;                       % LENGTH A
RATIO=1.0/1.0;
B=A/RATIO;                   % LENGTH B
%-------------------------------------------------------------------------------------
THEATA=input('Enter THEATA [Degree]: ');
VV=0.3;
TH=THEATA/180*pi;            % Skew angles, 0 for rectangular plate

D11=1.0;                     % Results are non-dimensional frequency parameters
D16=-sin(TH);
D22=1.0;
D26=D16;
D12=VV*cos(TH)^2+sin(TH)^2;
D66=(1.0+sin(TH)^2-VV*cos(TH)^2)/2.0;

D(1,1)=D11;
D(1,2)=D12;
D(1,3)=D16;
D(2,1)=D12;
D(2,2)=D22;
D(2,3)=D26;
D(3,1)=D16;
D(3,2)=D26;
D(3,3)=D66;
%-------------------------------------------------------------------------------------
```

```
FACTOR=cos(TH)^2;              % FOR NON-DIMENSIONAL FREQUENCIES
X=XX;                          % XX BELONG TO [-1,1]
Y=X;
%------------------------------------------------------------------------------
 [AX, BX]=DQMNEW2(N, X, A/2.0);              % MMWC-3
% [AY, BY]=DQMNEW2(N, Y, B/2.0);

%------------------------------------------------------------------------------
%    IF DQEM IS USED, ONLY MIXED WAY CAN BE USED SINCE THE FOUR WXY HAS
%    NOT BEEN CONSIDERED IN THIS PROGRAM.
%    AN EXAMPLE IS GIVEN. IN THE CURRENT VERSION OF PROGRAM, THE MIXED METHOD IS USED.
%    SEE THE FOUR CALL STATEMENTS FOR DETAILS.
%------------------------------------------------------------------------------
% [AX, BX]=DQELE2(N, X, A/2.0);
[AY, BY]=DQELE2(N, Y, B/2.0);

BXX=zeros(NW,MMM);
BYY=zeros(NW,MMM);
BXY=zeros(NW,MMM);

for I=1:NY                               % MIXED DERIVATIVES
    KI=NX*(I-1);
    for L=1:NY2
        KL=NX2*(L-1);
        for J=1:NX
            II=KI+J;
            for K=1:NX2
                JJ=KL+K;
                BXY(II,JJ)=2.0*AY(I,L)*AX(J,K);
            end
        end
    end
end

for I=1:NY
    KI=NX*(I-1);
    KL=NX2*(I-1);
    for J=1:NX
        II=KI+J;
        for K=1:NX2
            JJ=KL+K;
            BXX(II,JJ)=BY(J,K);                % WXX
        end
    end
end

for I=1:NY
    KI=NX*(I-1);
    for L=1:NY2
        KL=NX2*(L-1);
        for J=1:NX
            II=KI+J;
            JJ=KL+J;
            BYY(II,JJ)=BY(I,L);                % WYY
        end
    end
end

%------------------------------------------------------------------------------
%    REARRANGE NOT TO INCLUDE THE FOUR CORNER DEGREES
%    I.E., THE DOFS AT THE FOUR OPEN CIRCLES IN FIG. IV.1 ARE NOT USED.
%------------------------------------------------------------------------------
```

```
IJK=NX2*NY+NX;
for I=1:NW
        for J=1:NX
                BXX(I,IJK+J)=BXX(I,IJK+J+2);        % MOVE THE COLUMN
                BYY(I,IJK+J)=BYY(I,IJK+J+2);
                BXY(I,IJK+J)=BXY(I,IJK+J+2);
        end
end

STIFF=zeros(MMM-4,MMM-4);

%
%       NUMERICAL INTEGRATION (GLL QUADRATURE)
%
for I=1:NY
        for J=1:NX
                WIJ=WI(I)*WI(J);                % WEIGHT,THE DETERMINANT OF [J] IS NOT INCLUDED
                IJ=NX*(I-1)+J;

                for K=1:MMM-4
                        EX(1,K)=BXX(IJ,K);
                        EX(2,K)=BYY(IJ,K);
                        EX(3,K)=BXY(IJ,K);
                end

                DEX=D*EX;
                STIFF=STIFF+WIJ*EX'*DEX;

        end
end
%
%       THE END OF NUMERICAL INTEGRATION
%

%       Finding the NN non-zero deflections
%

IJK=0;
for I=1:NY                                      % RELATED TO NON-ZERO W ONLY
        KI=NX2*(I-1);
        for J=1:NX
                IJK=IJK+1;
                LI(IJK)=KI+J;
        end
end
%
%       The end of finding the NN non-zero deflections
%
for I=1:NN
        for J=1:NN
                GS(I,J)=STIFF(LI(I),LI(J));        % RETAINED
        end
end

%
%       Finding the NF non-zero first order derivatives at boundary points
%
```

```
IJK=0;                                      % TO BE ELIMINATED----RELATED TO WX OR WY
for I=1:NY
        KI=NX2*(I-1);
        for J=NX+1:NX2                      % WX(0,Y), WX(A,Y) ARE NOT ZERO
                IJK=IJK+1;
                LN(IJK)=KI+J;
        end
end

KI=NX2*NY;
for J=1:NX                                  % WY(X,0) IS NOT ZERO
        IJK=IJK+1;
        LN(IJK)=KI+J;
end

KI=NX2*NY+NX;
for J=1:NX                                  % WY(X,B) IS NOT ZERO
        IJK=IJK+1;
        LN(IJK)=KI+J;
end
%
%     The end of finding the NF non-zero first order derivatives at boundary points
%
for I=1:NF
        for J=1:NF
                FF(I,J)=STIFF(LN(I),LN(J));  % [KBB]-TO BE ELIMINATED
        end
end

for I=1:NN
        for J=1:NF
                GF(I,J)=STIFF(LI(I),LN(J));  % [KIB]
        end
end

for I=1:NF
        for J=1:NN
                FG(I,J)=STIFF(LN(I),LI(J));  % [KBI]
        end
end

%-----------------------------------------------------------------------------------------------
%   FUNCTION INV IS USED TO FIND INVERSE OF [FF].
%-----------------------------------------------------------------------------------------------

FF=inv(FF);                                 % FIND INVERSE OF [KBB]
GFF=-GF*FF;
GS=GS+GFF*FG;                               % FINAL [KII]

%-----------------------------------------------------------------------------------------------
%
%     Diagonal terms in mass matrix
%
IJK=0;
for I=1:NY                                  % RELATED TO W ONLY FOR THE DIAGONAL MASS MATRIX
```

```
      KI=NX*(I-1);
      for J=1:NX
            IJK=IJK+1;
            EM(IJK)=WI(I)*WI(J);
      end
end
%
%    The end of diagonal terms in mass matrix
%
for I=1:NN
      for J=1:NN
            GS(I,J)=GS(I,J)/EM(I);              % IN ORDER TO USE STANDARD SOLVER FOR EIGENVALUES
      end
end


EIG_GS=eig(GS);                              % Find all real eigen-values & sort them in ascending order
EIG_GS=EIG_GS(abs(imag(EIG_GS))<=1e-2);
EIG_GS=sort(real(EIG_GS));
EIG_GS=sqrt(EIG_GS)/FACTOR;

save EigenvalueData.txt -ascii -double EIG_GS         %Save the eigen-values in the file EigenvalueData.txt


%
%------------------------------------------------------------------------------------------------------------------------
%    The end of the converted MATLAB file Program3_Main
%------------------------------------------------------------------------------------------------------------------------
%
```

VI.2 ILLUSTRATION FOR APPLYING OTHER BOUNDARY CONDITIONS

For other combinations of boundary conditions, both NF and NN are needed to be modified. Besides, commenting or deleting some statements are needed in the section of finding the NN nonzero deflections and finding the NF nonzero first-order derivatives at boundary points. Four examples are given for illustrations, which are based on the program of "Free vibration of isotropic FFFF skew plate by the QEM" (Program 3 or Program3_Main). Slight difference between Program 3 (Program3_Main) and Program 2 (Program2_Main) exists on the DOFs at corners and will be discussed in detail for the case of SSSS skew plate. Note that NX = NY = *N*.

1. CCCC skew plate

The modification for this case is also simple. Besides setting NN = $(N - 2) \times (N - 2)$, some modifications in the section of "Finding the NN nonzero deflections" and in the section of "Diagonal terms in mass matrix" are needed. Detailed modifications are in bold letters and given below.

```
C------------------------------------------------------------------------------------------------------------------------
C       FORTRAN Codes
C       Finding the NN non-zero deflections
C------------------------------------------------------------------------------------------------------------------------
      IJK=0
      DO I=2,NY-1                               ! RELATED TO NON-ZERO W ONLY
      KI=NX2*(I-1)
```

```
      DO J=2,NX-1
      IJK=IJK+1
      LI(IJK)=KI+J
      ENDDO
      ENDDO
C--------------------------------------------------------------------------------
C    The end of finding the NN non-zero deflections
C--------------------------------------------------------------------------------


C--------------------------------------------------------------------------------
C
C    Diagonal terms in mass matrix
C--------------------------------------------------------------------------------
      IJK=0
      DO I=2,NY-1                          ! RELATED TO W ONLY FOR THE DIAGONAL MASS MATRIX
      KI=NX*(I-1)
      DO J=2,NX-1
      IJK=IJK+1
      EM(IJK)=WI(I)*WI(J)
      ENDDO
      ENDDO
C--------------------------------------------------------------------------------
C    The end of diagonal terms in mass matrix
C--------------------------------------------------------------------------------
```

In addition, put a letter C (the comment) at the beginning of the statement of GS(I,J) = GS(I,J) + GFF(I,K)*FG(K,J) since NF = 0 (modification of NF is not necessary), namely,

```
      DO I=1,NN
      DO J=1,NN
      DO K=1,NF                            ! Pay attention to the bold letter C
C     GS(I,J)=GS(I,J)+GFF(I,K)*FG(K,J)     ! This statement is not needed for the CCCC skew plate.
      ENDDO
      ENDDO
      ENDDO
```

```
%--------------------------------------------------------------------------------
%      Converted MATLAB codes
%--------------------------------------------------------------------------------
%
%      Finding the NN non-zero deflections
%--------------------------------------------------------------------------------
IJK=0;
for I=2:NY-1                % RELATED TO NON-ZERO W ONLY
      KI=NX2*(I-1);
      for J=2:NX-1
            IJK=IJK+1;
            LI(IJK)=KI+J;
      end
end
```

```
%------------------------------------------------------------------------------------------------
%      The end of finding the NN non-zero deflections
%------------------------------------------------------------------------------------------------


%------------------------------------------------------------------------------------------------
%      Diagonal terms in mass matrix
%------------------------------------------------------------------------------------------------

       IJK=0;
       for I=2:NY-1                          % RELATED TO W ONLY FOR THE DIAGONAL MASS MATRIX
              KI=NX*(I-1);
       for J=2:NX-1
              IJK=IJK+1;
              EM(IJK)=WI(I)*WI(J);
       end
end
%------------------------------------------------------------------------------------------------
%      The end of diagonal terms in mass matrix
%------------------------------------------------------------------------------------------------
```

In addition, put the symbol % (The comment) at the beginning of the statement of GS = GS + GFF*FG since NF = 0 (modification of NF is not necessary), namely,

%   GS=GS+GFF*FG;   % FINAL [KII], Pay attention to the bold symbol **%**.

2. SSSS skew plate

Set NN = $(N-2) \times (N-2)$ and NF = 4N. Note NF is different from the DQM (Program 2 or Program2_Main) and not equal to $4 \times N - 8$. The first-order derivatives, w_ξ, w_η, at corners are independent variables thus are different from the ones expressed in terms of differential quadrature. Therefore, the bending moments M_ξ, M_η should be zero. For isotropic rectangular plate; however, w_x, w_y and M_x, M_y are all zero at corners when they are expressed in terms of differential quadrature, thus whether NF = $4 \times N - 8$ or NF = $4 \times N$ does not matter and the same results will be obtained.

The modifications in the section of "Finding the NN nonzero deflections" and in the section of "Diagonal terms in mass matrix" are the same as the ones in CCCC skew plates, given previously. Thus, they are omitted.

Since NF is 4N, thus modifications are not necessary in the section of "Finding the NF nonzero first-order derivatives at boundary points." In other words, the zero bending moment conditions are the same as the FFFF skew plate.

3. SFSF skew plate

In this book, two different edge-numbering sequences are introduced for comparisons with existing results. The edge-numbering sequence shown in Fig. 8.2 is used here. Refer to Fig. IV.1, symbols SFSF stand for the plate with edges at $y = 0$ and b simply supported and the other two free.

Set NN = $N \times N - 2N$ and NF = 4N. Note that NF is different from the one in the DQM (Program 2 or Program2_Main) and is not equal to $4N - 4$. The first-order derivatives w_ξ, w_η at corners are independent variables. Thus, the corresponding bending moments M_ξ, M_η should be zero.

Since NF is 4*N*, thus modifications are not necessary in the section of "Finding the NF nonzero first-order derivatives at boundary points." In other words, the zero bending moment conditions are the same as the FFFF skew plate.

Some modifications are needed in the section of "Finding the NN nonzero deflections" and in the section of "Diagonal terms in mass matrix." Detailed modifications are in bold letters and given below.

```
C-----------------------------------------------------------------------------------------
C       FORTRAN Codes
C-----------------------------------------------------------------------------------------
C       Finding the NN non-zero deflections
C-----------------------------------------------------------------------------------------

        IJK=0
        DO I=2,NY-1                               ! RELATED TO NON-ZERO W ONLY
        KI=NX2*(I-1)
        DO J=1,NX
        IJK=IJK+1
        LI(IJK)=KI+J
        ENDDO
        ENDDO
C-----------------------------------------------------------------------------------------
C       The end of finding the NN non-zero deflections
C-----------------------------------------------------------------------------------------


C-----------------------------------------------------------------------------------------
C
C       Diagonal terms in mass matrix
C-----------------------------------------------------------------------------------------
        IJK=0
        DO I=2,NY-1                   ! RELATED TO W ONLY FOR THE DIAGONAL MASS MATRIX
        KI=NX*(I-1)
        DO J=1,NX
        IJK=IJK+1
        EM(IJK)=WI(I)*WI(J)
        ENDDO
        ENDDO
C-----------------------------------------------------------------------------------------
C       The end of diagonal terms in mass matrix
C-----------------------------------------------------------------------------------------



%-----------------------------------------------------------------------------------------
%       Converted MATLAB codes
%-----------------------------------------------------------------------------------------
%
```

```
%      Finding the NN non-zero deflections
%-------------------------------------------------------------------------------------
IJK=0;
for I=2:NY-1                              % RELATED TO NON-ZERO W ONLY
      KI=NX2*(I-1);
      for J=1:NX
             IJK=IJK+1;
             LI(IJK)=KI+J;
      end
end
%
%      The end of finding the NN non-zero deflections
%
%
%-------------------------------------------------------------------------------------
%
%      Diagonal terms in mass matrix
%-------------------------------------------------------------------------------------
IJK=0;
for I=2:NY-1                              % RELATED TO W ONLY FOR THE DIAGONAL MASS MATRIX
      KI=NX*(I-1);
      for J=1:NX
             IJK=IJK+1;
             EM(IJK)=WI(I)*WI(J);
      end
end

%-------------------------------------------------------------------------------------
%      The end of diagonal terms in mass matrix
%-------------------------------------------------------------------------------------
```

4. CSFF skew plate

Refer to Fig. IV.1, symbols CSFF stand for the plate with edge at $y = 0$ clamped, edge at $x = a$ simply supported and the other two edges free. In other words, the edge-numbering sequence is the same as the one shown in Fig. 8.2.

Set NN = $N \times N - 2N$ and NF = $3N - 2$. Some modifications in the section of "Finding the NN nonzero deflections" and in the section of "Diagonal terms in mass matrix" are needed. Detailed modifications are in bold letters and given below.

```
C-------------------------------------------------------------------------------------
C      FORTRAN Codes
C-------------------------------------------------------------------------------------
C      Finding the NN non-zero deflections
C-------------------------------------------------------------------------------------
      IJK=0
      DO I=2,NY                              ! RELATED TO NON-ZERO W ONLY
      KI=NX2*(I-1)
      DO J=1,NX-1
      IJK=IJK+1
      LI(IJK)=KI+J
      ENDDO
      ENDDO
```

```
C-------------------------------------------------------------------------------
C    The end of finding the NN non-zero deflections
C-------------------------------------------------------------------------------


C-------------------------------------------------------------------------------
C
C    Diagonal terms in mass matrix
C-------------------------------------------------------------------------------
     IJK=0
     DO I=2,NY                          ! RELATED TO W ONLY FOR THE DIAGONAL MASS MATRIX
     KI=NX*(I-1)
     DO J=1,NX-1
     IJK=IJK+1
     EM(IJK)=WI(I)*WI(J)
     ENDDO
     ENDDO
C-------------------------------------------------------------------------------
C    The end of diagonal terms in mass matrix
C-------------------------------------------------------------------------------
```

Since NF = 3N−2, some modifications in the section of "Finding the NF nonzero first-order derivatives at boundary points" are also needed. Detailed modifications are in bold letters and given below.

```
C-------------------------------------------------------------------------------
C    Finding the NF non-zero first order derivatives at boundary points
C-------------------------------------------------------------------------------
     IJK=0                             ! TO BE ELIMINATED----RELATED TO WX OR WY
     DO I=2,NY
     KI=NX2*(I-1)
     DO J=NX+1,NX2                     ! WX(0,Y), WX(A,Y) ARE NOT ZERO
     IJK=IJK+1
     LN(IJK)=KI+J
     ENDDO
     ENDDO

C    KI=NX2*NY
C    DO J=1,NX                         ! SINCE WY(X,0) IS ZERO, thus the five statements are not needed
C    IJK=IJK+1
C    LN(IJK)=KI+J
C    ENDDO

     KI=NX2*NY+NX
     DO J=1,NX                         ! WY(X,B) IS NOT ZERO
     IJK=IJK+1
     LN(IJK)=KI+J
     ENDDO
C-------------------------------------------------------------------------------
C    The end of finding the NF non-zero first order derivatives at boundary points
C-------------------------------------------------------------------------------
```

```
%-------------------------------------------------------------------------------
%       Converted MATLAB codes
%-------------------------------------------------------------------------------
%
%-------------------------------------------------------------------------------
%       Finding the NN non-zero deflections
%-------------------------------------------------------------------------------
IJK=0;
for I=2:NY                          % RELATED TO NON-ZERO W ONLY
      KI=NX2*(I-1);
      for J=1:NX-1
            IJK=IJK+1;
            LI(IJK)=KI+J;
      end
end
%-------------------------------------------------------------------------------
%       The end of finding the NN non-zero deflections
%-------------------------------------------------------------------------------
%
%-------------------------------------------------------------------------------
%
%       Diagonal terms in mass matrix
%-------------------------------------------------------------------------------
IJK=0;
for I=2:NY                          % RELATED TO W ONLY FOR THE DIAGONAL MASS MATRIX
      KI=NX*(I-1);
      for J=1:NX-1
            IJK=IJK+1;
            EM(IJK)=WI(I)*WI(J);
      end
end

%-------------------------------------------------------------------------------
%       The end of diagonal terms in mass matrix
%-------------------------------------------------------------------------------


%-------------------------------------------------------------------------------
%       Finding the NF non-zero first order derivatives at boundary points
%-------------------------------------------------------------------------------

IJK=0;                              % TO BE ELIMINATED----RELATED TO WX OR WY
for I=2:NY
      KI=NX2*(I-1);
      for J=NX+1:NX2                % WX(0,Y), WX(A,Y) ARE NOT ZERO
            IJK=IJK+1;
            LN(IJK)=KI+J;
      end
end

%     KI=NX2*NY;
%     for J=1:NX                    % WY(X,0) IS ZERO, Pay attention to the bold symbol %.
%           IJK=IJK+1;
%           LN(IJK)=KI+J;
%     end
```

```
KI=NX2*NY+NX;
for J=1:NX                        % WY(X,B) IS NOT ZERO
       IJK=IJK+1;
       LN(IJK)=KI+J;
End
```

```
%------------------------------------------------------------------------------------------------------------------
%      The end of finding the NF non-zero first order derivatives at boundary points
%------------------------------------------------------------------------------------------------------------------
```

It is demonstrated that only a few simple modifications on the program written for analyzing isotropic FFFF skew plate (Program 3 or Program3_Main) are needed. The modified program can solve the free vibration of the skew plate with any combinations of boundary conditions by using the QEM.

Appendix VII

VII PROGRAM 4 – BUCKLING OF ANISOTROPIC SSSS RECTANGULAR PLATE BY THE QEM

For readers' reference, Program 4 named as "Buckling of anisotropic SSSS rectangular plate by the QEM" is given. Note that subroutine INVMAT is called to find the inverse of the real matrix [FF] or [GS], subroutine TZZ is the standard eigenvalue solver to find all eigenvalues of the real matrix [PE]. Subroutine MTM is given in Section II.1. The three commonly used subroutines are not included in FORTRAN Program 4. Subroutines (functions) DQMNEW2 and DQELE2 should be included to run the program.

The program can be used to obtain the buckling load of isotropic and anisotropic rectangular plates under constant or linear variation edge loadings. In other words, the distribution of in-plane stress components are the same as the ones of the edge applied stresses.

For nonuniformly distributed edge loadings, the in-plane stress components at all grid points should be known first. This can be done by using the method of Airy stress function if all in-plane boundary conditions are stress conditions. A FORTRAN program, Program 5, will be given later in Section VIII. For other in-plane boundary conditions, a FORTRAN, Program 6, can be used to obtain the in-plane stress components at all grid points. Program 6 is also provided in Section VIII.

For other combinations of out-of-plane boundary conditions, the modifications are similar to the ones for Program 3. Details can be found in Section VI.2.

Similar to Program 3 (or Program3_Main), the element stiffness, geometric stiffness matrix, and mass matrices are numerically integrated by the GLL quadrature. Currently the mixed method is used to introduce the additional degree of freedom at boundary points and $N = 15$. If N is to be changed, XX(I) and WI(I) should be changed accordingly.

VII.1 FORTRAN AND MATLAB CODES

As in previous sections, FORTRAN codes are given first and then the converted MATLAB file is followed. The converted MATLAB file has the function of showing the buckling mode shape graphically.

Description of some variables in Program 4

$X(I)$: nondimensional x-coordinate of a grid point, $x_i [-1,1]$
$Y(I)$: nondimensional y-coordinate of a grid point, $y_i [-1,1]$

Input data

N: number of grid points in both directions
A: plate length in x-direction
B: plate length in y-direction
VV: Poisson's ratio
THICK: total thickness of the laminated plate
THEATA: the angle (degrees) between principal direction 1 and x-axis, the only input datum currently.
D11, D12, D22, D66, D16, D26: the flexural stiffness

XX(I): abscissa in GLL quadrature, xx_i $[-1,1]$
WI(I): weight in GLL quadrature
PX(I): in-plane stress N_x at all grid points
PY(I): in-plane stress N_y at all grid points
PXY(I): in-plane stress N_{xy} at all grid points
Note that except THEATA all other input data have been assigned currently.

Output data

1.0D0/BUCK: buckling load coefficient

```
C
C-------------------------------------------------------------------------------------
C       Program 4 (FORTRAN Codes)
C       Buckling of anisotropic SSSS rectangular plate by the QEM.
C       N—The number of grid points in both directions which is fixed! NF—the number of non-zero DOFs
C           associated to the first order derivatives at the boundary points (non zero Wx and Wy only),
C           the nonzero deflections at the boundary points are not included, different from the DQM.
C           It varies for different combinations of boundary conditions and LN (NF) should be changed accordingly.
C       NW—The number of non-zero deflections, including the ones on the free edges. LI (NW) and EM
C           should be changed accordingly, different from the DQM. For DQM, NW is fixed & equals to
C           the number of inner grid points.
C       THEATA—the only input datum to run the program, the angle of the principal direction of the material.
C       Only GLL points are used currently in this program.
C-------------------------------------------------------------------------------------
C
        IMPLICIT REAL*8 (A-H,O-Z)
        PARAMETER (N=15,MMM=(N+2)*(N+2),NW=N*N,NN=NW-4*N+4,NF=4*N)  ! NN- non-zero w only.
        CHARACTER FILE_MODE*80
        DIMENSION X(N),Y(N),AX(N,N+2),BX(N,N+2),AY(N,N+2),BY(N,N+2)
        DIMENSION XX(N),WI(N),Z(NW),LI(NW),LN(NF)
        DIMENSION GS(NN,NN),BK(NW),CK(NW),LM(NW)
        DIMENSION FF(NF,NF),GF(NN,NF),FG(NF,NN),FFG(NF,NN)
C
        DIMENSION D(3,3),BXX(NW,MMM),BYY(NW,MMM),BXY(NW,MMM)
        DIMENSION EM(NN),EX(3,MMM-4),DEX(3,MMM-4),STIFF(MMM-4,MMM-4)
        DIMENSION SS(2,2),AAX(NW,MMM),AAY(NW,MMM)
        DIMENSION GX(2,MMM-4),GEX(2,MMM-4),GGFF (MMM-4,MMM-4)
        DIMENSION PX(NW),PY(NW),PXY(NW),PF(NN,NF),PP(NN,NN),PE(NN,NN)
C-------------------------------------------------------------------------------------
C           SSSS --QEM   (3) – THE MIXED METHOD
C
C     INPUT FILENAME OF OUTPUT
C-------------------------------------------------------------------------------------
C
        XX(1)=-1.0D0
        XX(2)=-0.965245926503839D0
        XX(3)=-0.885082044222977D0
        XX(4)=-0.763519689951815D0
        XX(5)=-0.606253205469846D0
        XX(6)=-0.420638054713673D0
        XX(7)=-0.215353955363794D0
        XX(8)= 0.0D0
```

```
      WI(1)= 1.0D0/105.0D0
      WI(2)= 0.058029893034830D0
      WI(3)= 0.101660070340631D0
      WI(4)= 0.140511699803666D0
      WI(5)= 0.172789647252341D0
      WI(6)= 0.196987235964736D0
      WI(7)= 0.211973585926826D0
      WI(8)= 0.217048116348816D0
      DO I=9,15
      XX(I)=-XX(16-I)
      WI(I)= WI(16-I)
      ENDDO
C
      WRITE(*,'  (A)  ') ' ENTER OUTPUT FILE NAME'          ! File to store output data
      READ(*,'  (A)  ') FILE_MODE
      OPEN(UNIT=9,STATUS='UNKNOWN',FILE=FILE_MODE)
C
C-------------------------------------------------------------------------------
C
      NX=N                        ! TOTAL POINTS IN THE X-DIRECTION
      NY=N                        ! TOTAL POINTS IN THE Y-DIRECTION
      NY2=NY+2
      NX2=NX+2

      PI=DATAN(1.0D0)*4.0D0
      B=1.D0                    !  LENGTH B,    KEPT UNCHANGED FOR BUCKLING COEFFICIENT K
      RATIO=1.0D0/1.0D0
      A=B/RATIO                 !  LENGTH A,   CHANGE FOR DIFFERENT ASPECT RATIOS !
C
C-------------------------------------------------------------------------------
C
      WRITE (*,*) 'ENTER THEATA'
      READ (*,*) THEATA        ! IN DEGREES
      TH=THEATA/180.D0*PI   ! (TH—ANGLE OF THE PRINCIPAL DIRECTION IN RADIUS)
C
C-------------------------------------------------------------------------------
C   SETTING MATERIAL PARAMETERS
C-------------------------------------------------------------------------------
C
C-------------------------------------------------------------------------------
C   SINGLE LAYER
C-------------------------------------------------------------------------------
C
      Q11=25.0D0
      Q22=1.0D0
      Q12=0.25D0
      Q66=0.5D0
      THICK=1.0D0

      QQ11=Q11*(DCOS(TH)**4)+2.D0*(Q12+2.D0*Q66)*(DSIN(TH)**2)*
     $     (DCOS(TH)**2)+Q22*(DSIN(TH)**4)

      QQ12=(Q11+Q22-4.D0*Q66)*(DSIN(TH)**2)*(DCOS(TH)**2)+Q12*
     $     (DSIN(TH)**4+DCOS(TH)**4)

      QQ22=Q11*(DSIN(TH)**4)+2.D0*(Q12+2.D0*Q66)*(DSIN(TH)**2)*
     $     (DCOS(TH)**2)+Q22*(DCOS(TH)**4)
```

```
      QQ16=(Q11-Q12-2.D0*Q66)*(DSIN(TH))*(DCOS(TH)**3)+
   $        (Q12-Q22+2.D0*Q66)*(DSIN(TH)**3)*(DCOS(TH))

      QQ26=(Q11-Q12-2.D0*Q66)*(DCOS(TH))*(DSIN(TH)**3)+
   $        (Q12-Q22+2.D0*Q66)*(DCOS(TH)**3)*(DSIN(TH))

      QQ66=(Q11+Q22-2.D0*Q12-2.D0*Q66)*(DSIN(TH)**2)*(DCOS(TH)**2)
   $        +Q66*(DSIN(TH)**4+DCOS(TH)**4)


      D11=QQ11*(THICK**3)/12.D0
      D12=QQ12*(THICK**3)/12.D0
      D16=QQ16*(THICK**3)/12.D0
      D22=QQ22*(THICK**3)/12.D0
      D26=QQ26*(THICK**3)/12.D0
      D66=QQ66*(THICK**3)/12.D0
C
C-----------------------------------------------------------------------------------
C     END OF THE SETTING MATERIAL PARAMETERS
C-----------------------------------------------------------------------------------
C
      D(1,1)=D11
      D(1,2)=D12
      D(1,3)=D16
      D(2,1)=D12
      D(2,2)=D22
      D(2,3)=D26
      D(3,1)=D16
      D(3,2)=D26
      D(3,3)=D66
C
C-----------------------------------------------------------------------------------
C
      DO I=1,N
        X(I)=XX(I)        !XX [-1,1]
      ENDDO

      DO I=1,N
        Y(I)=X(I)
      ENDDO
C-----------------------------------------------------------------------------------
C     QEM (3)  -  MIXED
C-----------------------------------------------------------------------------------
C
      CALL DQMNEW2(N,X,AX,BX,A/2.0D0)      ! DQM
C     CALL DQMNEW2(N,Y,AY,BY,B/2.0D0)      ! DQM
      CALL DQELE2(N,Y,AY,BY,B/2.0D0)       ! DQEM
C     CALL DQELE2(N,X,AX,BX,B/2.0D0)       ! DQEM

C     CALL HDQMNEW2(N,X,AX,BX,A/2.0D0)     ! HDQM
C     CALL HDQMNEW2(N,Y,AY,BY,B/2.0D0)     ! HDQM

C-----------------------------------------------------------------------------------
C  THE STRESS COMPONENTS CAN BE OBTAINED BY AIRY STRESS FUNCTION IF THE STRESSES ARE
C  UNKNOWN WITHIN THE PLATE, SUCH AS PLATE UNDER NON-UNIFORM EDGE LOADINGS
C-----------------------------------------------------------------------------------
```

```fortran
C
      DO I=1,NW
      PX (I)=0.0D0
      PY (I)=0.0D0
      PXY(I)=0.0D0
      ENDDO

      DO I=1,NY
      KI=NX*(I-1)
      DO J=1,NX
      PX (KI+J)= 1.0D0                      ! UNI-AXIAL COMPRESSION IN X DIRECTION
C     PXY(KI+J)= 1.0D0                      ! PURE SHEAR
      PY (KI+J)= 1.0D0                      ! UNI-AXIAL COMPRESSION IN Y DIRECTION
      ENDDO
      ENDDO

C-------------------------------------------------------------------------------------------------------------------
C         END OF STRESS COMPONENTS
C-------------------------------------------------------------------------------------------------------------------
      DO I=1,NW
      DO J=1,MMM
      AAX(I,J)=0.0D0
      AAY(I,J)=0.0D0
      BXX(I,J)=0.0D0
      BYY(I,J)=0.0D0
      BXY(I,J)=0.0D0
      ENDDO
      ENDDO
C
      DO I=1,NY                          ! MIXED DERIVATIVES
      KI=NX*(I-1)
      DO L=1,NY2
      KL=NX2*(L-1)
      DO J=1,NX
      II=KI+J
      DO K=1,NX2
      JJ=KL+K
      BXY(II,JJ)=2.D0*AY(I,L)*AX(J,K)
      ENDDO
      ENDDO
      ENDDO
      ENDDO

      DO I=1,NY                          !   WXX, WX
      KI=NX*(I-1)
      KL=NX2*(I-1)
      DO J=1,NX
      II=KI+J
      DO K=1,NX2
      JJ=KL+K
      BXX(II,JJ)=BX(J,K)
      AAX(II,JJ)=AX(J,K)
      ENDDO
      ENDDO
      ENDDO
```

```
      DO I=1,NY                              !  WYY, WY
      KI=NX*(I-1)
      DO L=1,NY2
      KL=NX2*(L-1)
      DO J=1,NX
      II=KI+J
      JJ=KL+J
      AAY(II,JJ)=AY(I,L)
      BYY(II,JJ)=BY(I,L)
      ENDDO
      ENDDO
      ENDDO
C-------------------------------------------------------------------------------------------------------
C      DELETE THE FOUR CORNER DEGREES
C      I.E., THE DOFS AT THE FOUR OPEN CIRCLES IN FIG. IV.1 ARE NOT USED.

C-------------------------------------------------------------------------------------------------------
      IJK=NX2*NY+NX

      DO I=1,NW
      DO J=1,NX
      BXX(I,IJK+J)=BXX(I,IJK+J+2)     ! MOVE THE COLUMN
      BYY(I,IJK+J)=BYY(I,IJK+J+2)
      BXY(I,IJK+J)=BXY(I,IJK+J+2)
      AAX(I,IJK+J)=AAX(I,IJK+J+2)     ! MOVE THE COLUMN
      AAY(I,IJK+J)=AAY(I,IJK+J+2)
      ENDDO
      ENDDO

      DO I=1,MMM-4
      DO J=1,MMM-4
      STIFF(I,J)=0.0D0
      GGFF (I,J)=0.0D0
      ENDDO
      ENDDO

      DO I=1,NY
      KI=NX*(I-1)
      DO J=1,NX
      WIJ=WI(I)*WI(J)                          ! WEIGHT
      IJ=NX*(I-1)+J

      DO K=1,MMM-4
      EX(1,K)=BXX(IJ,K)
      EX(2,K)=BYY(IJ,K)
      EX(3,K)=BXY(IJ,K)
      GX(1,K)=AAX(IJ,K)
      GX(2,K)=AAY(IJ,K)
      ENDDO

      SS(1,1)=PX (KI+J)
      SS(1,2)=PXY(KI+J)
      SS(2,1)=PXY(KI+J)
      SS(2,2)=PY (KI+J)

      CALL MTM (D, EX,DEX,3,3,MMM-4)
      CALL MTM (SS,GX,GEX,2,2,MMM-4)
```

```
      DO I1=1,MMM-4
      DO J1=1,MMM-4
      DO K=1,3
       STIFF(I1,J1)=STIFF(I1,J1)+WIJ*EX(K,I1)*DEX(K,J1)     ! The determinant of Jacobian matrix J is not included.
      ENDDO
      ENDDO
      ENDDO
      DO I1=1,MMM-4
      DO J1=1,MMM-4
      DO K=1,2
       GGFF(I1,J1)=GGFF(I1,J1)+WIJ*GX(K,I1)*GEX(K,J1)   ! The determinant of Jacobian matrix J is not included.
      ENDDO
      ENDDO
      ENDDO
      ENDDO
      ENDDO
C
C     Finding the NN non-zero deflections
C
      IJK=0
      DO I=2,NY-1                              ! RELATED TO NON-ZERO W ONLY
      KI=NX2*(I-1)
      DO J=2,NX-1
      IJK=IJK+1
      LI(IJK)=KI+J
      ENDDO
      ENDDO
C
C     The end of finding the NN non-zero deflections
C
      DO I=1,NN
      DO J=1,NN
      GS(I,J)=STIFF(LI(I),LI(J))                    ! RETAINED
      PP(I,J)=GGFF (LI(I),LI(J))
      ENDDO
      ENDDO
C
C     Finding the NF non-zero first order derivatives at boundary points
C
      IJK=0                                   ! TO BE ELIMINATED-----RELATED TO WX OR WY

      DO I=1,NY
      KI=NX2*(I-1)
      DO J=NX+1,NX2                           !WX1, WXN ARE NOT ZERO
      IJK=IJK+1
      LN(IJK)=KI+J
      ENDDO
      ENDDO

      KI=NX2*NY                               ! WY1 IS NOT ZERO
      DO J=1,NX
      IJK=IJK+1

      LN(IJK)=KI+J
      ENDDO
```

```
        KI=NX2*NY+NX                            ! Wyn is not zero (the four Wxys have been removed)
        DO J=1,NX
        IJK=IJK+1
        LN(IJK)=KI+J
        ENDDO
C
C    The end of finding the NF non-zero first order derivatives at boundary points
C
        DO I=1,NF
        DO J=1,NF
        FF(I,J)=STIFF(LN(I),LN(J))              ! TO BE ELIMINATED
        ENDDO
        ENDDO
C
        DO I=1,NN
        DO J=1,NF
        GF(I,J)=STIFF(LI(I),LN(J))
        PF(I,J)=GGFF (LI(I),LN(J))
        ENDDO
        ENDDO

        DO I=1,NF
        DO J=1,NN
        FG(I,J)=STIFF(LN(I),LI(J))
        ENDDO
        ENDDO

        CALL INVMAT(NF,FF,ZZ,1.D-30,ISW,BK,CK,LM)
C
C--------------------------------------------------------------------------------------------------------------------------
C
        DO I=1,NF
        DO J=1,NN
        FFG(I,J)=0.0D0
        DO K=1,NF
        FFG(I,J)=FFG(I,J)-FF(I,K)*FG(K,J)
        ENDDO
        ENDDO
        ENDDO

         DO I=1,NN
        DO J=1,NN
        DO K=1,NF
        GS(I,J)=GS(I,J)+GF(I,K)*FFG(K,J)         ! COMMENT THIS SENTENCE (CCCC)
        PP(I,J)=PP(I,J)+PF(I,K)*FFG(K,J)         ! COMMENT THIS SENTENCE (CCCC)
        ENDDO
        ENDDO
        ENDDO

        CALL INVMAT(NN,GS,ZZ,1.D-30,ISW,BK,CK,LM)

        DO I=1,NN
        DO J=1,NN
        PE(I,J)=0.0D0
        DO K=1,NN
        PE(I,J)=PE(I,J)+GS(I,K)*PP(K,J)
        ENDDO
        ENDDO
        ENDDO
```

```
          CALL TZZ(NN,NN,PE,Z,BK,CK,LM)

          BUCK=0.0D0
          DO 400 I=1,NN
          IF (BK(I) .LT. BUCK)GO TO 400
          BUCK=BK(I)
   400    CONTINUE
          WRITE (*,551) I,1.0D0/BUCK/PI/PI,1.0D0/BUCK
          WRITE (9,552)     1.0D0/BUCK
   551    FORMAT (1X,I5,2G14.6)
   552    FORMAT (1X,2G14.6)

          STOP
          END
C
C-----------------------------------------------------------------------------------------------------
C        THE END OF FORTRAN PROGRAM 4.
C-----------------------------------------------------------------------------------------------------
C


%
%-----------------------------------------------------------------------------------------------------
%     The converted MATLAB file named Program4_Main
%
%-----------------------------------------------------------------------------------------------------
%        PROGRAM 4
%        BUCKLING OF AN ISOTROPIC SSSS RECTANGULAR PLATE BY THE QEM.
%        N—THE NUMBER OF GRID POINTS IN BOTH DIRECTIONS WHICH IS FIXED
%        NF—THE NUMBER OF NON-ZERO DOFS ASSOCIATED TO THE FIRST ORDER DERIVATIVES AT
%        THE BOUNDARY POINTS (NON ZERO WX AND WY ONLY),
%        THE NONZERO DEFLECTIONS AT THE BOUNDARY POINTS ARE NOT INCLUDED,DIFFERENT
% FROM THE DQM.
%        IT VARIES FOR DIFFERENT COMBINATIONS OF BOUNDARY CONDITIONS AND LN (NF) SHOULD
% BE CHANGED ACCORDINGLY.
%        NW—THE NUMBER OF NON-ZERO DEFLECTIONS, INCLUDING THE ONES ON THE FREE EDGES.
%        LI (NW) AND EM SHOULD BE CHANGED ACCORDINGLY, DIFFERENT FROM THE DQM. FOR DQM,
%        NW IS FIXED & EQUALS TO THE NUMBER OF INNER GRID POINTS.
%        THEATA—THE ONLY INPUT DATUM TO RUN THE PROGRAM, THE ANGLE OF THE PRINCIPAL
%        DIRECTION OF THE MATERIAL.
%        ONLY GLL POINTS ARE USED CURRENTLY IN THIS PROGRAM.
%-----------------------------------------------------------------------------------------------------
%
%
          CLC; CLEAR; CLOSE ALL;
          N=15;
          MMM=(N+2)*(N+2);     NW=N*N;   NN=NW-4*N+4;     NF=4*N;
%-----------------------------------------------------------------------------------------------------
%          SSSS --QEM   (3) – THE MIXED METHOD
%
%     INPUT FILENAME OF OUTPUT
%-----------------------------------------------------------------------------------------------------
```

```
%
      XX=ZEROS(1,N);
      WI=ZEROS(1,N);

      XX(1)=-1.0D0;
      XX(2)=-0.965245926503839D0;
      XX(3)=-0.885082044222977D0;
      XX(4)=-0.763519689951815D0;
      XX(5)=-0.606253205469846D0;
      XX(6)=-0.420638054713673D0;
      XX(7)=-0.215353955363794D0;
      XX(8)= 0.0D0;

      WI(1)= 1.0D0/105.0D0;
      WI(2)= 0.058029993034830D0;
      WI(3)= 0.101660070340631D0;
      WI(4)= 0.140511699803666D0;
      WI(5)= 0.172789647252341D0;
      WI(6)= 0.196987235964736D0;
      WI(7)= 0.211973585926826D0;
      WI(8)= 0.217048116348816D0;

      FOR I=9:15
          XX(I)=-XX(16-I);
          WI(I)= WI(16-I);
      END
%
%-----------------------------------------------------------------------------------------
%
      NX=N;                          % TOTAL POINTS IN THE X-DIRECTION
      NY=N;                          % TOTAL POINTS IN THE Y-DIRECTION
      NY2=NY+2;
      NX2=NX+2;
      PI=ATAN(1.0D0)*4.0D0;
      B=1.D0;                        %  LENGTH B,    KEPT UNCHANGED FOR BUCKLING COEFFICIENT K
      RATIO=1.0D0/1.0D0;
      A=B/RATIO;                     % LENGTH A,   CHANGE FOR DIFFERENT ASPECT RATIOS !
%
%-----------------------------------------------------------------------------------------
%
      THEATA=INPUT('ENTER THEATA (DEGREES) :');
      TH=THEATA/180.D0*PI;   % (TH—ANGLE OF THE PRINCIPAL DIRECTION)
%
%-----------------------------------------------------------------------------------------
%    SETTING MATERIAL PARAMETERS
%-----------------------------------------------------------------------------------------
%
%-----------------------------------------------------------------------------------------
%    SINGLE   LAYER
%-----------------------------------------------------------------------------------------
%
      Q11=25.0D0;
      Q22=1.0D0;
      Q12=0.25D0;
      Q66=0.5D0;
      THICK=1.0D0;
```

```
QQ11=Q11*(COS(TH)^4)+2.D0*(Q12+2.D0*Q66)*(SIN(TH)^2)*...
        (COS(TH)^2)+Q22*(SIN(TH)^4);

QQ12=(Q11+Q22-4.D0*Q66)*(SIN(TH)^2)*(COS(TH)^2)+Q12*...
        (SIN(TH)^4+COS(TH)^4);

QQ22=Q11*(SIN(TH)^4)+2.D0*(Q12+2.D0*Q66)*(SIN(TH)^2)*...
        (COS(TH)^2)+Q22*(COS(TH)^4);

QQ16=(Q11-Q12-2.D0*Q66)*(SIN(TH))*(COS(TH)^3)+...
        (Q12-Q22+2.D0*Q66)*(SIN(TH)^3)*(COS(TH));

QQ26=(Q11-Q12-2.D0*Q66)*(COS(TH))*(SIN(TH)^3)+...
        (Q12-Q22+2.D0*Q66)*(COS(TH)^3)*(SIN(TH));

QQ66=(Q11+Q22-2.D0*Q12-2.D0*Q66)*(SIN(TH)^2)*(COS(TH)^2)...
        +Q66*(SIN(TH)^4+COS(TH)^4);


D11=QQ11*(THICK^3)/12.D0;
D12=QQ12*(THICK^3)/12.D0;
D16=QQ16*(THICK^3)/12.D0;
D22=QQ22*(THICK^3)/12.D0;
D26=QQ26*(THICK^3)/12.D0;
D66=QQ66*(THICK^3)/12.D0;
%
%-----------------------------------------------------------------------------
%    END OF THE SETTING MATERIAL PARAMETERS
%-----------------------------------------------------------------------------
%

D(1,1)=D11;
D(1,2)=D12;
D(1,3)=D16;
D(2,1)=D12;
D(2,2)=D22;
D(2,3)=D26;
D(3,1)=D16;
D(3,2)=D26;
D(3,3)=D66;
%
%-----------------------------------------------------------------------------
%
X=XX;
Y=X;


%-----------------------------------------------------------------------------
%    QEM (3)   -   THE MIXED METHOD
%-----------------------------------------------------------------------------
    [AX,BX]=DQMNEW2(N,X,A/2.0D0);          % DQM

%   [AY,BY]=DQMNEW2(N,Y,B/2.0D0);          % DQM
    [AY,BY]=DQELE2(N,Y,B/2.0D0);           % DQEM
%   [AX,BX]=DQELE2(N,X,B/2.0D0);           % DQEM

%   [AX,BX]=HDQMNEW2(N,X,A/2.0D0);         % HDQM
%   [AY,BY]=HDQMNEW2(N,Y,B/2.0D0);         % HDQM
```

```
%-------------------------------------------------------------------------------
%   THE STRESS COMPONENTS CAN BE OBTAINED BY AIRY STRESS FUNCTION IF THE STRESSES ARE
%   UNKNOWN WITHIN THE PLATE, SUCH AS PLATE UNDER NON-UNIFORM EDGE LOADINGS
%-------------------------------------------------------------------------------
        PX=ZEROS(1,NW);
        PY=ZEROS(1,NW);
        PXY=ZEROS(1,NW);

        FOR I=1:NY
            KI=NX*(I-1);
            FOR J=1:NX
                PX (KI+J)= 1.0D0;               % UNI-AXIAL COMPRESSION IN X DIRECTION
            %   PXY(KI+J)= 1.0D0;               % PURE SHEAR
                PY (KI+J)= 1.0D0;               % UNI-AXIAL COMPRESSION IN Y DIRECTION
            END
        END
%-------------------------------------------------------------------------------
%       END OF STRESS COMPONENTS
%-------------------------------------------------------------------------------

        AAX=ZEROS(NW,MMM);
        AAY=ZEROS(NW,MMM);
        BXX=ZEROS(NW,MMM);
        BYY=ZEROS(NW,MMM);
        BXY=ZEROS(NW,MMM);


%
        FOR I=1:NY                              % MIXED DERIVATIVES
            KI=NX*(I-1);
            FOR L=1:NY2
                KL=NX2*(L-1);
                FOR J=1:NX
                    II=KI+J;
                    FOR K=1:NX2
                        JJ=KL+K;
                        BXY(II,JJ)=2.D0*AY(I,L)*AX(J,K);
                    END
                END
            END
        END

        FOR I=1:NY                              %   WXX, WX
            KI=NX*(I-1);
            KL=NX2*(I-1);
            FOR J=1:NX
                II=KI+J;
                FOR K=1:NX2
                    JJ=KL+K;
                    BXX(II,JJ)=BX(J,K);
                    AAX(II,JJ)=AX(J,K);
                END
            END
        END
```

```
    FOR I=1:NY                              %   WYY, WY
        KI=NX*(I-1);
        FOR L=1:NY2
            KL=NX2*(L-1);
            FOR J=1:NX
                II=KI+J;
                JJ=KL+J;
                AAY(II,JJ)=AY(I,L);
                BYY(II,JJ)=BY(I,L);
            END
        END
    END
%----------------------------------------------------------------------------------------------------------
%      DELETE THE FOUR CORNER DEGREES
%      I.E., THE DOFS AT THE FOUR OPEN CIRCLES IN FIG. IV.1 ARE NOT USED.
%----------------------------------------------------------------------------------------------------------
    IJK=NX2*NY+NX;


    FOR J=1:NX
        BXX(:,IJK+J)=BXX(:,IJK+J+2);    %MOVE THE COLUMN
        BYY(:,IJK+J)=BYY(:,IJK+J+2);
        BXY(:,IJK+J)=BXY(:,IJK+J+2);
        AAX(:,IJK+J)=AAX(:,IJK+J+2);    %MOVE THE COLUMN
        AAY(:,IJK+J)=AAY(:,IJK+J+2);
    END


    STIFF=ZEROS(MMM-4,MMM-4);
    GGFF=ZEROS(MMM-4,MMM-4);

    FOR I=1:NY
        KI=NX*(I-1);
        FOR J=1:NX
            WIJ=WI(I)*WI(J);                        % WEIGHT
            IJ=NX*(I-1)+J;

            EX(1,:)=BXX(IJ,:);
            EX(2,:)=BYY(IJ,:);
            EX(3,:)=BXY(IJ,:);
            GX(1,:)=AAX(IJ,:);
            GX(2,:)=AAY(IJ,:);


            SS(1,1)=PX (KI+J);
            SS(1,2)=PXY(KI+J);
            SS(2,1)=PXY(KI+J);
            SS(2,2)=PY (KI+J);

            %-------------------------------------------------------------------------------------------

            DEX=D*EX;
            GEX=SS*GX;
            %-------------------------------------------------------------------------------------------
```

```
%------

                STIFF=STIFF+WIJ*EX(1:3,1:MMM-4)'*DEX(1:3,1:MMM-4);
                GGFF=GGFF+WIJ*GX(1:2,1:MMM-4)'*GEX(1:2,1:MMM-4);

        END
    END
%
%    FINDING THE NN NON-ZERO DEFLECTIONS
%

    IJK=0;
    FOR I=2:NY-1                          % RELATED TO NON-ZERO W ONLY
        KI=NX2*(I-1);
        FOR J=2:NX-1
            IJK=IJK+1;
            LI(IJK)=KI+J;
        END
    END
%
%    THE END OF FINDING THE NN NON-ZERO DEFLECTIONS
%

    GS=STIFF(LI,LI);                % RETAINED
    PP=GGFF(LI,LI);


%
%    FINDING THE NF NON-ZERO FIRST ORDER DERIVATIVES AT BOUNDARY POINTS
%
    IJK=0;                              % TO BE ELIMINATED-----RELATED TO WX OR WY

    FOR I=1:NY
        KI=NX2*(I-1);
        FOR J=NX+1:NX2                  %WX1, WXN ARE NOT ZERO
            IJK=IJK+1;
            LN(IJK)=KI+J;
        END
    END

    KI=NX2*NY;                      % WY1 IS NOT ZERO
    FOR J=1:NX
        IJK=IJK+1;
        LN(IJK)=KI+J;
    END

    KI=NX2*NY+NX;              % WYN IS NOT ZERO (THE FOUR WXYS HAVE BEEN REMOVED)
    FOR J=1:NX
        IJK=IJK+1;
        LN(IJK)=KI+J;
    END
%
%    THE END OF FINDING THE NF NON-ZERO FIRST ORDER DERIVATIVES AT BOUNDARY POINTS

    FF=STIFF(LN,LN);          % TO BE ELIMINATED
%
    GF=STIFF(LI,LN);
    PF=GGFF(LI,LN);
```

```
    FG=STIFF(LN,LI);

    FF=INV(FF);
%
%------------------------------------------------------------------------------------------------------------------
%

    FFG=-FF*FG;

    GS=GS+GF*FFG;          % COMMENT THIS SENTENCE (CCCC)
    PP=PP+PF*FFG;          % COMMENT THIS SENTENCE (CCCC)
    PE=GS\PP;

    [EIGENVECTOR, EIGENVALUE]=EIG(PE);
    BUCKLOAD=DIAG(EIGENVALUE);
    BUCKLOAD=REAL(BUCKLOAD);
    [CRITICALLOAD, EIGENNO] = MAX(BUCKLOAD);
    CRITICALLOAD = 1.0D0 /CRITICALLOAD;
    DISP(['CRITICAL LOAD = ',NUM2STR(CRITICALLOAD)])

%
%   PLOT THE MODE SHAPE CORRESPOD TO THE CRETICAL LOAD
%
    MODEELIMINATED = EIGENVECTOR(:,EIGENNO);
    MODEELIMINATED=REAL(MODEELIMINATED);

    [MAXDIS,MAXLOC]=MAX(ABS(MODEELIMINATED));
    MODEELIMINATED = MODEELIMINATED / MODEELIMINATED(MAXLOC);

    MODE=ZEROS(NX,NY);
    [XCOORD,YCOORD] = MESHGRID(X,Y);
    MODE(2:NX-1,2:NY-1) = RESHAPE(MODEELIMINATED,NX-2,NY-2);
    SURF(XCOORD,YCOORD,MODE)

    XLABEL('X')
    YLABEL('Y')
    ZLABEL('Z')
    TITLE({ ['BUCKLING   MODE   SHAPE   (\THETA=',NUM2STR(THEATA),'^O)'];[ 'CRITICAL  LOAD  =
',NUM2STR(CRITICALLOAD)] })

%
%------------------------------------------------------------------------------------------------------------------
%   The end of the converted MATLAB file Program4_Main
%------------------------------------------------------------------------------------------------------------------
%
```

VII.2 ILLUSTRATION FOR DIFFERENT MATERIALS

For different materials, simply make changes in the Section of "SETTING MATERIAL PARAMETERS."
Some examples are given below for illustrations.

1. Isotropic materials

VV is Poisson's ratio and TH can be set to any real number.

```
C------------------------------------------------------------------------------------------------
C      FORTRAN Codes
C      SETTING MATERIAL PARAMETERS
C------------------------------------------------------------------------------------------------
C
C------------------------------------------------------------------------------------------------
C    1,   ISOTROPIC MATERIALS
C------------------------------------------------------------------------------------------------
C
       VV=0.3D0
       D11=1.0D0
       D12=VV*D11
       D16=0.0D0
       D22=D11
       D26=0.0D0
       D66=(1.0D0-VV)/2.0D0
C
C------------------------------------------------------------------------------------------------
C      END OF THE SETTING MATERIAL PARAMETERS
C------------------------------------------------------------------------------------------------
C
```

```
%
%------------------------------------------------------------------------------------------------
%        Converted MATLAB codes
%------------------------------------------------------------------------------------------------
%
%------------------------------------------------------------------------------------------------
%    SETTING MATERIAL PARAMETERS
%------------------------------------------------------------------------------------------------
%
%------------------------------------------------------------------------------------------------
%    1. ISOTROPIC MATERIALS
%------------------------------------------------------------------------------------------------
%
       VV=0.3D0;
       D11=1.0D0;
       D12=VV*D11;
       D16=0.0D0;
       D22=D11;
       D26=0.0D0;
       D66=(1.0D0-VV)/2.0D0;
%
%------------------------------------------------------------------------------------------------
%    THE END OF THE SETTING MATERIAL PARAMETERS
%------------------------------------------------------------------------------------------------
%
```

2. Three-layer symmetric angle-ply laminated plate

TH is the angle (in degrees) and should be input. The total thickness is 1.0 and the thickness of each layer is the same. Currently, Q_{ij} are set and should be changed according to the real material properties of the lamina. Pay attention to the bold number 12.96.

```
C
C-----------------------------------------------------------------------------------------------------
C     FORTAN Codes
C     SETTING MATERIAL PARAMETERS
C-----------------------------------------------------------------------------------------------------
C
C-----------------------------------------------------------------------------------------------------
C     Three-layer (A/-A/A) symmetric angle ply laminated material, TH–angle (in degrees)
C-----------------------------------------------------------------------------------------------------
C
      Q11=25.0D0
      Q22=1.0D0
      Q12=0.25D0
      Q66=0.5D0
      THICK=1.0D0      ! TOTAL THICKNESS

      QQ11=Q11*(DCOS(TH)**4)+2.D0*(Q12+2.D0*Q66)*(DSIN(TH)**2)*
     $       (DCOS(TH)**2)+Q22*(DSIN(TH)**4)

      QQ12=(Q11+Q22-4.D0*Q66)*(DSIN(TH)**2)*(DCOS(TH)**2)+Q12*
     $       (DSIN(TH)**4+DCOS(TH)**4)

      QQ22=Q11*(DSIN(TH)**4)+2.D0*(Q12+2.D0*Q66)*(DSIN(TH)**2)*
     $       (DCOS(TH)**2)+Q22*(DCOS(TH)**4)

      QQ16=(Q11-Q12-2.D0*Q66)*(DSIN(TH))*(DCOS(TH)**3)+
     $       (Q12-Q22+2.D0*Q66)*(DSIN(TH)**3)*(DCOS(TH))

      QQ26=(Q11-Q12-2.D0*Q66)*(DCOS(TH))*(DSIN(TH)**3)+
     $       (Q12-Q22+2.D0*Q66)*(DCOS(TH)**3)*(DSIN(TH))

      QQ66=(Q11+Q22-2.D0*Q12-2.D0*Q66)*(DSIN(TH)**2)*(DCOS(TH)**2)
     $       +Q66*(DSIN(TH)**4+DCOS(TH)**4)


      D11=QQ11*(THICK**3)/12.D0
      D12=QQ12*(THICK**3)/12.D0
      D16=QQ16*(THICK**3)/12.96D0      ! 3 LAYERS
      D22=QQ22*(THICK**3)/12.D0
      D26=QQ26*(THICK**3)/12.96D0      ! 3 LAYERS
      D66=QQ66*(THICK**3)/12.D0


C-----------------------------------------------------------------------------------------------------
C     THE END OF THE SETTING MATERIAL PARAMETERS
C-----------------------------------------------------------------------------------------------------
C
```

```
%
%-----------------------------------------------------------------------------------------------------
%         Converted MATLAB codes
%-----------------------------------------------------------------------------------------------------
%
%-----------------------------------------------------------------------------------------------------
%         SETTING MATERIAL PARAMETERS
%-----------------------------------------------------------------------------------------------------
%
%-----------------------------------------------------------------------------------------------------
%         Three-layer (A/-A/A) symmetric angle ply laminated material, TH –angle (in degrees)
%-----------------------------------------------------------------------------------------------------
%
      Q11=25.0D0;
      Q22=1.0D0;
      Q12=0.25D0;
      Q66=0.5D0;
      THICK=1.0D0;       % TOTAL THICKNESS
      QQ11=Q11*(cos(TH)^4)+2.D0*(Q12+2.D0*Q66)*(sin(TH)^2)*...
          (cos(TH)^2)+Q22*(sin(TH)^4);
      QQ12=(Q11+Q22-4.D0*Q66)*(sin(TH)^2)*(cos(TH)^2)+Q12*...
          (sin(TH)^4+cos(TH)^4);
      QQ22=Q11*(sin(TH)^4)+2.D0*(Q12+2.D0*Q66)*(sin(TH)^2)*...
          (cos(TH)^2)+Q22*(cos(TH)^4);
      QQ16=(Q11-Q12-2.D0*Q66)*(sin(TH))*(cos(TH)^3)+...
          (Q12-Q22+2.D0*Q66)*(sin(TH)^3)*(cos(TH));
      QQ26=(Q11-Q12-2.D0*Q66)*(cos(TH))*(sin(TH)^3)+...
          (Q12-Q22+2.D0*Q66)*(cos(TH)^3)*(sin(TH));
      QQ66=(Q11+Q22-2.D0*Q12-2.D0*Q66)*(sin(TH)^2)*(cos(TH)^2)...
          +Q66*(sin(TH)^4+cos(TH)^4);
      D11=QQ11*(THICK^3)/12.D0;
      D12=QQ12*(THICK^3)/12.D0;
      D16=QQ16*(THICK^3)/12.96D0; % 3 LAYERS
      D22=QQ22*(THICK^3)/12.D0;
      D26=QQ26*(THICK^3)/12.96D0; % 3 LAYERS
      D66=QQ66*(THICK^3)/12.D0;
%
%-----------------------------------------------------------------------------------------------------
%   THE END OF THE SETTING MATERIAL PARAMETERS
%-----------------------------------------------------------------------------------------------------
```

3. Isotropic skew plate
TH is the skew angle (in degrees) and should be input. VV is Poisson's ratio.

```fortran
C-----------------------------------------------------------------------
C      FORTRAN Codes
C      SETTING MATERIAL PARAMETERS
C-----------------------------------------------------------------------
C
C-----------------------------------------------------------------------
C      ISOTROPIC SKEW PLATE, TH—SKEW ANGLE (IN DEGREES)
C-----------------------------------------------------------------------
C
       VV=0.3D0
       D11=1.0D0                         ! RESULTS ARE BUCKLING LOAD COEFFICIENT
       D16=-DSIN(TH)
       D22=1.0D0
       D26=D16
       D12=VV*DCOS(TH)**2+DSIN(TH)**2
       D66=(1.0D0+DSIN(TH)**2-VV*DCOS(TH)**2)/2.0D0
C
C-----------------------------------------------------------------------
C      THE END OF THE SETTING MATERIAL PARAMETERS
C-----------------------------------------------------------------------
C
```

```matlab
%
%-----------------------------------------------------------------------
%      Converted MATLAB codes
%-----------------------------------------------------------------------
%
%-----------------------------------------------------------------------
%      SETTING MATERIAL PARAMETERS
%-----------------------------------------------------------------------
%
%-----------------------------------------------------------------------
%      ISOTROPIC SKEW PLATE, TH—SKEW ANGLE(in degrees)
%-----------------------------------------------------------------------
%
       VV=0.3D0;
       D11=1.0D0;     % RESULTS ARE BUCKLING LOAD COEFFICIENT
       D16=-sin(TH);
       D22=1.0D0;
       D26=D16;
       D12=VV*cos(TH)^2+sin(TH)^2;
       D66=(1.0D0+sin(TH)^2-VV*cos(TH)^2)/2.0D0;
%
%-----------------------------------------------------------------------
%      THE END OF THE SETTING MATERIAL PARAMETERS
%-----------------------------------------------------------------------
%
```

Appendix VIII

VIII PROGRAMS 5 AND 6: IN-PLANE STRESS ANALYSIS OF RECTANGULAR PLATE BY THE DQM

Two programs, called Program 5 and Program 6, are given in this section for readers' references. Program 5 uses the method of Airy stress function thus can only be used for rectangular plates with all edges having stress boundary conditions. Program 6 uses the displacement components u and v as the unknowns thus can be used for rectangular plates with any combinations of boundary conditions. The DQM with MMWC-3, called the modified DQM, is used in Program 5. Grid III is used in both programs. If the results are to be used in the buckling analysis by using the QEM, i.e., by Program 4 or Program4_Main, then GLL points should be used as the grid points for the modified DQM.

Note that subroutine INVMAT is called to find the inverse of the real matrix [GG] in Program 5 and [GK] in Program 6. This commonly used subroutine is not included in Programs 5 and 6.

VIII.1 PROGRAM 5: METHOD OF AIRY STRESS FUNCTION

The program can be used to obtain the in-plane stress components at all grid points. Currently, the rectangular plate is under uniaxial cosine distributed loading given by Eq. (5.18). Since only stress boundary conditions are encountered, thus the method of Airy stress function is used. The fourth-order compatibility equation is solved by using the modified DQM.

If different distributed loads are considered, one only needs to change the given values at the boundary points and will be demonstrated at the end of this section.

As in previous sections, FORTRAN codes are given first and then the converted MATLAB file is followed. The converted MATLAB file has the function of showing the stress distributions graphically.

Description of some variables in Program 5

$X(I)$: x coordinate of a grid point, x_i [$-$A/2,A/2]
$Y(I)$: y coordinate of a grid point, y_i [$-$B/2,B/2]

Input data

N: number of grid points in both directions
A: plate length in x direction
B: plate length in y direction
VV: Poisson's ratio
SIGMAX0: maximum applied edge stress for parabolic or cosine distributed stress in x direction.

Note that all input data have been assigned currently.

Output data

$SX(I)$: in-plane stress σ_x at all grid points
$SY(I)$: in-plane stress σ_y at all grid points
$SXY(I)$: in-plane stress τ_{xy} at all grid points

Differential Quadrature and Differential Quadrature Based Element Methods. 978-0-12-803081-3

355

```
C
C-----------------------------------------------------------------------------------
C       Program 5 (FORTRAN Codes)
C       In plane stress analysis of rectangular plates by the modified DQM (Use the method of Airy stress function)
C       N—The number of grid points in both directions.
C       Currently Grid III is used in the program.
C-----------------------------------------------------------------------------------
C
        IMPLICIT REAL*8 (A-H,O-Z)
        PARAMETER (N=15,NN=N*N,MMM=(N+2)*(N+2))
        CHARACTER FILE_MODE*80
        DIMENSION X(N),Y(N),AX(N,N+2),BX(N,N+2),CX(N,N+2),DX(N,N+2)
        DIMENSION AY(N,N+2),BY(N,N+2),CY(N,N+2),DY(N,N+2)
        DIMENSION BK(MMM),CK(MMM),LM(MMM)
        DIMENSION AXY(NN,NN),BXX(NN,NN),BYY(NN,NN)
        DIMENSION GG(MMM,MMM),BBX(N,N),BBY(N,N), LI(NN)
        DIMENSION F(MMM),V(MMM),SX(NN),SY(NN),SXY(NN)
C
C-----------------------------------------------------------------------------------
C       INPUT FILENAME OF OUTPUT
C-----------------------------------------------------------------------------------
C
        WRITE(*,'   (A)   ')' ENTER OUTPUT (STRESS) FILE NAME'
        READ(*,'   (A)   ') FILE_MODE
        OPEN(UNIT=9,STATUS='UNKNOWN',FILE=FILE_MODE)
C
C-----------------------------------------------------------------------------------
C
        PI=DATAN(1.0D0)*4.0D0
        A=1.D0                    !   LENGTH A
        RATIO=1.0D0/1.0D0
        B=A/RATIO                 !   LENGTH B
C
C-----------------------------------------------------------------------------------
C
        VV=0.3D0
        SIGMAX0=1.0D0
        NX2=N+2
        NY2=N+2

        DO I=1,N
         X(I)=-DCOS((I-1)*PI/(N-1))/2.0D0*A            ! GRID III [-A/2, A/2]
        ENDDO

        DO I=1,N
         Y(I)=X(I)*B/A
        ENDDO
C
C-----------------------------------------------------------------------------------
C
        CALL DQMNEW(N,X,AX,BX,CX,DX,BBX)            ! MODIFIED DQM
        CALL DQMNEW(N,Y,AY,BY,CY,DY,BBY)
```

```
!       CALL HDQMNEW(N,X,AX,BX,CX,DX,BBX)       ! HARMONIC DQM, NOT USED CURRENTLY
!       CALL HDQMNEW(N,Y,AY,BY,CY,DY,BBY)

        DO I=1,MMM
        F(I)=0.0D0
        DO J=1,MMM
        GG(I,J)=0.0D0
        ENDDO
        ENDDO

        DO I=1,NN
        DO J=1,NN
        BXX(I,J)=0.0D0
        AXY(I,J)=0.0D0
        BYY(I,J)=0.0D0
        ENDDO
        ENDDO
C
C    DQ EQUATIONS AT ALL INNER GRID POINTS
C
        DO I=2,N-1
        KI=NX2*(I-1)
        DO L=1,NY2
        KL=NX2*(L-1)
        DO J=2,N-1
        II=KI+J
        DO K=1,NX2
        JJ=KL+K
        GG(II,JJ)=2.D0*BY(I,L)*BX(J,K)    ! MIXED DERIVATIVES
        ENDDO
        ENDDO
        ENDDO
        ENDDO

        DO I=2,N-1
        KI=NX2*(I-1)
        DO J=2,N-1
        II=KI+J
        DO K=1,NX2
        JJ=KI+K
        GG(II,JJ)=GG(II,JJ)+DX(J,K)     ! D4W/DX4
        ENDDO
        ENDDO
        ENDDO

        DO I=2,N-1
        KI=NX2*(I-1)
        DO L=1,NY2
        KL=NX2*(L-1)
        DO J=2,N-1
        II=KI+J
        JJ=KL+J
        GG(II,JJ)=GG(II,JJ)+DY(I,L)    ! D4W/DY4
```

```
        ENDDO
        ENDDO
        ENDDO
C
C---------------------------------------------------------------------------------
C
C       APPLYING THE EQUIVALENT STRESS BOUNDARY CONDITIONS
C
C---------------------------------------------------------------------------------
C
        DO J=1,N                ! Y=-B/2 (PHI), EQ. (5.19)-1ST EQ.
        GG(J,J)=1.0D0
        F(J)=0.0D0
        ENDDO

        KI=NX2*(N-1)            !Y=+B/2 (PHI) , EQ. (5.21) -1ST EQ.
        DO J=1,N
        II=KI+J
        GG(II,II)=1.0D0
        F(II)=-SIGMAX0*B*B/PI
        ENDDO

        KI=NX2*N                ! Y=-B/2, EQ. (5.19)-2ND EQ.
        DO J=1,N
        II=KI+J
        GG(II,II)=1.0D0
        F(II)=0.0D0
        ENDDO

        KI=NX2*(N+1)            ! Y=+B/2, EQ. (5.21)-2ND EQ.
        DO J=1,N
        II=KI+J
        GG(II,II)=1.0D0
        F(II)=-2.0D0*SIGMAX0*B/PI
        ENDDO

        DO I=1,N                ! X=-A/2 , EQ. (5.22)- 2ED EQ.
        KI=NX2*(I-1)
        II=KI+N+1
        GG(II,II)=1.0D0
        F(II)=0.0D0
        ENDDO

        DO I=1,N                ! X=+A/2, EQ. (5.20)-2ED EQ
        KI=NX2*(I-1)
        II=KI+NX2
        GG(II,II)=1.0D0
        F(II)=0.0D0
        ENDDO

        DO I=2,N-1              ! X=-A/2 (PHI) , EQ. (5.22) -1ST EQ.
        KI=NX2*(I-1)
        II=KI+1
        GG(II,II)=1.0D0
        F(II)=SIGMAX0*B/2.0D0/PI/PI*
```

```
     *       (2.0D0*B*DCOS(PI*Y(I)/B)-2.0D0*PI*Y(I)-PI*B)
            ENDDO

            DO I=2,N-1                  ! X=+A/2 (PHI) , EQ. (5.20) -1ST EQ.
            KI=NX2*(I-1)
            II=KI+N
            GG(II,II)=1.0D0
            F(II)=SIGMAX0*B/2.0D0/PI/PI*
     *       (2.0D0*B*DCOS(PI*Y(I)/B)-2.0D0*PI*Y(I)-PI*B)
            ENDDO
C
C     ADD    (THE FOUR OPEN CIRCLES IN FIG. IV.1)---JUST MAKE INVERSE OF [GG] EASY
C
            DO I=N+1,NY2
            KI=NX2*(I-1)
            DO J=N+1,NX2
            II=KI+J
            GG(II,II)=1.0D0
            F(II)=0.0D0
            ENDDO
            ENDDO
C--------------------------------------------------------------------------------------------------------------------------------
C
C     THE END OF APPLYING THE EQUIVALENT STRESS BOUNDARY CONDITIONS
C
C--------------------------------------------------------------------------------------------------------------------------------
C
            CALL INVMAT(MMM,GG,ZZ,1.D-30,ISW,BK,CK,LM)
C
            DO I=1,MMM
            V(I)=0.0D0
            DO J=1,MMM
            V(I)=V(I)+GG(I,J)*F(J)
            ENDDO
            ENDDO
C
C     COMPUTE STRESS COMPONENTS AT ALL GRID POINTS
C
            DO I=1,N
            KI=N*(I-1)
            DO L=1,N
            KL=N*(L-1)
            DO J=1,N
            II=KI+J
            DO K=1,N
            JJ=KL+K
            AXY(II,JJ)=-AY(I,L)*AX(J,K)   ! MIXED DERIVATIVES
            ENDDO
            ENDDO
            ENDDO
            ENDDO

            DO I=1,N
            KI=N*(I-1)
```

```
          DO J=1,N
          II=KI+J
          DO K=1,N
          JJ=KI+K
          BXX(II,JJ)=BBX(J,K)              ! D2W/DX2
          ENDDO
          ENDDO
          ENDDO
          DO I=1,N
          KI=N*(I-1)
          DO L=1,N
          KL=N*(L-1)
          DO J=1,N
          II=KI+J
          JJ=KL+J
          BYY(II,JJ)=BBY(I,L)              ! D2W/DY2
          ENDDO
          ENDDO
          ENDDO

          IJK=0
          DO I=1,N
          KI=NX2*(I-1)
          DO J=1,N
          IJK=IJK+1
          LI(IJK)=KI+J
          ENDDO
          ENDDO
          DO I=1,NN
          SX (I)=0.0D0
          SY (I)=0.0D0
          SXY(I)=0.0D0
          DO J=1,NN
          SX (I)=SX (I)+BYY(I,J)*V(LI(J))
          SY (I)=SY (I)+BXX(I,J)*V(LI(J))
          SXY(I)=SXY(I)+AXY(I,J)*V(LI(J))
          ENDDO
          ENDDO
C
C         PRINT OUT RESULTS
C
          WRITE(9,553)
          DO I=1,NN
          WRITE(9,550) SX(I)
          ENDDO
          WRITE(9,554)
          DO I=1,NN
          WRITE(9,550) SY(I)
          ENDDO
          WRITE(9,555)
          DO I=1,NN
          WRITE(9,550) SXY(I)
          ENDDO
550       FORMAT (1X,G20.10)
```

```
553    FORMAT (//5X,'SIGMA_X'/)
554    FORMAT (//5X,'SIGMA_Y'/)
555    FORMAT (//5X,'TAU_XY'/)

       STOP
       END
C
C------------------------------------------------------------------------------------------------------
C                        THE END OF FORTRAN PROGRAM5
C------------------------------------------------------------------------------------------------------
C
```

```
%
%------------------------------------------------------------------------------------------------------
%     The converted MATLAB file named Program5_Main
%------------------------------------------------------------------------------------------------------
%
%
%------------------------------------------------------------------------------------------------------
%     Program 5
%     In plane stress analysis of rectangular plates by the modified DQM (Use the method of Airy stress function)
%     N—The number of grid points in both directions.
%     Currently Grid III is used in this program.
%------------------------------------------------------------------------------------------------------
%
       clc; clear; close all;
       N=15; NN=N*N; MMM=(N+2)*(N+2);


%------------------------------------------------------------------------------------------------------
%
       PI=atan(1.0D0)*4.0D0;
       A=1.D0;                  %    LENGTH A
       RATIO=1.0D0/1.0D0;
       B=A/RATIO;               %    LENGTH B
%
%------------------------------------------------------------------------------------------------------
%
       VV=0.3D0;
       SIGMAX0=1.0D0;
       NX2=N+2;
       NY2=N+2;

       X=zeros(1,N);
        for I=1:N
           X(I)=-cos((I-1)*PI/(N-1))/2.0D0*A;         % GRID III [-A/2, A/2]
       end

       Y=X*B/A;

       [AX,BX,CX,DX,BBX]=DQMNEW(N,X);
       [AY,BY,CY,DY,BBY]=DQMNEW(N,Y);
```

```
%       [AX,BX,CX,DX,BBX]=HDQMNEW(N,X);
%       [AY,BY,CY,DY,BBY]=HDQMNEW(N,Y);

        F=zeros(MMM,1);
        GG=zeros(MMM,MMM);

        BXX=zeros(NN,NN);
        AXY=zeros(NN,NN);
        BYY=zeros(NN,NN);
%
%   DQ EQUATIONS AT ALL INNER GRID POINTS
%
        for I=2:N-1
           KI=NX2*(I-1);
           for L=1:NY2
               KL=NX2*(L-1);
               for J=2:N-1
                   II=KI+J;
                   for K=1:NX2
                       JJ=KL+K;
                       GG(II,JJ)=2.D0*BY(I,L)*BX(J,K);   % MIXED DERIVATIVES
                   end
               end
           end
        end

        for I=2:N-1
           KI=NX2*(I-1);
           for J=2:N-1
               II=KI+J;
               for K=1:NX2
                   JJ=KI+K;
                   GG(II,JJ)=GG(II,JJ)+DX(J,K);       % D4W/DX4
               end
           end
        end

        for I=2:N-1
           KI=NX2*(I-1);
           for L=1:NY2
               KL=NX2*(L-1);
               for J=2:N-1
                   II=KI+J;
                   JJ=KL+J;
                   GG(II,JJ)=GG(II,JJ)+DY(I,L);   % D4W/DY4
               end
           end
        end
%
%-------------------------------------------------------------------------------------------------------------------
%
%      APPLYING THE EQUIVALENT STRESS BOUNDARY CONDITIONS
%
%-------------------------------------------------------------------------------------------------------------------
%
```

```
    for J=1:N                   %Y=-B/2 (PHI), EQ. (5.19) -1ST EQ.
        GG(J,J)=1.0D0;
        F(J)=0.0D0;
    end

    KI=NX2*(N-1);               %Y=+B/2 (PHI) , EQ. (5.21) -1ST EQ.
    for J=1:N
        II=KI+J;
        GG(II,II)=1.0D0;
        F(II)=-SIGMAX0*B*B/PI;
    end

    KI=NX2*N;                   % Y=-B/2, EQ. (5.19)-2ND EQ.
    for J=1:N
        II=KI+J;
        GG(II,II)=1.0D0;
        F(II)=0.0D0;
    end

    KI=NX2*(N+1);               % Y=+B/2, EQ. (5.21)-2ND EQ.
    for J=1:N
        II=KI+J;
        GG(II,II)=1.0D0;
        F(II)=-2.0D0*SIGMAX0*B/PI;
    end

    for I=1:N                   % X=-A/2 , EQ. (5.22)- 2ED EQ.
        KI=NX2*(I-1);
        II=KI+N+1;
        GG(II,II)=1.0D0;
        F(II)=0.0D0;
    end

    for I=1:N                   % X=+A/2, EQ. (5.20)- 2ED EQ
        KI=NX2*(I-1);
        II=KI+NX2;
        GG(II,II)=1.0D0;
        F(II)=0.0D0;
    end

    for I=2:N-1                 % X=-A/2 (PHI) , EQ. (5.22) -1ST EQ.
        KI=NX2*(I-1);
        II=KI+1;
        GG(II,II)=1.0D0;
        F(II)=SIGMAX0*B/2.0D0/PI/PI*...
            (2.0D0*B*cos(PI*Y(I)/B)-2.0D0*PI*Y(I)-PI*B);
    end

    for I=2:N-1                 % X=+A/2 (PHI) , EQ. (5.20) -1ST EQ.
        KI=NX2*(I-1);
        II=KI+N;
        GG(II,II)=1.0D0;
        F(II)=SIGMAX0*B/2.0D0/PI/PI*...
```

```
                    (2.0D0*B*cos(PI*Y(I)/B)-2.0D0*PI*Y(I)-PI*B);
        end
%
%      ADD    (THE FOUR OPEN CIRCLES IN FIG. IV.1)---JUST MAKE INVERSE OF [GG] EASY
%
        for I=N+1:NY2
            KI=NX2*(I-1);
            for J=N+1:NX2
                II=KI+J;
                GG(II,II)=1.0D0;
                F(II)=0.0D0;
            end
        end
%-------------------------------------------------------------------------------------------------------------------------
%
%      THE END OF APPLYING THE EQUIVALENT STRESS BOUNDARY CONDITIONS
%
%-------------------------------------------------------------------------------------------------------------------------
%
        V=GG\F;
%
%      COMPUTE STRESS COMPONENTS AT ALL GRID POINTS
%
        for I=1:N
            KI=N*(I-1);
            for L=1:N
                KL=N*(L-1);
                for J=1:N
                    II=KI+J;
                    for K=1:N
                        JJ=KL+K;
                        AXY(II,JJ)=-AY(I,L)*AX(J,K);   % MIXED DERIVATIVES
                    end
                end
            end
        end
         for I=1:N
            KI=N*(I-1);
            for J=1:N
                II=KI+J;
                for K=1:N
                    JJ=KI+K;
                    BXX(II,JJ)=BBX(J,K);               % D2W/DX2
                end
            end
        end
         for I=1:N
            KI=N*(I-1);
            for L=1:N
                KL=N*(L-1);
                for J=1:N
                    II=KI+J;
                    JJ=KL+J;
                    BYY(II,JJ)=BBY(I,L);               % D2W/DY2
                end
            end
```

```
    end

  IJK=0;
  for I=1:N
      KI=NX2*(I-1);
      for J=1:N
             IJK=IJK+1;
             LI(IJK)=KI+J;
        end
  end

  SX =BYY*V(LI);
  SY =BXX*V(LI);
  SXY=AXY*V(LI);


  [XCoord,YCoord]=meshgrid(X,Y);
  XCoord = ( XCoord - X(1) ) * RATIO;
  YCoord = ( YCoord -Y(1) ) * RATIO;


  Sigma_X=reshape(SX,N,N)';
  Sigma_Y=reshape(SY,N,N)';
  Tau_XY =reshape(SXY,N,N)';

  figure
  surfc(XCoord,YCoord,Sigma_X)
  xlabel('x')
  ylabel('y')
  zlabel('\sigma_x/\sigma_0')

  figure
  surfc(XCoord,YCoord,Sigma_Y)
  xlabel('x')
  ylabel('y')
  zlabel('\sigma_y/\sigma_0')

  figure
  surfc(XCoord,YCoord,Tau_XY)
  xlabel('x')
  ylabel('y')
  zlabel('\tau_x_y/\sigma_0')

%
%-------------------------------------------------------------------------------------------------------------
%    The end of the converted MATLAB file Program5_Main
%-------------------------------------------------------------------------------------------------------------
%
```

For other nonlinear distributed edge loading, simply change the expression in Program 5 or Program5_Main after "APPLYING THE EQUIVALENT STRESS BOUNDARY CONDITIONS". An example, namely, the rectangular plate under parabolic distributed edge loading, is given for illustrations. Pay attention to the bold letters in the program given below.

```fortran
C
C--------------------------------------------------------------------------------
C    FORTRAN Codes
C    APPLYING THE EQUIVALENT STRESS BOUNDARY CONDITIONS
C
C--------------------------------------------------------------------------------
C
      DO J=1,N                 ! Y=-B/2 (PHI), EQ. (5.41)-1ST EQ.
      GG(J,J)=1.0D0
      F(J)=0.0D0
      ENDDO

      KI=NX2*(N-1)             !Y=+B/2 (PHI) , EQ. (5.43) -1ST EQ.
      DO J=1,N
      II=KI+J
      GG(II,II)=1.0D0
      F(II)=-SIGMAX0*B*B/ 3.0D0
      ENDDO

      KI=NX2*N                 ! Y=-B/2, EQ. (5.41)-2ND EQ.
      DO J=1,N
      II=KI+J
      GG(II,II)=1.0D0
      F(II)=0.0D0
      ENDDO

      KI=NX2*(N+1)             ! Y=+B/2, EQ. (5.43)-2ND EQ.
      DO J=1,N
      II=KI+J
      GG(II,II)=1.0D0
      F(II)=-2.0D0*SIGMAX0*B/ 3.0D0
      ENDDO

      DO I=1,N                 ! X=-A/2 , EQ. (5.44)- 2ED EQ.
      KI=NX2*(I-1)
      II=KI+N+1
      GG(II,II)=1.0D0
      F(II)=0.0D0
      ENDDO

      DO I=1,N                 ! X=+A/2, EQ. (5.42)-2ED EQ
      KI=NX2*(I-1)
      II=KI+NX2
      GG(II,II)=1.0D0
      F(II)=0.0D0
      ENDDO

      DO I=2,N-1               ! X=-A/2 (PHI) , EQ. (5.44) -1ST EQ.
      KI=NX2*(I-1)
      II=KI+1
      GG(II,II)=1.0D0
      F(II)=SIGMAX0 /48.0D0/B/B*(16.0D0* Y(I)**4 -3.0D0*B**4
     *      -16.0D0*B**3*Y(I)-24.0D0*B*B*Y(I)*Y(I))
      ENDDO
```

```
      DO I=2,N-1                ! X=+A/2 (PHI) , EQ. (5.42) -1ST EQ.
      KI=NX2*(I-1)
      II=KI+N
      GG(II,II)=1.0D0
      F(II)=SIGMAX0 /48.0D0/B/B*(16.0D0* Y(I)**4 -3.0D0*B**4
     *        -16.0D0*B**3*Y(I)-24.0D0*B*B*Y(I)*Y(I))
      ENDDO
C
C     ADD   (THE FOUR OPEN CIRCLES IN FIG. IV.1.) ---JUST MAKE INVERSE OF [GG] EASY
C
      DO I=N+1,NY2
      KI=NX2*(I-1)
      DO J=N+1,NX2
      II=KI+J
      GG(II,II)=1.0D0
      F(II)=0.0D0
      ENDDO
      ENDDO
C-------------------------------------------------------------------------------
C
C     THE END OF APPLYING THE EQUIVALENT STRESS BOUNDARY CONDITIONS
C
C-------------------------------------------------------------------------------
C


%
%-------------------------------------------------------------------------------
%     Converted MATLAB codes
%-------------------------------------------------------------------------------
%
%-------------------------------------------------------------------------------
%
%     APPLYING THE EQUIVALENT STRESS BOUNDARY CONDITIONS
%
%-------------------------------------------------------------------------------
%
      for J=1:N      % Y=-B/2 (PHI), EQ. (5.41) -1ST EQ.
      GG(J,J)=1.0D0;
      F(J)=0.0D0;
      end

      KI=NX2*(N-1);       %Y=+B/2 (PHI) , EQ. (5.43) -1ST EQ.
        for J=1:N
          II=KI+J;
          GG(II,II)=1.0D0;
          F(II)=-SIGMAX0*B*B/ 3.0D0;
      end

      KI=NX2*N;      % Y=-B/2, EQ. (5.41)-2ND EQ.
```

```
    for J=1:N
        II=KI+J;
        GG(II,II)=1.0D0;
        F(II)=0.0D0;
    end

    KI=NX2*(N+1);   % Y=+B/2, EQ. (5.43)-2ND EQ.
      for J=1:N
        II=KI+J;
        GG(II,II)=1.0D0;
        F(II)=-2.0D0*SIGMAX0*B/ 3.0D0;
    end

      for I=1:N    % X=-A/2 , EQ. (5.44)- 2ED EQ.
        KI=NX2*(I-1);
        II=KI+N+1;
        GG(II,II)=1.0D0;
        F(II)=0.0D0;
    end

    for I=1:N     % X=+A/2, EQ. (5.42)- 2ED EQ
        KI=NX2*(I-1);
        II=KI+NX2;
        GG(II,II)=1.0D0;
        F(II)=0.0D0;
    end

      for I=2:N-1      % X=-A/2 (PHI) , EQ. (5.44) -1ST EQ.
        KI=NX2*(I-1);
        II=KI+1;
        GG(II,II)=1.0D0;
        F(II)=SIGMAX0 /48.0D0/B/B*(16.0D0* Y(I)^4 -3.0D0*B^4-16.0D0*B^3*Y(I)-24.0D0*B*B*Y(I)*Y(I));
    end

      for I=2:N-1       % X=+A/2 (PHI) , EQ. (5.42) -1ST EQ.
        KI=NX2*(I-1);
        II=KI+N;
        GG(II,II)=1.0D0;
        F(II)=SIGMAX0 /48.0D0/B/B*(16.0D0* Y(I)^4 -3.0D0*B^4-16.0D0*B^3*Y(I)-24.0D0*B*B*Y(I)*Y(I));
    end
%
%   ADD (THE FOUR OPEN CIRCLES IN FIG. IV.1.) ---JUST MAKE INVERSE OF [GG] EASY
%
    for I=N+1:NY2
        KI=NX2*(I-1);
        for J=N+1:NX2
            II=KI+J;
            GG(II,II)=1.0D0;
            F(II)=0.0D0;
        end
    end
%-------------------------------------------------------------------------------------------------------------------------
%
%   THE END OF APPLYING THE EQUIVALENT STRESS BOUNDARY CONDITIONS
%
%-------------------------------------------------------------------------------------------------------------------------
.%
```

VIII.2 PROGRAM 6: DIRECT SOLVE THE EQUILIBRIUM EQUATIONS

The program can be used to obtain the in-plane stress and displacement components at all grid points. Currently, the rectangular plate is under uniaxial cosine distributed loading, given by Eq. (5.18). Since the differential equations in terms of u and v are solved by the DQM, thus any combinations of boundary conditions can be applied by using Program 6.

As in previous sections, FORTRAN codes are given first and then the converted MATLAB file is followed. The converted MATLAB file has the function of showing the stress as well as the displacement distributions graphically.

Description of some variables in Program 6

$X(I)$: x coordinate of a grid point, x_i [$-A/2,A/2$]
$Y(I)$: y coordinate of a grid point, y_i [$-B/2,B/2$]

Input data

N: number of grid points in both directions, an odd number only
A: plate length in x direction
B: plate length in y direction
VV: Poisson's ratio (μ)
EE: modulus of elasticity (E)
SIGMAX0: maximum applied edge stress for parabolic or cosine distributed stress in x direction.

Note that all input data have been assigned currently.

Output data

$SX(I)$: in-plane stress σ_x at all grid points
$SY(I)$: in-plane stress σ_y at all grid points
$SXY(I)$: in-plane stress τ_{xy} at all grid points
$V(I)$: in-plane displacement u at all grid points
$V(I + MM)$: in-plane displacement v at all grid points (MM $= N^2$)

```
C
C-----------------------------------------------------------------------
C      Program 6   (FORTRAN Codes)
C      In-plane stress analysis of rectangular plates by the DQM (Solve the equilibrium equations in terms of u & v).
C      N—The number of grid points in both directions. It should be an odd number!
C      NI—The number of inner grid points.
C      Currently Grid III is used in the program.
C      The applied edge load should be symmetric about x and y axes since symmetric conditions are used.
C-----------------------------------------------------------------------
C
       IMPLICIT REAL*8 (A-H,O-Z)
       PARAMETER (N=15,MM=N*N,NN=2*MM,NI=(N-2)*(N-2),N2=2*N)
       CHARACTER FILE_MODE*80
       DIMENSION X(N),Y(N),AX(N,N),BX(N,N),AY(N,N),BY(N,N)
```

```
      DIMENSION GK(NN,NN),BK(NN),CK(NN),Z(NN),LM(NN)
      DIMENSION F(NN),V(NN),SX(MM),SY(MM),SXY(MM)
      DIMENSION AXX(MM,MM),BXX(MM,MM),AXY(MM,MM)
      DIMENSION AYY(MM,MM),BYY(MM,MM)
      DIMENSION LI(NI),LX(N2),LY(N2)
C
C-------------------------------------------------------------------------------------------------------
C
C    UNKNOWN-DISPLACEMENTS
C    U—N*N, V--N*N
C
C-------------------------------------------------------------------------------------------------------
C    INPUT FILENAME OF OUTPUT
C-------------------------------------------------------------------------------------------------------
C
      WRITE(*,'   (A)   ') ' ENTER OUTPUT FILE NAME'
      READ(*,'   (A)   ') FILE_MODE
      OPEN(UNIT=9,STATUS='UNKNOWN',FILE=FILE_MODE)        ! FILE TO STORE OUTPUT DATA
C
C-------------------------------------------------------------------------------------------------------
C
C
      PI=DATAN(1.0D0)*4.0D0
      NX=N                       ! TOTAL POINTS IN THE X-DIRECTION
      NY=N                       ! TOTAL POINTS IN THE Y-DIRECTION
      A=1.D0                     ! LENGTH A CHANGE FOR DIFFERENT ASPECT RATIOS A/B ONLY !
      RATIO=1.0D0/1.0D0
      B=A/RATIO                  ! LENGTH B
C
C-------------------------------------------------------------------------------------------------------
C
      VV=0.3D0                   ! POISSON'S RATIO
      EE=1.0D0
      C11=EE/(1.0D0-VV*VV)
      C12=(1.0D0-VV)/2.0D0*C11
      C13=(1.0D0+VV)/2.0D0*C11
      GG=EE/2.0D0/(1.0D0+VV)
      SIGMA0=1.0D0
C
C-------------------------------------------------------------------------------------------------------
C
      DO I=1,N
      X(I)=-DCOS((I-1)*PI/(N-1))/2.0D0*A          ! GRID III [-A/2, A/2]
      ENDDO

      DO I=1,N
      Y(I)=X(I)*B/A
      ENDDO
C
C-------------------------------------------------------------------------------------------------------
C
      CALL DQM(N,X,AX,BX)
      CALL DQM(N,Y,AY,BY)
```

```
C
      DO I=1,MM
      DO J=1,MM
      AXX(I,J)=0.0D0
      BXX(I,J)=0.0D0
      AXY(I,J)=0.0D0
      AYY(I,J)=0.0D0
      BYY(I,J)=0.0D0
      ENDDO
      ENDDO
C
C     DQ EQUATIONS AT ALL GRID POINTS
C
      DO I=1,NY                              ! MIXED DERIVATIVES
      KI=NX*(I-1)
      DO L=1,NY
      KL=NX*(L-1)
      DO J=1,NX
      II=KI+J
      DO K=1,NX
      JJ=KL+K
      AXY(II,JJ)=AY(I,L)*AX(J,K)
      ENDDO
      ENDDO
      ENDDO
      ENDDO
      DO I=1,NY                              ! D/DX
      KI=NX*(I-1)
      DO J=1,NX
      II=KI+J
      DO K=1,NX
      JJ=KI+K
      AXX(II,JJ)=AX(J,K)
      BXX(II,JJ)=BX(J,K)
      ENDDO
      ENDDO
      ENDDO

      DO I=1,NY                              ! D/DY
      KI=NX*(I-1)
      DO L=1,NY
      KL=NX*(L-1)
      DO J=1,NX
      II=KI+J
      JJ=KL+J
      AYY(II,JJ)=AY(I,L)
      BYY(II,JJ)=BY(I,L)
      ENDDO
      ENDDO
      ENDDO
C
C     EQUATIONS AT INNER NODES
C
      DO I=1,NN
      F(I)=0.0D0
      DO J=1,NN
```

```
      GK(I,J)=0.0D0
      ENDDO
      ENDDO
C
C     FIND THE POSITIONS OF INNER NODES
C
      DO I=2,NY-1
      KI=NX*(I-1)
      IJ= (NX-2)*(I-2)
      DO J=2,NX-1
      IJK=IJ+J-1
      LI(IJK)=KI+J
      ENDDO
      ENDDO

      DO I=1,NI
      DO J=1,MM
      GK(LI(I),J    )=C11*BXX(LI(I),J)+C12*BYY(LI(I),J)          ! X-DIRECTION
      GK(LI(I),J+MM)=C13*AXY(LI(I),J)                           ! X-DIRECTION
      GK(LI(I)+MM,J+MM)=C11*BYY(LI(I),J)+C12*BXX(LI(I),J) ! Y-DIRECTION
      GK(LI(I)+MM,J    )=C13*AXY(LI(I),J)                       ! Y-DIRECTION
      ENDDO
      ENDDO
C
C     EQUATIONS AT BOUNDARY NODES (X=0,A)
C
      IX=0
      DO I=1,N
      KI=N*(I-1)
      DO J=1,N,N-1
      IX      =IX+1
      LX(IX)=KI+J
      ENDDO
      ENDDO
      DO I=1,IX
      DO J=1,MM
      GK(LX(I),J    )=C11*AXX(LX(I),J)            ! SIGMA_X (AT U)
      GK(LX(I),J+MM)=C11*AYY(LX(I),J)*VV
      ENDDO
      ENDDO

      DO I=3,IX-2
      DO J=1,MM
      GK(LX(I)+MM,J    )=GG*AYY(LX(I),J)          ! TAU_XY   (AT V)
      GK(LX(I)+MM,J+MM)=GG*AXX(LX(I),J)
      ENDDO
      ENDDO
C
C     EQUATIONS AT BOUNDARY NODES (Y=0,B)
C
      IY=0
      DO I=1,N,N-1
      KI=N*(I-1)
      DO J=1,N
```

```
      IY     =IY+1
      LY(IY)=KI+J
      ENDDO
      ENDDO

      DO I=1,IY
      DO J=1,MM
      GK(LY(I)+MM,J    )=C11*AXX(LY(I),J)*VV   ! SIGMA_Y (AT V)
      GK(LY(I)+MM,J+MM)=C11*AYY(LY(I),J)
      ENDDO
      ENDDO

      DO I=2,N-1                              ! Y=0
      DO J=1,MM
      GK(LY(I),J    )=GG*AYY(LY(I),J)          ! TAU_XY (AT U)
      GK(LY(I),J+MM)=GG*AXX(LY(I),J)
      ENDDO
      ENDDO

      DO I=N+2,N2-1                           ! Y=B
      DO J=1,MM
      GK(LY(I),J    )=GG*AYY(LY(I),J)          ! TAU_XY   (AT U)
      GK(LY(I),J+MM)=GG*AXX(LY(I),J)
      ENDDO
      ENDDO
C
C     EDGE LOAD AY X=0 AND A
C
      K=0
      DO I=1,N
      K=K+1
C     F(LX(K))=-SIGMA0                        ! FOR CONSTANT STRESS
      F(LX(K))=-SIGMA0*DCOS(PI*Y(I)/B)
      K=K+1
C     F(LX(K))=-SIGMA0                        ! FOR CONSTANT STRESS
      F(LX(K))=-SIGMA0*DCOS(PI*Y(I)/B)
      ENDDO
C
C     APPLY THE ZERO DISPLACEMENT CONDITION (U=0 AT X=0, V=0 AT Y=0)
C     NOTE: N IS ODD NUMBER SINCE THE ENTIRE PLATE IS MODELLED
C
      II=N*(N/2)+MM
      DO I=1,N
      J=II+I
      GK(J,J)=1.0D30                ! V=0 AT Y=0 (SYMMETRIC)
      ENDDO

      MD=(N+1)/2
      DO I=1,N
      J=N*(I-1)+MD
      GK(J,J)=1.0D30                ! U=0 AT X=0 (SYMMETRIC)
      ENDDO

      CALL INVMAT(NN,GK,DE,1.0D-30,ISW,BK,CK,LM)
C
C     FIND DISPLACEMENT
```

```
C
      DO I=1,NN
      V(I)=0.0D0
      DO J=1,NN
      V(I)=V(I)+GK(I,J)*F(J)
      ENDDO
      ENDDO
C
C     FIND STRESS COMPONENTS
C
      DO I=1,MM
      SX (I)=0.0D0
      SY (I)=0.0D0
      SXY(I)=0.0D0
      DO K=1,MM
      SX (I)=SX (I)+C11*(AXX(I,K)*V(K)+VV*AYY(I,K)*V(K+MM))
      SY (I)=SY (I)+C11*(AXX(I,K)*V(K)*VV+AYY(I,K)*V(K+MM))
      SXY(I)=SXY(I)+GG *(AYY(I,K)*V(K)+ AXX(I,K)*V(K+MM))
      ENDDO
      ENDDO
C
C     PRINT OUT RESULTS
C
      WRITE(9,553)
      DO I=1,MM
      WRITE(9,550) SX(I)
      ENDDO
      WRITE(9,554)
      DO I=1,MM
      WRITE(9,550) SY(I)
      ENDDO
      WRITE(9,555)
      DO I=1,MM
      WRITE(9,550) SXY(I)
      ENDDO
      WRITE(9,551)
      DO I=1,MM
      WRITE(9,550) V(I)            ! DISPLACEMENTU
      ENDDO
      WRITE(9,552)
      DO I=1,MM
      WRITE(9,550) V(I+MM)   ! DISPLACEMENTV
      ENDDO

550   FORMAT (1X,G20.10)
551   FORMAT (  5X,' DISPLACEMENT_U'/)
552   FORMAT (//5X,'DISPLACEMENT_V'/)
553   FORMAT (//5X,'SIGMA_X'/)
554   FORMAT (//5X,'SIGMA_Y'/)
555   FORMAT (//5X,'TAU_XY'/)

      STOP
      END
C
C-------------------------------------------------------------------------------
C                       THE END OF FORTRAN PROGRAM 6
C-------------------------------------------------------------------------------
```

```
%
%-----------------------------------------------------------------------------
%     The converted MATLAB file named Program 6_Main
%-----------------------------------------------------------------------------
%

%
%-----------------------------------------------------------------------------
%     Program 6
%     In-plane stress analysis of rectangular plates by the DQM (Solve the equilibrium equations in terms of u & v).
%     N—The number of grid points in both directions. It should be an odd number!
%     NI—The number of inner grid points.
%     Currently Grid III is used.
%     The applied edge load should be symmetric about x and y axes since symmetric conditions are used.
%-----------------------------------------------------------------------------
%
      clc; clear; close all;
      N=15; MM=N*N; NN=2*MM; NI=(N-2)*(N-2); N2=2*N;
%
%-----------------------------------------------------------------------------
%
%     UNKNOWN-DISPLACEMENTS
%     U-N*N, V--N*N
%
%-----------------------------------------------------------------------------
%

      PI=atan(1.0D0)*4.0D0;
      NX=N;                      % TOTAL POINTS IN THE X-DIRECTION
      NY=N;                      % TOTAL POINTS IN THE Y-DIRECTION
      A=1.D0;             % LENGTH A
      RATIO=1.0D0/1.0D0;
      B=A/RATIO;          %   LENGTH B
%
%-----------------------------------------------------------------------------
%
      VV=0.3D0;          %   POISSON'S RATIO
      EE=1.0D0;
      C11=EE/(1.0D0-VV*VV);
      C12=(1.0D0-VV)/2.0D0*C11;
      C13=(1.0D0+VV)/2.0D0*C11;
      GG=EE/2.0D0/(1.0D0+VV);
      SIGMA0=1.0D0;
%
%-----------------------------------------------------------------------------
%
      X=zeros(1,N);
       for I=1:N
        X(I)=-cos((I-1)*PI/(N-1))/2.0D0*A;        % GRID III [-A/2, A/2]
       end

      Y=X*B/A;
%
%-----------------------------------------------------------------------------
```

```
%
      [AX,BX]=DQM(N,X);
      [AY,BY]=DQM(N,Y);
%
      AXX=zeros(MM,MM);
      BXX=zeros(MM,MM);
      AXY=zeros(MM,MM);
      AYY=zeros(MM,MM);
      BYY=zeros(MM,MM);
%
%     DQ EQUATIONS AT ALL GRID POINTS
%
      for I=1:NY                              % MIXED DERIVATIVES
         KI=NX*(I-1);
         for L=1:NY
             KL=NX*(L-1);
             for J=1:NX
                 II=KI+J;
                 for K=1:NX
                     JJ=KL+K;
                     AXY(II,JJ)=AY(I,L)*AX(J,K);
                 end
             end
         end
      end

      for I=1:NY                              % D/DX
         KI=NX*(I-1);
         for J=1:NX
             II=KI+J;
             for K=1:NX
                 JJ=KI+K;
                 AXX(II,JJ)=AX(J,K);
                 BXX(II,JJ)=BX(J,K);
             end
         end
      end

      for I=1:NY                              %D/DY
         KI=NX*(I-1);
         for L=1:NY
             KL=NX*(L-1);
             for J=1:NX
                 II=KI+J;
                 JJ=KL+J;
                 AYY(II,JJ)=AY(I,L);
                 BYY(II,JJ)=BY(I,L);
             end
         end
      end
%
%     EQUATIONS AT INNER NODES
%
      F=zeros(NN,1);
      GK=zeros(NN,NN);
```

```
%
%      FIND THE POSITIONS OF INNER NODES
%
    LI=zeros(NI,1);

      for I=2:NY-1
         KI=NX*(I-1);
         IJ= (NX-2)*(I-2);
         for J=2:NX-1
              IJK=IJ+J-1;
              LI(IJK)=KI+J;
         end
      end

      GK(LI    ,1:MM          )=C11*BXX(LI,:)+C12*BYY(LI,:);          % X-DIRECTION
      GK(LI    ,1+MM:MM+MM)=C13*AXY(LI,:);                           % X-DIRECTION
      GK(LI+MM,1+MM:MM+MM)=C11*BYY(LI,:)+C12*BXX(LI,:);             % Y-DIRECTION
      GK(LI+MM,1:MM          )=C13*AXY(LI,:);                       % Y-DIRECTION
%
%      EQUATIONS AT BOUNDARY NODES (X=0,A)
%
    IX=0;
    LX=zeros(1,N2);
     for I=1:N
        KI=N*(I-1);
        for J=1:N-1:N
             IX    =IX+1;
             LX(IX)=KI+J;
        end
     end

     I=1:IX;
       GK( LX(I), 1   :MM    )=C11*AXX( LX(I),:);                % SIGMA_X (AT U)
       GK( LX(I), 1+MM:MM+MM)=C11*AYY( LX(I),:)*VV;

     I=3:IX-2;
       GK( LX(I)+MM, 1    :MM    )=GG*AYY( LX(I),:);             % TAU_XY   (AT V)
       GK( LX(I)+MM, 1+MM:MM+MM)=GG*AXX( LX(I),:);

%
%      EQUATIONS AT BOUNDARY NODES (Y=0,B)
%
    IY=0;
    LY=zeros(1,N2);
     for I=1:N-1:N
        KI=N*(I-1);
        for J=1:N
             IY    =IY+1;
             LY(IY)=KI+J;
        end
     end

     I=1:IY;
       GK( LY(I)+MM, 1   :MM    )=C11*AXX(LY(I),:)*VV;      % SIGMA_Y (AT V)
       GK( LY(I)+MM, 1+MM:MM+MM)=C11*AYY(LY(I),:);
```

```
     I=2:N-1;      % Y=0
     GK(LY(I), 1   :MM   )=GG*AYY(LY(I),:);              % TAU_XY (AT U)
     GK(LY(I), 1+MM:MM+MM)=GG*AXX(LY(I),:);


     I=N+2:N2-1; % Y=B
     GK(LY(I),1    :MM   )=GG*AYY(LY(I),:) ;             % TAU_XY   (AT U)
     GK(LY(I),1+MM:MM+MM)=GG*AXX(LY(I),:)  ;

%
%     EDGE LOAD AY X=0 AND A
%
     K=0;
     for I=1:N
        K=K+1;
%        F(LX(K))=-SIGMA0;                       ! CONSTANT STRESS
         F(LX(K))=-SIGMA0*cos(PI*Y(I)/B);
         K=K+1;
%        F(LX(K))=-SIGMA0;                       ! CONSTANT STRESS
         F(LX(K))=-SIGMA0*cos(PI*Y(I)/B);
     end
%
%    APPLY THE ZEOR DISPLACEMENT CONDITION (U=0 AT X=0, V=0 AT Y=0)
%    NOTE: N IS ODD NUMBER SINCE THE ENTIRE PLATE IS MODELLED
%
     II=N*fix(N/2)+MM;
     for I=1:N
        J=II+I;
        GK(J,J)=1.0D30;        % V=0 AT Y=0 (SYMMETRIC)
     end

     MD=(N+1)/2;
     for I=1:N
        J=N*(I-1)+MD;
        GK(J,J)=1.0D30;                    % U=0 AT X=0 (SYMMETRIC)
     end

%
%     FIND DISPLACEMENT
%
     V=GK\F;
%
%     FIND STRESS COMPONENTS
%
     SX =C11*(AXX*V(1:MM)+VV*AYY(:,1:MM)*V(1+MM:MM+MM));
     SY =C11*(AXX*V(1:MM)*VV+AYY(:,1:MM)*V(1+MM:MM+MM));
     SXY=GG *(AYY*V(1:MM)+    AXX(:,1:MM)*V(1+MM:MM+MM));
%
%     PRINT OUT RESULTS
%
     [XCoord,YCoord]=meshgrid(X,Y);
     XCoord = ( XCoord - X(1) ) * RATIO;
     YCoord = ( YCoord -Y(1) ) * RATIO;
     Sigma_X=reshape(SX,N,N)';
```

```
        Sigma_Y=reshape(SY,N,N)';
        Tau_XY =reshape(SXY,N,N)';
        XDis=reshape(V(1    :MM     ),N,N)';
        YDis=reshape(V(1+MM:MM+MM),N,N)';

% Plot the normalized u
        figure
        pcolor(XCoord,YCoord,XDis)
        xlabel('x')
        ylabel('y')
        title('Eu/a\theta_0','fontsize',12)
        colorbar
        shading interp

% Plot the normalized v
        figure
        pcolor(XCoord,YCoord,YDis)
        xlabel('x')
        ylabel('y')
        title('Ev/a\theta_0','fontsize',12)
        colorbar
        shading interp

% Plot the normalized Sigma_X
        figure
        surf(XCoord,YCoord,Sigma_X)
        xlabel('x')
        ylabel('y')
        zlabel('\sigma_x/\sigma_0')

% Plot the normalized Sigma_Y
        figure
        surf(XCoord,YCoord,Sigma_Y)
        xlabel('x')
        ylabel('y')
        zlabel('\sigma_y/\sigma_0')

% Plot the normalized shear stress Tau_XY
        figure
        surf(XCoord,YCoord,Tau_XY)
        xlabel('x')
        ylabel('y')
        zlabel('\tau_x_y/\sigma_0')


%
%------------------------------------------------------------------------------------------------------
%     The end of the converted MATLAB file Program6_Main
%------------------------------------------------------------------------------------------------------
%
```

Appendix IX

IX SUMMARY OF THE RESEARCH WORK RELATED TO THE DQM

For readers' reference, the publications written by the author, his colleagues and his former graduate students are summarized and briefly discussed. The contribution to the development of the differential quadrature method (DQM) or the differential quadrature element-based method (the DQEM and the QEM) is mentioned. The publications listed at the end of Appendix IX are arranged according to the order of published date and not according to the date when the work is completed.

As was mentioned in the preface, the DQM was introduced to the author by Professor Charles W. Bert in 1991. Ref. [1] is the first published paper which the author is contributed to. To overcome the difficulty in applying multiple boundary conditions, a new approach called the method of modification of weighting coefficient-1 or simply MMWC-1 is proposed [1]. The essence of the method is that one of the two boundary conditions is applied during formulation of the weighting coefficients of higher order derivatives. The method works very well for beams and isotropic/orthotropic plates, accurate results can be obtained by the DQM for problems of deflection, buckling and free vibration of beams and rectangular or annular plates [2–4]. However, the method is only applicable for simply supported boundary and cannot be used for clamped and free boundaries. Thus, the problem of applying multiple boundary conditions has not been solved completely.

It is recognized that the DQM is similar to the mixed collocation method, thus nonuniform grid points, the roots of $(N-2)$th Chebyshev polynomial together with two end points, are proposed [5], since the DQM with uniform grid distribution cannot yield reliable and accurate solutions for problems of anisotropic rectangular plates. The proposed grid spacing has been successfully applied to analyze problems of anisotropic rectangular plate [5] and isotropic skew plate [6]. The DQM with the proposed nonuniform grid spacing can yield reliable and accurate solutions. Convergence of the DQM with the proposed nonuniform grid spacing is investigated and good convergence rate is observed [7].

In the early version of the DQM, polynomials are used to determine the weighting coefficients. The harmonic function is proposed to determine the weighting coefficients [5,8] and the method is named as the harmonic differential quadrature method (HDQM) by Professor Alfred G. Striz, or simply the HDQM. With the increase of the number of grid points, results obtained by the DQM and the HDQM are almost the same and the difference is negligible. Thus, the HDQM is not used very often. Later an explicit formulation is given to compute the weighting coefficients for the HDQM [9] and the method is called HDQMNEW in this book.

In Ref. [10], several nonuniform grid spacing is given and the error is analyzed for the DQM with different grid distributions. If N, the number of grid points, is an odd number and uniform grid spacing is used, the accuracy of the DQM at the middle point is almost an order higher than the one at all other grid points. This may be one of the reasons why the discrete singular convolution (DSC) algorithm with the nonregularized Lagrange's delta sequence kernel (called the DSC-LK) is an efficient and accurate numerical method. However, Runge's phenomenon is observed for the DQM with uniform grid spacing. Thus, nonuniform grid spacing should be used in the DQ analysis for reliability and computational efficiency considerations.

Differential Quadrature and Differential Quadrature Based Element Methods. 978-0-12-803081-3

The DQM with three nonuniform grid distributions, different from the one presented in [5], is used to analyze the buckling of laminated composite plates [11]. Although the convergence rate is quite different for smaller N; however, when the number of grid points is large enough, say $N = 17$, the accuracy of results obtained by the DQM with different nonuniform grid distributions is similar. Accurate and reliable solutions are obtained by the DQM for the free vibration of the circular annular plates with nonuniform thickness [12]. It is emphasized that the DQM with nonuniform grid distributions should be used in practical applications.

Ref. [13] is a review paper on the DQM and its applications. The objective of the paper is to introduce the DQM as well as its new development to the Chinese research community. The paper has been paid attention and the translated name of the DQM is accepted by the Chinese research community. The paper has been cited more than 100 times.

Regarding the difficulty in applying the multiple boundary conditions in the ordinary DQM, an additional degree of freedom (DOF), the first-order derivative with respect to x, is introduced at each boundary point if two boundary conditions are to be satisfied at each boundary point. Hermite interpolation, instead of the Lagrange interpolation, is used to determine the weighting coefficients. The method is called the differential quadrature element method or simply the DQEM [14–17]. Since, the number of DOFs at each boundary point is the same as the number of boundary conditions, thus the difficulty in applying the multiple boundary conditions has been completely removed. The DQEM extends the application range of the ordinary DQM, since it can be used to solve problems with discontinuous load and geometry; such as the concentrated load, stepped beam and plate structures, and even frame structures. The assemblage procedures are similar to the conventional finite element method.

Meanwhile, the applications of the DQM have been extended to shell structures [18] and nonlinear analysis [19]. In Ref. [20], the principle and applications of the DQM and the DQEM are investigated in details. The explicit formulations to compute the weighting coefficients of the DQM and the DQEM are derived independently. The analogy between the DQM and the mixed collocation method and between the DQEM and subdomain together with the mixed collocation method is discussed. A number of numerical examples are given for straight and curved beams; rectangular, circular, annular, and sector plates; conical and cylindrical shells. Some results have also been published in domestic and international journals [18,19,21–23].

Although the DQEM [14–16] has several advantages over the DQM; however, it has one drawback that one more condition should be found at the corner point if a rectangular plate is considered. In the DQEM, each corner has four DOFs, i.e., w, w_x, w_y, and w_{xy}; however, only three boundary conditions are available. For isotropic or orthotropic rectangular plate, the condition applied at the DOF w_{xy} may affect the results only slightly; however, this is not the case when anisotropic materials are involved. The added condition may affect the final results a lot when the corner is free. Therefore, to find an appropriate missing condition is important, especially if more DQ elements are to be used in the analysis. On the other hand, MMWC-1 [1] is simple and efficient, but limited to simply supported boundary conditions only. Thus, a new version of the DQM [24–28] is proposed to overcome the difficulty existing in the DQEM and the drawback existing in MMWC-1. The method is called the method of modification of weighting coefficient-3 and the method of modification of weighting coefficient-4, or simply MMWC-3 and MMWC-4.

Similar to the DQEM, an additional DOF, the first-order derivative with respect to x, is introduced at each boundary point if two boundary conditions are to be satisfied at each boundary point. Instead of Hermite interpolation, Lagrange interpolation is adopted to compute the weighting coefficient of the first-order derivative, exactly the same as the ordinary DQM. The additional DOF is introduced during formulations of the weighting coefficients of the second, third, and fourth-order derivatives. The idea is similar to MMWC-1. Although the DQ equations in one dimension are similar to the DQEM in form; however, the new version of the DQM has three DOFs, i.e., w, w_x, and w_y at each corner point for the rectangular plate. Since the number of DOFs at all boundary points are the same as the number of the boundary conditions, thus the difficulty existing in the DQEM has been overcome and the limitation existing in MMWC-1 has been completely removed. With the proposed method, any combinations of boundary condition can be applied without any difficulties. Accurate results for problems of static, buckling and free vibration of plates are obtained by the new version of the DQM [24–28]. Besides the MMWC-4 can be employed if the La-DQM is to be used.

Various ways to apply the multiple boundary conditions are summarized and illustrated in Ref. [29]. According to the book reviewers' suggestion, the names of some methods in this book are different from the ones in Ref. [29]. For the rectangular plate, only MMWC-3 (MMWC-4) or the mixed method of MMWC-3 (MMWC-4) with DQEM is recommended. If the isotropic or orthotropic rectangular plate without a free edge, it is recommended that MMWC-1 is used to deal with the simply supported boundary conditions and MMWC-3 or MMWC-4 is used to deal with the clamped boundary conditions. Since, one of the boundary condition has been built in during formulation of weighting coefficients and the deflection at all boundary points is zero, therefore, only the DQ equations at inner points are needed. The implementation is very simple and the results are very accurate [30].

The DQM and the DQEM have been successfully applied to problems with geometrical discontinuous, such as the buckling of drill-string with connector [31], stiffened rectangular plate [32], and stiffened cylindrical shell [33,34], the buckling of rectangular plate under nonuniformly distributed edge compressive load [35–37]. The fourth-order partial differential equation; namely, the compatibility equation in terms of Airy stress function, is successfully solved by the DQM with MMWC-3, then the buckling load of the rectangular plate is obtained by the same method. No difficulty is encountered and accurate results are obtained. The DQM is successfully used to obtain solutions to the nonlinear stability problem of thin doubly curved orthotropic shallow shells [38].

The suitability of the existing DQ-based time integration scheme for analyzing dynamic problems has been investigated [39]. An assessment of the DQ-based time integration scheme for nonlinear dynamic equations has been made. It is concluded that overall speaking the existing DQ-based time integration scheme is reliable, computationally efficient and also suitable for time integrations over long time durations. Care should be taken; however, in choosing a time step when applying the scheme to nonlinear dynamic systems. Although it is an unconditionally stable time integration scheme, it may yield inaccurate results for nonlinear dynamic systems if an inappropriate large time step is used [40], this is similar to the other existing unconditionally stable time integration schemes.

The DQEM has been tried to solve problems in the area of fracture mechanics [41]. Plane stress DQ element is formulated in terms of displacements. The stress intensity factor (SIF) is gained by using the displacement extrapolation technique. Comparisons to the analytical solutions show that the DQEM can calculate the SIF conveniently and accurately.

The DQM and DQEM have been used to solve practical problems, such as the nonlinear buckling analysis of long drill-strings in straight and curved wellbores [42]. For uniform long drill-strings, the DQEM or the LaDQM should be used even the strings are uniform in cross section, more grid points are needed to describe the post buckling deformation, since the DQM with larger number of grid points may cause numerical instability.

The modified DQM is successfully used to obtain elastoplastic buckling load of rectangular plates under various combinations of edge loadings. As is expected, accurate results are obtained for both thin and thick plates [43–45]. To solve the difficulty in programming for the DQM with MMWC-3, a simple way is proposed [44,46]. FORTRAN programs and MATLAB files included in this book adopt this simple method. It is seen that any combinations of boundary conditions are easily applied by the provided programs with only a few simple modifications.

The DQ-based time integration scheme is used to solve nonlinear boundary value problems. The buckling and post-buckling behavior of an extensible beam is obtained by the DQ-based time integration scheme [47]. Since the problem is not an initial value problem, thus a way to satisfy the other boundary condition is proposed.

The weak form quadrature element method or simply the QEM is different from the strong form DQEM. In principle, the QEM is the same as higher order finite element method (FEM) or the time-domain spectral element method (SEM). Due to using differential quadrature rule to compute the weighting coefficients at integration point, the QEM is more flexible and simpler in implementation than the FEM and SEM, since explicit expressions of the derivatives of shape functions are not needed. The QEM is used to obtain the dynamic response of a flexible rod hit by rigid ball [48]. After removing the computational noise by a low pass filter, accurate impact force history is obtained. More recently, the QEM is successfully used to analyze the static problem of higher order sandwich beams [49] and dynamic response FGM bars [50]; accurate static and dynamic responses are obtained.

The DQM, DQEM, and QEM have been successfully used to analyze the free vibration of multiple-stepped beams [51] and skew plates [46,52]. The important finding is that the way to apply the multiple boundary conditions may affect the solution accuracy and reliability for the case of skew plates with large skew angles. The DQM with MMWC-3 and Grid III can yield accurate frequencies for the skew plate without a free edge [46]. It is, however, only the DQM with MMWC-3 and Grid V can yield accurate frequencies for the skew plate with free edges [52]. In other words, the DQM with other nonuniform grid distributions, successfully used in literature, cannot yield reliable solutions for the skew plate with free edges and large skew angles [52]. Thus, care should be taken in using the DQM and attention should be paid on the way to applying the multiple boundary conditions accurately and on the choice of a proper grid distribution. Otherwise, wrong results may be obtained. For the quadrature element method; however, accurate results can always be obtained for skew plates with free edges and large skew angles [52].

Currently only GLL points are used as the quadrature element nodes in literature. Otherwise the way to compute the weighting coefficients explicitly at integration points cannot be used, thus the QEM reduces to the time-domain SEM. To overcome this difficulty, a method is proposed [53]. With the proposed method, the quadrature element can have any type of nodes, such as Chebyshev nodes or extended Chebyshev nodes.

It should be mentioned that the content presented in Sections 6.2.3, 6.3.3, and 6.3.4 is on-going and unpublished research results. With the proposed method to deal with the Dirac-delta function, the DQM can analyze concentrated and line distributed loads conveniently. The DQM has been successfully used to analyze the moving load problem. For analyzing dynamic problem by the DQM, care should be taken on the method of application of multiple boundary conditions as well as the choice of proper grid spacing. If spurious eigenvalues exist, then numerical instability would occur during step-by-step time integration.

REFERENCES

[1] X. Wang, C.W. Bert, A new approach in applying differential quadrature to static and free vibration analyses of beams and plates, J. Sound Vib. 162 (3) (1993) 566–572.

[2] X. Wang, A.G. Striz, C.W. Bert, Free vibration analysis of annular plates by the DQ method, J. Sound Vib. 164 (1993) 173–175.

[3] X. Wang, C.W. Bert, A.G. Striz, Differential quadrature analysis of deflection, buckling, and free vibration of beams and rectangular plates, Comput. Struct. 48 (1993) 473–479.

[4] C.W. Bert, X. Wang, A.G. Striz, Static and free vibrational analysis of beams and plates by differential quadrature method, Acta Mech. 102 (1994) 11–24.

[5] C.W. Bert, X. Wang, A.G. Striz, Differential quadrature for static and free vibration analyses of anisotropic plates, Int. J. Solids Struct. 30 (1993) 1737–1744.

[6] X. Wang, A.G. Striz, C.W. Bert, Buckling and vibration analysis of skew plates by the DQ method, AIAA J. 32 (1994) 886–889.

[7] C.W. Bert, X. Wang, A.G. Striz, Convergence of the DQ method in the analysis of anisotropic plates, J. Sound Vib. 170 (1994) 140–144.

[8] A.G. Striz, X. Wang, C.W. Bert, Harmonic differential quadrature method and applications to structural components, Acta Mech. 111 (1995) 85–94.

[9] X. Wang, B. He, An explicit formulation for weighting coefficients of harmonic differential quadrature, J. NUAA 27 (1995) 496–501 (in Chinese).

[10] B. He, X. Wang, Error analysis in differential quadrature method, T. NUAA 11 (2) (1994) 194–200.

[11] X. Wang, Differential quadrature for buckling analysis of laminated plates, Comput. Struct. 57 (4) (1995) 715–719.

[12] X. Wang, J. Yang, J. Xiao, On free vibration analysis of isotropic circular annular plates with non-uniform thickness by the differential quadrature method, J. Sound Vib. 184 (1995) 547–551.

[13] X. Wang, Differential quadrature in the analysis of structural components, Adv. Mech. 25 (2) (1995) 232–240 (in Chinese).

[14] X. Wang, H. Gu, B. Liu, On buckling analysis of beams and frame structures by the differential quadrature element method, Proc. Eng. Mech. 1 (1996) 382–385.

[15] X. Wang, H. Gu, Static analysis of frame structures by the differential quadrature element method, Int. J. Numer. Meth. Eng. 40 (1997) 759–772.

[16] H. Gu, X. Wang, On free vibration analysis of circular plates with stepped thickness over a concentric region by the DQEM, J. Sound Vib. 202 (1997) 452–459.

[17] X. Wang, Y. Wang, R. Chen, Static and free vibrational analysis of rectangular plates by the differential quadrature element method, Commun. Numer. Meth. Eng. 14 (1998) 1133–1141.

[18] Y. Wang, R. Liu, X. Wang, Free vibration analysis of truncated conical shells by the differential quadrature method, J. Sound Vib. 224 (1999) 387–394.

[19] Y. Wang, R. Liu, X. Wang, On free vibration analysis of nonlinear piezoelectric circular shallow spherical shells by the differential quadrature element method, J. Sound Vib. 245 (2001) 179–185.

[20] Y. Wang, Differential quadrature method and differential quadrature element method – theory and applications, Ph.D. Dissertation, Namjing University of Aeronautics and Astronautics, China, 2001 (in Chinese).

[21] Y. Wang, X. Wang, On a high-accuracy curved differential quadrature beam element and its applications, J. NUAA 33 (6) (2001) 516–520 (in Chinese).

[22] Y. Wang, X. Wang, Analysis of nonlinear piezoelectric circular shallow spherical shells by the differential quadrature element method, T. NUAA 18 (2) (2001) 130–136.

[23] X. Wang, Y. Wang, On nonlinear behavior of spherical shallow shells bonded with piezoelectric actuators by the differential quadrature element method (DQEM), Int. J. Numer. Meth. Eng. 53 (2002) 1477–1490.

[24] X. Wang, M. Tan, Y. Zhou, Buckling analyses of anisotropic plates and isotropic skew plates by the new version differential quadrature method, Thin Wall Struct. 41 (2003) 15–29.

[25] X. Wang, Y. Wang, Y. Zhou, Application of a new differential quadrature element method for free vibrational analysis of beams and frame structures, J. Sound Vib. 269 (2004) 1133–1141.

[26] Y. Wang, X. Wang, Y. Zhou, Static and free vibration analyses of rectangular plates by the new version of differential quadrature element method, Int. J. Numer. Meth. Eng. 59 (9) (2004) 1207–1226.

[27] X. Wang, Y. Wang, Free vibration analyses of thin sector plates by the new version of differential quadrature method, Comput. Meth. Appl. M. 193 (2004) 3957–3971.

[28] X. Wang, Y. Wang, Re-analysis of free vibration of annular plates by the new version of differential quadrature method, J. Sound Vib. 278 (2004) 685–689.

[29] X. Wang, F. Liu, X. Wang, et al. New approaches in application of differential quadrature method for fourth-order differential equations, Commun. Numer. Meth. Eng. 21 (2) (2005) 61–71.

[30] X. Wang, L. Gan, Y. Wang, A differential quadrature analysis for vibration and buckling of an SS-C-SS-C rectangular plate loaded by linearly varying in-plane stresses, J. Sound Vib. 298 (2006) 420–431.

[31] M.S., Wang., Buckling of drill-strings with connectors in inclined well-bores, Master Thesis, Nanjing University of Aeronautics and Astronautics, China, 2006 (in Chinese).

[32] C.W., Dai, Buckling analysis of stiffened plates with differential quadrature element method, Master Thesis, Nanjing University of Aeronautics and Astronautics, China, 2006 (in Chinese).

[33] L.H., Jiang, Buckling analysis of stiffened circular cylindrical panels with differential quadrature element method, Master Thesis, Nanjing University of Aeronautics and Astronautics, China, 2007 (in Chinese).

[34] L. Jiang, Y. Wang, X. Wang, Buckling analysis of stiffened circular cylindrical panels using differential quadrature element method, Thin Wall Struct. 46 (4) (2008) 390–398.

[35] X. Wang, X. Wang, X. Shi, Differential qudrature buckling analyses of rectangular plates subjected to non-uniform distributed in-plane loadings, Thin Wall Struct. 44 (2006) 837–843.

[36] X. Wang, X. Wang, X. Shi, Accurate buckling loads of thin rectangular plates under parabolic edge compressions by differential quadrature method, Int. J. Mech. Sci. 49 (2007) 447–453.

[37] X. Wang, L. Gan, Y. Zhang, Differential quadrature analysis of the buckling of thin rectangular plates with cosine-distributed compressive loads on two opposite sides, Adv. Eng. Softw. 39 (2008) 497–504.

[38] X. Wang, Nonlinear stability analysis of thin doubly curved orthotropic shallow shells by the differential quadrature method, Comput. Meth. Appl. M. 196 (2007) 2242–2251.

[39] J. Liu, On the differential quadrature method for analyzing dynamic problems, Master Thesis, Nanjing University of Aeronautics and Astronautics, China, 2007 (in Chinese).

[40] J. Liu, X. Wang, An assessment of the differential quadrature time integration scheme for non-linear dynamic equations, J. Sound Vib. 314 (2008) 246–253.

[41] Z.B., Zhou, Application of differential quadrature element method to fracture analysis, Master Thesis, Nanjing University of Aeronautics and Astronautics, China, 2008 (in Chinese).

[42] L.F., Gan, Nonlinear analysis of tubular buckling in straight and curved wells, Ph.D. Dissertation, Nanjing University of Aeronautics and Astronautics, China, 2008 (in Chinese).

[43] X. Wang, J. Huang, Elastoplastic buckling analyses of rectangular plates under biaxial loadings by the differential quadrature method, Thin Wall Struct. 47 (2009) 12–20.

[44] W. Zhang, Elastoplastic buckling analysis of rectangular plates by using the differential quadrature method, Master Thesis, Nanjing University of Aeronautics and Astronautics, China, 2010 (in Chinese).

[45] W. Zhang, X. Wang, Elastoplastic buckling analysis of thick rectangular plates by using the differential quadrature method, Comput. Math. Appl. 61 (2011) 44–61.

[46] X. Wang, Y. Wang, Z. Yuan, Accurate vibration analysis of skew plates by the new version of the differential quadrature method, Appl. Math. Model 38 (2014) 926–937.

[47] Z. Yuan, X. Wang, Buckling and post-buckling analysis of extensible beams by using the differential quadrature method, Comput. Math. Appl. 62 (2011) 4499–4513.

[48] C. Xu, X. Wang, Efficient numerical method for dynamic analysis of flexible rod hit by rigid ball, T. NUAA 29 (4) (2012) 338–344.

[49] Y. Wang, X. Wang, Static analysis of higher order sandwich beams by weak form quadrature element method, Compos. Struct. 116 (2014) 841–848.

[50] C. Jin, X. Wang, Dynamic analysis of functionally graded material bars by using novel weak form quadrature element method, J. Vibroeng. 16 (6) (2014) 2790–2799.

[51] X. Wang, Y. Wang, Free vibration analysis of multiple-stepped beams by the differential quadrature element method, Appl. Math. Comput. 219 (2013) 5802–5810.

[52] X. Wang, Z. Wu, Differential quadrature analysis of free vibration of rhombic plates with free edges, Appl. Math. Comput. 225 (2013) 171–183.

[53] C. Jin, X. Wang, L. Ge, Novel weak form quadrature element method with expanded Chebyshev nodes, Appl. Math. Lett. 34 (2014) 51–59.

Index